DRINKING WATER

DRINKING WATER

Editor

Lucio Piovene

scitus academics

Drinking Water

Edited by **Lucio Piovene**

This Book is Printed on Acid-Free Paper. ∞

ISBN: 9781681179674

Library of Congress Control Number: 2017949108

SCITUS Academics LLC,

616, Corporate Way, Suite 2, 4766,

Valley Cottage, NY 10989

www.scitusacademics.com

Notice

Printed in United States of America.

Preface

Optimum management of global water resources presents one of the most crucial challenges of the 21st century. Global population will increase by three billion or more over the next 50–75 years, and the number of people living in urban areas wills more than double. Most of the world's population growth will occur in developing countries where water is already critically short and many of the residents are impoverished. Even today, more than 1 billion people do not have access to safe and affordable drinking water and perhaps twice that many lack adequate sanitation services. In fact, inadequate drinking water quality is a leading cause of infant mortality worldwide. This requires proper management during planning, construction and installation, operation and maintenance of the entire system from catchment to consumer. Also, good understanding of these processes facilitates early identification of potentially present weaknesses. Projections of global water needs are worrisome enough when the water demands arising from future population and economic growth are compared with current estimates of developed and developable supplies. However, the reliability of current supplies is also in question. The fact is that there are trends and circumstances which will almost certainly reduce available supplies in the face of sharply escalating water demands world-wide.

This volume 'Drinking Water' focuses on fundamental and applied research in water sources, substances, drinking water treatment processes, distribution systems, and residual management. The book covers a wide range of current issues, reflecting on current problems and demonstrating the complexity of water management. This book will be of valuable for scientists and researchers and engineers from water supply companies and engineering consulting firms.

Table of Contents

CHAPTER 1

New Perspectives in Monitoring Drinking Water Microbial Quality

Ma José Figueras [1,*] and Juan J. Borrego [2]

[1]*Unit of Microbiology, School of Medicine, IISPV, Universitat Rovira i Virigili, Sant Llorenç 21, 43201-Reus, Spain*

[2]*Department of Microbiology, Faculty of Sciences, University of Malaga, Campus Universitario Teatinos, 29071-Malaga, Spain*

ABSTRACT

The safety of drinking water is evaluated by the results obtained from faecal indicators during the stipulated controls fixed by the legislation. However, drinking-water related illness outbreaks are still occurring worldwide. The failures that lead to these outbreaks are relatively common and typically involve preceding heavy rain and inadequate disinfection processes. The role that classical faecal indicators have played in the protection of public health is reviewed and the turning points expected for the future explored. The legislation for protecting the quality of drinking water in Europe is under revision, and the planned modifications include an update of current indicators and methods as well as the introduction of Water Safety Plans (WSPs), in line with WHO recommendations. The principles of the WSP approach and the advances signified by the introduction of these preventive measures in the future improvement of dinking water quality are presented. The expected impact that climate change will have in the quality of drinking water is also critically evaluated.

Keywords: indicator microorganisms; drinking water quality; WSP; European drinking water directive; climate change

1. INTRODUCTION

The evaluation of the microbiological quality of drinking water aims to protect consumers from illness due to consumption of water that may contain pathogens such as bacteria, viruses and protozoa, and thus to prevent drinking-water related illness outbreaks. This has been, and still is nowadays an important challenge. For the past century, this evaluation has been performed through the analysis in finished drinking water of faecal pollution indicators, which are expected to predict the potential presence of pathogenic microorganisms in the water. However, scientists, engineers, public health officials and water pollution control agencies have faced cases in which the quality of water have showed the presence of indicators when water was already served to the consumers. In addition, drinking water outbreaks have occurred both in presence or absence of indicator organisms and involved pathogenic microorganisms that have contaminated the drinking water, and that either were not eliminated during treatment, or the latter failed at the time of the outbreak. The United States Centre for Disease Control has reported 780 disease outbreaks associated with the consumption of contaminated drinking waters from 1971 to 2006, which affected 577,094 persons [1]. A number of outbreaks have also occurred in Europe. For instance in Spain, in the 1999–2006 period, 413 outbreaks were recorded that involved 23,642 cases [2]. These outbreaks occurred despite specific legislations designed to prevent them, and the associated microbial control measures being carried out. The World Health Organization (WHO) has been very active in this field developing important guidelines of universal application and has promoted, in recent years, a more preventive approach than only checking the quality of the finished drinking water [3,4]. This "Water Safety Plans" (WSPs) approach takes into account all factors that endanger the quality of drinking water from the source to the final tap water at the consumer's home. The WHO alone, and in collaboration with the International Water Association (IWA), has developed several guideline documents that are freely accessible through Internet, the most recent of which is the *Water Safety Plan Manual* [4]. Furthermore, both, in collaboration with the Organization for Economic Co-operation and Development (OECD), published a book entitled Assessment of the Safety of Drinking Water, underlying the challenges for the 21st century [5]. Another WHO document is *Emerging Issues in Water and Infectious Diseases* that reviews the problem of emerging pathogens and other aspects that endanger water safety [6]. All of them are important reference manuals associated with water quality [3–6]. In 2003 the European Union initiated an extensive revision of the existing Drinking Water Directive (98/83/EC) [7], and is currently deciding what modifications will be included in the new and updated Directive in order to increase the quality of drinking water and protect public health.

In previous studies, we and other authors, have reviewed the definitions of index and indicator organisms used to evaluate the microbiological quality of water [3,8–12], as well as the relevance of some of them, e.g., faecal streptococci, or their relationship with pathogenic bacteria, e.g., *Salmonella*, also involved in drinking water outbreaks [9,10,13]. The specific methods used for the analysis of indicators have also been reviewed in detail [8,12,14], so in this update we will review recent advances in relation to indicator/index organisms. Furthermore, this overview aims to introduce the current proposed modifications for the new EU Drinking Water Directive, among which the WSPs are a key element. The WSPs principles will be presented in order to raise awareness of water quality professionals so that they can prepare for future developments. The expected influence of climate change in the quality of drinking water will be discussed.

2. FAECAL INDICATORS IN DRINKING WATER CONTROL

There are hundreds of different enteric microorganisms that are known to infect humans. Enteric microorganisms are excreted in the faeces of infected individuals or animals, and may directly or indirectly contaminate water intended for human consumption [3,8,14–17]. Since the adoption of disinfection practices by drinking water utilities, the incidence of waterborne diseases has decreased drastically. However, the WHO estimates that in developing countries, some two million children die each year of infectious diseases associated with contaminated water [17,18]. According to the American Society for Microbiology "many serious health problems could be eliminated if more countries adopted water quality practices, including the simple steps of source water protection and disinfection to ensure safe water supplies" (http://www.epa-gov./OWOW/watershed/statewidn/table.htm).
Therefore, the control of microbial pathogens must be carried out by the use of a multibarrier approach, including source protection, proper treatment and disinfection, and optimal distribution maintenance [19,20]. This approach has been adopted in the form of the WSPs, by the WHO [3,4] and will be discussed in the present review.

The presence of enteric pathogens in drinking waters is of great concern, and thus, legislation either in Europe, USA and other countries requires analysis of indicators to determine the microbiological quality of these waters. Ideally we would like to analyze the waters for the presence and quantification of specific enteric pathogens. However, many waterborne pathogens are still difficult to detect and/or quantify in waters and for most of the newly recognized agents, easy methods to detect them in water samples have still to be developed [12,14]. The introduction of molecular methods has advanced the recognition of these new agents and their benefits were recently reviewed [21]. However, the routine application of these methods for the analysis of pathogens is not a reality yet and is restricted to research studies or to cases of suspected outbreaks. Nowadays, new approaches based on virulence factor-activity relationships to discover and detect emerging waterborne pathogens are being explored [22]. Therefore, the most useful tool to determine the potential presence of pathogenic microorganisms in waters is the analysis of several microorganisms classed as either "indicator, or, index" organisms [8,9]. These indicators must fulfil the requirements indicated in a previous study [8].

To avoid the ambiguity in the term "microbial indicator", the following three groups are now recognized: process microbial indicators, faecal indicators and index and model organisms. Process indicators comprise a group of organisms that demonstrate the efficacy of a process; faecal indicators are those organisms that indicate the presence of faecal contamination, hence, they only infer that pathogens may be present; index and model organisms include a group or species indicative of pathogenic presence and behaviour, respectively [9]. The use of index and indicator organisms to assess the microbiological and sanitary quality of waters is well established and has been practiced for almost a century. The most widely used indicators are coliforms (total coliforms), faecal or thermotolerant coliforms, *Escherichia coli*, enterococci (faecal streptococci or intestinal enterococci) and bacteriophages.

2.1. Coliforms

By definition, coliform bacteria are facultative anaerobes, Gram-negative, non-spore-forming, rod-shaped bacteria that ferment lactose with acid production in 24 to 48 h at 36 ºC, and indole-negative. Coliforms belong to the family *Enterobacteriaceae* and include *Escherichia, Enterobacter, Klebsiella* and *Citrobacter, Kluyvera, Leclercia* genera, and some members of the genus *Serratia*. These bacteria were classically used as indicators of faecal contamination of waters because they were considered to be inhabitants of the intestinal tracts of homeothermic animals [8,11]. However, the ability of some coliforms to grow in natural waters, the lack of correlation between the number of coliforms and those of pathogenic microorganisms, and the detection of atypical strains has led them to become unsuitable faecal indicators [8,11,23,24]. Furthermore, several studies have demonstrated presence of coliforms in drinking water distribution systems associated with biofilm growth problems [25,26]. The coliform bacteria, traditionally termed the "total coliform" group, have been the primary standard for potable water in most of the world. However, many regulatory agencies have questioned its utility as an indicator. For these reasons, this is one of the parametres that has been eliminated from the European legislation for the management of the quality of bathing waters (2006/7/EC) and most probably will also disappear in the modification of the current drinking water legislation (98/83/CE) [7]. Nowadays, coliforms are typically associated with treatment effectiveness, and should be absent from adequately treated plant effluents [25,26]. The presence of coliforms in the distribution system, while possibly due to inadequate treatment, could also be due to cross-connections or failure to maintain an adequate disinfectant residual [26,27].

2.2. Faecal Coliforms

These bacteria conform to all the criteria used to define total coliforms plus the requirement that they grow and ferment lactose with the production of acid at 44.5 ºC. For this reason, "thermotolerant coliform" would be the scientifically accurate term for this group [8,11,23,24]. Bacteria in this coliform subgroup have been found to have a positive correlation with faecal contamination of warm-blooded animals [8,11,15,23,24]. However, some thermotolerant coliform bacteria that conform to this definition also belong to the genus *Klebsiella* and have been isolated from environmental samples in the apparent absence of faecal pollution [8,11,15]. Similarly, other members of the thermotolerant coliform group, including *Escherichia coli*, have been detected in some pristine areas [28], and associated with regowth events in potable water distribution systems [25]. Faecal coliforms display a survival pattern similar to these of bacterial pathogens but their usefulness as indicators of protozoan and viral contamination is limited, therefore, tended to be replaced by *E. coli* in several legislations [26].

2.3. Escherichia coli

E. coli is a member of faecal coliform group, being a more specific indicator for the presence of faecal contamination. In addition, *E. coli* conforms to taxonomic as well as functional identification criteria and is enzymatically distinguished by the lack of urease and presence of β-glucuronidase. One disadvantage associated with this organism as an indicator is that it has been consistently found in pristine tropical rain forest aquatic and plant systems as well as soils [28,29]. Additionally, it seems to survive for short periods in aquatic temperate environments [23,30]. *E. coli* is the faecal indicator of choice used in WHO Guidelines for Drinking-water Quality [3,26], and several countries include this organism in their regulations as the primary indicator of faecal pollution (*i.e.*, Europe, USA). Although it has long been known that *E. coli* can cause disease in humans, the bacteria naturally occurs in the lower part of the gut of warm-blooded animals [11,31]. Its role as an enteric pathogen has been reinforced with the discovery of *E. coli* O157:H7 associated with haemorrhagic enteritis and haemolytic uremic syndrome, that was responsible of producing several drinking water outbreaks, and some of them lack of β-glucuronidase activity [32,33].

2.4. Faecal Streptococci, Enterococci or Intestinal Enterococci

This group of microorganisms has received widespread acceptance as useful indicators of microbiological water quality, since: (i) they show a high and close relationship with health hazards associated with the water use, mainly for gastrointestinal symptoms; (ii) they are always present in faeces of warm-blooded animals; (iii) they unable to multiply in sewage-contaminated waters; and (iv) their die-off is less rapid than those of coliforms in water, and persistence patterns are similar to those of potential waterborne pathogenic bacteria [8,11,35–39].

Faecal streptococci, enterococci and intestinal enterococci are three synonyms used to refer to species described as members of the genus *Enterococcus,* which also fulfil Sherman's criteria (growth at 10 °C and 45 °C, resistance to 60 °C, growth at pH 9.6 and at 6.5% NaCl, and reduction of 0.1% methylene blue) [8,10,11]. They comprise species of different sanitary significance and survival characteristics and, in addition, the proportions of the species of this group are not the same in animal and human faeces [8,10,11]. *Enterococcus faecalis* and *Ent. faecium,* are the predominant species in human faeces and sewage [10,11,37]. In a European study that investigated enterococcal populations in animals, humans, and the environment the most common species detected were *Ent. faecium* (33%), *Ent. faecalis* (29%), and *Ent. hirae* (24%) [38]. This *Enterococcus* species distribution in human and animal hosts has been recently confirmed using a molecular multiplex PCR technique [39].

Despite the definitions provided above for the indicators (total coliforms, faecal coliforms, *E. coli* and enterococci) in practical terms these are determined on the basis of the biochemical reactions evaluated in culture media that are recognized either by the appearance of characteristic colonies (with a specific colour as response to this reaction in chromogenic substrates) and/or by the emission of fluorescence. Colour and fluorescence are also the responses expected in presence/absence tests in liquid media either in bottles or in a Most Probable Number approach designed as blisters or microplate systems that enable quantification [8,11]. Microbiological methods for indicators are far from perfect because they can produce false positive and negative results [8,11,35–37,40–42].

Molecular methods are useful both to monitor natural communities of bacteria, and to track specific bacterial markers in complex environments. Length-heterogeneity polymerase chain reaction (LHPCR) and terminal restriction fragment length polymorphism (T-RFLP) of 16S rDNAs of anaerobic bacteria have been used by Field *et al.* [43] to develop an alternative indicator that distinguishes the source of faecal pollution in water.

2.5. Bacteriophages

Several bacteriophage groups have also been classically used as faecal and viral indicators, and as models to evaluate the efficiency of the chlorination of drinking waters [44–46]. The proposed groups are somatic coliphages, F (male)-specific RNA bacteriophages (FRNA phages) and phages of *Bacteroides fragilis* [47–49].

Somatic coliphages are specific viruses of *E. coli* and have been commonly used as indicators of faecal and/or sewage pollution in several water types and as biotracers to identify pollution sources in surface waters and aquifers [50,51]. In addition, they may also serve as indicators for assessing the removal efficiency during the treatment of water and wastewater treatment plants [52]. On the basis of the differences in origin and ecology between enteric viruses and somatic coliphages, it is doubtful to conclude that this phage group could successfully be used in all situations as enteric viruses indicators [47], and they may not be a useful indicator of a distribution system integrity problem, even when the problem involves the introduction of faecal contamination [53].

The use of FRNA phages was proposed as faecal pollution indicators and as model viruses in water hygiene on the basis of: (i) their similar sizes and shapes to human enteric viruses; (ii) their correlation with the sewage contamination degree; and (iii) their inability to replicate in the water ecosystem [46]. However, the low incidence of this phage group in human faeces and its low specificity for its bacterial host, suggest that they would multiply in the sewerage system [48,54,55]. Hence, the presence of FRNA phages in water should be primarily used as an index of sewage pollution rather than faecal pollution [56].

Bacteroides fragilis is a strict anaerobe found in high concentrations in the human intestinal tract and dies rapidly when discharged into environmental waters. A phage of the strain HSP 40 of *B. fragilis* (isolated from Hospital San Pablo, Barcelona, Spain) has been proposed as a specific index of human faecal pollution of waters [49], because: (i) phages against this bacterial strain are human specific and are not isolated from the faeces of other homoeothermic animals; (ii) *B. fragilis* HSP 40 phages are consistently isolated from sewage, faecally-polluted waters, and their sediments but not from unpolluted samples; (iii) the levels of phages are related to the degree of pollution; (iv) *B. fragilis* phages always outnumber human enteric viruses; and (iv) in model experiments, no replication of these phages has been observed under simulated environmental conditions [57]. The low prevalence of these phages in waters with low and moderate levels of faecal pollution and the complex methodology for their recovery are the main drawbacks for the general use of these viruses as an indicator group [58,59].

3. DRINKING WATER OUTBREAKS

The total number of drinking water-related illness in the USA has been estimated at 19 million/year; however, this figure depends upon the approach considered [60]. The detected water-borne outbreaks are considered to be just the tip of the iceberg of the total drinking-water-related illness. In fact the actual disease burden in Europe, as in other parts of the world, is difficult to estimate and is, most likely underestimated [61]. Outbreaks have the potential to be rather large as in the case of the Milwaukee (USA) *Cryptosporidium* outbreak that affected over 400,000 people in 1993 [1]. At least 325 drinking water-associated outbreaks of parasitic protozoan diseases have been reported all over the world between 1954 and 2003, over 30% (106) of all outbreaks were documented from Europe, *Giardia duodenalis* and *Cryptosporidium parvum* accounting for the majority of these outbreaks [62]. The first outbreak that provided evidence that *E. coli* O157:H7 was transmitted by drinking water occurred in a small rural town in Missouri (USA) that had an unchlorinated water supply [63]. There were 243 cases, of whom 86 presented bloody stools, 32 were hospitalized, four died and two developed haemolytic ureamic syndrome (HUS). HUS is a severe disease that may cause an acute renal failure, which may require dialysis or kidney transplantation.

Other outbreaks like the one of Walkerton, Ontario (Canada) in 2000, which affected over 2,300 cases, revealed a mixed aetiology by *Cryptosporidium* and *Escherichia coli* 0157:H7 [60]. In our view, outbreaks of mixed aetiologies should be more common than what is detected because the sewage contamination drinking water contains many potential pathogenic microorganisms. A good example of this is the outbreak that occurred in South Bass Island (Ohio, USA) in 2004, which revealed a massive microbiological contamination (with total coliforms, *E. coli*, enterococci, *Campylobacter*, *Arcobacter*, coliphages and adenoviruses) of the ground water used for drinking water producing 1,450 cases of gastroenteritis [64,65]. This outbreak also revealed that the deterioration of the water occurred over years, and that a poorly known microorganism *Arcobacter* was implicated [64]. The latter is often confounded with *Campylobacter* if inappropriate molecular identification methods are applied. Furthermore, *Arcobacter* is frequently present in human sewage showing a good correlation with indicators of faecal pollution [66–68]. However, despite source water showing a high prevalence of *Arcobacter* spp. appropriate treatment can remove these microorganisms as well as noroviruses from the finished drinking water [68].

In Europe, monographic water outbreak reports (e.g., those produced by CDC in the USA [1]) are not available, because drinking water is defined as food, and therefore reporting is included with food-borne outbreaks [69]. In 2007, only 17 water-borne outbreaks were reported by eight countries [69], clearly indicating an under-reporting. They involved 10,912 cases, with 232 hospitalizations. The main microorganisms involved were *Campylobacter*, norovirus, *Giardia* and *Cyptosporidium*. Interestingly, the biggest outbreaks had multiple aetiologies, one involving 453 registered cases in Denmark and a large outbreak with 8,000 cases in Finland of which approximately 1,000 sought medical attention and 200 were hospitalized [69]. In the latter three major (*Campylobacter*, norovirus, *Giardia*), and three minor causative agents (*Salmonella* Enteritidis, *Clostridium difficile* and rotavirus) were isolated from the patients, and all the causative agents were also isolated from water samples [69]. Multiple microorganisms (enteroviruses, *Giardia*, *Cryptosporidium*, *Campylobacter* and *Arcobacter*) were recovered from the patients in an outbreak that occurred in Slovenia in 2008 [70]. This reinforces our idea that multiple microorganism aetiologies maybe more common than as well as the prevalence of *Arcobacter*.

3.1. Principal Failures Associated with Outbreaks and Lessons Learned

Prevention and containment of outbreaks requires examination of the causative events responsible for their occurrence. As indicated by Risebro *et al.* [71], retrospective analysis of outbreaks of enteric diseases can be used to inform outbreak investigators, facilitate corrective measures, and further develop multi-barrier approaches. In this sense these authors developed an outbreak fault tree that was applied to 61 enteric outbreaks related to public drinking water supplies in the EU. The approach found that failures in the source water and in water treatment, independently or together, were the cause of more than 50% (34/61) of the outbreaks. Faults at the distribution system occurred less frequently (19/61 outbreaks) but were often solitary events contributing heavily towards the outbreak (a mean % score of 87.42). Livestock and rainfall in the catchment with none or inadequate filtration of the water sources contributed to *Cryptosporidium* outbreaks. Of the 23 protozoan parasite outbreaks that showed one treatment causative event, 90% of these events were filtration deficiencies. However, by contrast, for bacterial, viral, and mixed pathogen outbreaks, disinfection deficiencies were associated with 75% of the outbreaks [71].

Excessive rainfall has been an important contributor to historical waterborne disease outbreaks [72–77]. In fact, most of the bacteriological parametres (heterotrophic bacteria, *E. coli*, total coliforms, faecal streptococci, and *Clostridium perfringens* counts) increased considerably during extreme runoff events as do the concentrations of *Giardia* and *Cryptosporidium* [74–76]. Another important lesson learned, after an outbreak, is that once *Cryptosporidium* has colonized a drinking water system, it can persist for a long time despite vigorous and repetitive flushing of the system [77]. The investigators suggested that oocysts were being trapped in the biofilm in the distribution network and then were being released back into the supply. A very long persistence of norovirus in the water distribution system was also observed in the outbreak that occurred in Finland mentioned previously, and it required advising people in the affected areas to boil the water before use for a ten-week period [69].

3.2. Further Health Consequences of Gastroenteritis Outbreaks

Nowadays, it is known that gastroenteritis may have other important health sequels, like reactive arthritis, irritable bowel syndrome, cancer predisposition, to name a few [17]. In this sense, a study carried out with the patients of the Walkerton outbreak in Canada, showed that 15.7% of the asymptomatic patients during the outbreak, and in 17.6 and 21.6% of those who had moderate and severe symptoms of acute gastroenteritis respectively, showed problems with arthritis 4.5 years later [78]. Gastroenteritis is also associated with subsequent post-infectious irritable bowel syndrome [79], and HUS [63].

Such outbreaks can generate high societal alarm, which can result in the introduction of new Drinking Water Regulations. An example is the case of the biggest worldwide outbreak of *Legionella* that motivated the first Spanish legislation in relation to the control of this microorganism. Another example is the incorporation of significant requirements for drinking water providers, following the Walkerton outbreak [80]. In fact the latter outbreak has also influenced the EU legislation for drinking water, which had incorporated the control of *Cryptosporidium* in specific circumstances. These interventions lead to a significant decline in cryptosporidiosis [81].

4. ROLE OF FAECAL INDICATORS IN THE PROTECTION OF PUBLIC HEALTH: ALTERNATIVE INDICATORS AND RECOVERY OF INJURED BACTERIA

The failure of measurements of single indicator organisms to correlate with pathogens suggests that public health is not adequately protected by simple monitoring schemes based on detection of a single indicator, particularly at the detection limits routinely employed. In addition, the classical microbial indicators proposed present several shortcomings and they cannot be used in all water types. In this instance, other indicators, named "Alternative", should be used to determine the possible threat to public health [3,10,11,26].

4.1. Alternative Indicators

The use of the sulphite-reducing members of the genus *Clostridium* (*C. perfringens*) as indicators of faecal pollution is based on: (i) the presence of these microorganisms in the faeces of all warmblooded animals as well as in sewage, (ii) more stability in environmental waters and greater resistance to the disinfection processes than most pathogens, and (iii) successful use in monitoring sewage-contaminated waters [3,10, 11,26,82]. Nevertheless, sulphite-reducing clostridia are considered ubiquitous in aquatic sediments and the spore form explains their persistence, although they can be used as indicators of remote or non-point faecal pollution or to evaluate the virus and cyst inactivation in the drinking water disinfection processes [10,11,83]. However, the WHO [3,24,82] does not recommend clostridia for routine distribution system monitoring because, due to their length of survival, they may be detected long after (and far from) the pollution event, leading to possible false alarms.

Heterotrophic plate count (HPC) or total aerobic bacteria were among the first parametres used to monitor the safety of finished drinking water. However, presently they have become an indicator of general water quality within distribution systems [3,26,84]. It provides a good operational monitoring parametre that measures the deterioration of water quality through distribution systems. It is considered that the bacteriological content of drinking-water leaving treatment plants should contain only very low levels of heterotrophic and aerobic spore-forming microorganisms [26]. In fact this parametre, evaluated both at 22° and 37 °C, is included in the EU drinking water legislation, and is also required to be evaluated monthly for the control of *Legionella* in the Spanish and other specific legislations to control the latter microorganism. A series of review papers appeared in 2004, including one by the WHO, which evaluated the role of this parametre in water as a control measure in drinking water safety management [83].

Members of the genus *Pseudomonas* are possibly the microorganisms most often isolated from bodies of water. However, contrary to the previously discussed indicators, their presence does not necessarily indicate a possible risk to public health. *P. aeruginosa* was found to be more resistant than acid-fast bacteria during ozonation processes, demonstrating its resistance to chemical disinfection and thus its usefulness in the analysis of waters that receive chemical disinfection, including drinking waters [10,83,86–88]. Their role and significance in water has recently been reviewed by Mena and Gerba [88]. These authors estimated the health risks associated with the exposure to *P. aeruginosa*, and conclude that the risk derived from drinking water ingestion is low; however, it is slightly higher if the subject is taking an antibiotic to which this microorganism is resistant.

The mycobacteria belong to a group of microorganisms considered emerging pathogens of increasing importance. Their role as aetiological agents of waterborne disease is still not completely understood, although this group includes pathogenic species such as *Mycobacterium tuberculosis* or *M. bovis* and other atypical mycobacteria, for example, *M. intracellulare* and *M. avium* [4,89]. According to the WHO these bacterial species are relatively resistant to treatment and disinfection and have been detected in well operated and maintained drinking-water supplies with HPC less than 500/mL and total chlorine residuals of up to 2.8 mg/L. Furthermore, the growth of these organisms in biofilms reduces the effectiveness of disinfection [4].

Aeromonas are considered autochthonous microorganisms of water and are responsible of producing several diseases in humans, some of them related to water exposure or consumption of contaminated water [4,90–92]. Considerable new knowledge has been accumulated in recent years about the taxonomy, virulence properties and disease presentations of the species included in this genus [92–108]. The introduction of new molecular approaches, e.g., housekeeping genes enabled to recognise new species from freshwater (*A. fluvialis* and *A. rivuli*), tap water (*A. tecta*) and new and/or relevant clinical species like *A. taiwanensis*, *A. saranelli* and *A. aquariorum* [93–96]. The latter three species have been associated mainly with extraintestinal infections [94,95]. These clinical species together with the previously known species associated with different human disease (*A. hydrophila*, *A. caviae*, *A. veronii* bt. sobria, *A. veronii* bt. veronii, *A. jandaei* and *A. schubertii*) [90–92] should nowadays be considered relevant for public health. *Aeromonas* can be readily isolated from drinking water distribution systems, where they appear to survive well, to proliferate at low temperatures and to be associated with pipe biofilms where populations may survive at high chlorine levels [4,91,97]; and therefore, they may be considered as potential indicators of disinfection efficacy and biofilm development [4,91]. In The Netherlands, the public health authorities defined maximum values for *Aeromonas* densities, *i.e.*, 20 CFU/100 mL for finished water, and 200 CFU/100 mL for drinking water in the distribution system [91]. The factors that influence the occurrence and population sizes of *Aeromonas* spp. in water distribution systems include organic content, temperature, the residence time of water in the distribution network, and the presence of residual chlorine [4]. Strains isolated from drinking water contain virulence factors [100–105]. In one of the latter studies, it was demonstrated a clonal relationship between the isolates recovered from patients with diarrhoea and those recovered from the drinking water [103]. However, the role of *Aeromonas* in gastroenteritis has been questioned, but many arguments support its true association with diarrheal disease [98]. Recently a new case of *Aeromonas* HUS has been published and previous described cases were reviewed [66]. Furthermore, shigatoxin genes homologous to those of *E. coli* O157:H7 responsible of HUS were found to be present in some *Aeromonas* strains, reinforcing its role of as an aetiological agent of HUS [105]. The public health importance of this finding, together with that derived from the recent isolation of the clinical relevant species *A. aquariorum* [94], in chironomid egg masses, which may infest drinking water systems, needs to be further clarified [106].

Some groups of human viruses have also been proposed as alternative indicators for the control of drinking water quality, such as adenoviruses and polyomaviruses [107,108], on the basis that adenoviruses have been found to be significantly more stable than faecal indicator bacteria and other enteric viruses during UV treatment. Some researchers have suggested enteroviruses or noroviruses as indicators of other enteric viruses [109,110]. However, these viruses exhibit seasonal fluctuations and epidemic spikes [111]. Griffin *et al.* [112] have proposed Torque teno virus as a more appropriate indicator of viral pathogens in drinking waters due to its characteristics. Torque teno virus is a small, non-enveloped DNA virus that is likely to exhibit similar transport characteristics to pathogenic enteric viruses. Torque teno virus is unique among enteric viral pathogens in that it appears to be ubiquitous in humans, elicits seemingly innocuous infections, and does not exhibit seasonal fluctuations or epidemic spikes. Torque teno virus is transmitted primarily via the faecal-oral route and can be assayed using rapid molecular techniques.

4.2. Recovery of Injured Bacteria

Indicator bacteria become injured in water and wastewater following sublethal exposure to a wide variety of chemical and physical agents [15,113–115]. Such bacteria are unable to form colonies on most selective media, and between 10% and 90% of indicator bacteria in treated drinking water may be injured [116]. As a consequence, injured cells are undetected in water, leading to an underestimation of the faecal pollution level of the finished drinking water in distribution networks [117]. The advantages of several methods to detect injured bacteria have been reviewed earlier [3,11,26,114]. The detection of injured bacteria in treated waters may be indicative of the potential regrowth in the distribution system, due especially to the presence of high levels of nutrients, and thus may provide guidance in the diagnosis of problems within water distribution systems [15].

5. EUROPEAN LEGISLATION ON DRINKING WATER

At present, the Council Directive 98/83/EC [7] is the legislation that is applied for the protection of the quality of water intended for human consumption. Within this Directive it was indicated that the standards included in this legislation were meant to be reviewed by the European

Commission (EC) every five years to adapt them to the latest scientific state of the art. In this sense the Commission initiated a process in 2003 in which a wide range of stakeholders participated to discuss the key elements that could be modified in light of current knowledge and advances in technology. The agreed elements were: (i) the inclusion of a more preventative approach for improving the quality of drinking water based on the evaluation of risk to contamination from the source water to the tap through a risk assessment approach in line with the defined WSPs developed by the WHO [3,4,26]; (ii) to update and review both chemical and microbiological parametres and to include standard methods for monitoring, sampling and analysis; (iii) to pay special attention to small water supply systems, which are now known to be those at higher risk globally; and (iv) to introduce criteria for construction products in contact with drinking water. In this process, the EC engaged the WHO to advise on how the WSP concept could be incorporated in the revised drinking water legislation and EC has set up expert groups and employed consultants to address all the key elements mentioned: http://ec.europa.eu/environment/water/water-drink/revision_en.html. In relation to the microbiological parametres it has been agreed that *E. coli* and enterococci have proven to be useful and therefore both parametres will remain in the new proposal; however, it has been recommended that the sampling frequency for enterococci is increased to one similar to that required for monitoring *E. coli*. Another point raised is that the current reference Membrane Filtration method ISO 9308-1 for *E. coli* is not suitable, and the proposal is to change it, and include more than one method. It has also been agreed that the coliform bacteria parametre is not suitable as a faecal indicator but may serve other purposes (*i.e.*, operational monitoring). It is considered that if coliforms are finally kept in the new Directive, there are issues concerning their method of analysis and their definition that should be reviewed. There is also a general agreement about removing *C. perfringens* from the list of parametres for routine compliance monitoring. In relation to total colony count, where the existing Directive the standard is "no abnormal change", it is recognized that this needs to be rephrased, and that more clear guidance value is needed. The report can be consulted at:http://circa.europa.eu/Public/irc/env/ drinking_water_rev/library?l=/microbiological/17102007_28022008/_EN_1.0_&a=d.

An important interlink should exist in the new proposed Directive with other already existing or planned EU legislation related to the quality of water, e.g., Water Framework Directive 2000/60/EC (WFD) or the Groundwater Directive 2006/118/EC (GWD). The WFD has a major role in the protection and the prevention of pollution of water resources and therefore may be an important driving force to improve the quality of raw water. An important integration between responsible authorities dealing with the obligations arising from these legislations will be required.

6. WHO GUIDELINES AND WATER SAFETY PLANS

The WHO recommend WSPs as the most effective approach for consistently ensuring the safety of a drinking-water supply, because this approach manages the risk from the catchment or water sources to the consumer's tap [3,4,26]. The WSP approach is based on the hazard analysis and critical control point (HACCP) system, used classically in the food industry for controlling food quality. The risk assessment of the complete water system (catchment to tap), included in the WSP, should provide a better understanding of the risks of contamination by pathogens at each step along the system. Then preventative strategies should be designed (a multi-barrier protection) in order to correctly manage these risks to efficiently and effectively protect public health. This approach does not rely solely on end point testing, but on the establishment of critical control points that will be subject to on-line monitoring. The parameters that can be measured on-line and in real-time are: free chlorine, water pressure, dissolved oxygen and turbidity, for which safety critical limits are established. Any sudden anomalous changes in any of these parameters may indicate a problem within the system that can be managed before water is supplied to the consumer. The introduction of these early warning or control parametres from source to tap, that can predict or alert of a possible deterioration of the drinking water quality before it is distributed to the population, are the key elements of the WSPs. The microbiological analysis for the identification and enumeration of indicator microorganisms are too slow (require 24–48 h), and therefore are not suitable for this, however, they have an important role as validation tools because they verify that the barriers work properly and that the complete process is under control.

In reality many big water companies have long been adopting the principles of risk assessment and risk management (mostly in the form of operational procedures) for their treatment works and distribution networks, and therefore adaptation to these approaches will not be difficult but beneficial as already reported [118,http://www.gov.ns.ca/nse/water/docs/NSWaterStrategy.pdf]. Within an EU research financed project (Healthy Water) one of the objectives was to train water companies on the principles of the WSPs, the experience demonstrated that the companies acknowledged the benefit of the new approach, and some of the largest water companies have incorporated it already.

However, small water companies would only be implementing WSPs when the approach becomes a mandatory requirement under the new EU legislation. In reality legislations are good driving forces for such improvements as outlined previously. For instance the national Spanish legislation on *Legionella* promulgated after the world largest outbreak, requires a part of the microbiological controls two other obligations: (i) to pass a training course for those responsible of handling installations at risk of propagating the legionellosis; and (ii) to implant control plans based on the methodology of HACCP. According to Bartram *et al.* [85] health care facilities should have general water safety plans as part of their infection control strategy. Such plans may be generic (e.g., applicable to health centres in general) or specific for larger buildings (*i.e.*, hospitals and nursing homes) and should address microbial growth in addition to control of external contamination by *P. aeruginosa*, and *Legionella*. The WSPs have to be developed by a team and require:

- Specific measures to protect raw water used to produce drinking water (*i.e.*, fencing).
- Ensuring the appropriate level of treatment at the water company and during storage and distribution pipe networks to customer's tap is maintained to guarantee the water quality.
- Ensuring that customers are aware of their role and responsibility for keeping water wholesome in their properties—it includes public buildings as well as private homes

Protection of the entire catchment areas is the first step of the multiple-barrier protection concept. Modelling can be used for establishing microbial risks in drinking water catchments and can be an excellent management tool [118] in the development of the WSP. There is much evidence that inappropriate water handling is one of the main sources of water contamination at the consumer's homes. Considering this, WHO prepared a specific manual for *Managing Microbial Water Quality in Piped Distribution Systems* [26, http://www.who.int/water_sanitation_health/dwq/924156251X/en/].

Further information, of how to implement a WSP, can be found in the WHO technical guidance documents (http://www.who.int/water_sanitation_health/dwq/wsp170805.pdf), including the WSP Manual [4,http://whqlibdoc.who.int/publications/2009/9789241562638_eng.pdf] in which specific case studies are presented. A dedicated web site on the WSPs has been developed by the International Water Association (IWA) (http://www.wsportal.org/ibis/water-safety-portal/eng/ we-lcome). In addition, WHO and IWA developed guiding documents to initiate such process considering all levels of resources available, so that it can be implemented all over the world, even at poorly developed countries: (www.unwater.org/worldwaterday/.../WSP_ RoadMap_Final_ 3_19_10.pdf).

7. QUALITY OF DRINKING WATER AND CLIMATE CHANGE

The main impacts of climate change on water availability are flooding and droughts. However, besides these quantitative impacts, the climate change will affect the surface water quality [4,120]. The climate change determinants affecting water quality are mainly the air temperature, the increase of extreme hydrological events, soil drying-rewetting cycles and solar radiation. First of all, temperature is the main factor affecting almost all physico-chemical equilibriums and biological reactions. Consequently, several transformations or effects related to water will be favoured by water temperature increase such as dissolution, solubilisation, complexation, degradation, evaporation, *etc.* This phenomenon globally leads to the concentration increase of dissolved substances in water but also to the concentration decrease of dissolved gasses, such as oxygen. Floods and droughts will also modify water quality by direct effects of dilution or concentration of dissolved substances [4,121]. A positive effect is the concentration decrease of some pollutants due to a low water velocity, which allow the assimilation of nutrients by aquatic plants and the adsorption/complexation of heavy metals on suspended matter and settling [122]. Runoff and solid material transportation are the main consequences of heavy rainfalls; for example, in the temperate zone, climate change will decrease the number of rainy days but increase the average volume of each rainfall event [73,123]. As a consequence, drought–rewetting cycles may impact water quality as it enhances decomposition of organic matter into streams [124]. A study performed by Nichols *et al.* [125] provided evidence that both low rainfall and heavy rain precede many drinking water outbreaks, and therefore both should be considered when assessing the health impacts of climate change. Solar irradiation increase could also alter water quality and especially characteristics of natural organic matter in freshwater systems both by warming and UVB radiation (increasing photolysis) [126].

Waterborne pathogens could be spread within the freshwater after a contamination by animal or human waste due to heavy rainfall discharge in combined sewer systems (CSS) [4,74,127]. When the flow exceeds the CSS capacity, the sewers overflow directly into surface water body [127]. Pednekar *et al.* [128] have studied coliform load in a tidal embayment and shown that storm-water coming from the surrounding watershed is a primary source of coliform. Moreover, higher water temperatures will probably lead to a pathogen survival increase in the environment, although there is still no clear evidence [129]. Floods often led to a contamination of groundwater and additional disease outbreaks [4,130]. Even though the risk of diseases outbreaks linked to drinking waters is low in developed countries, private supplies would be at risk [4,26,129]. In addition, an increase in temperature threats water quality with regard to waterborne diseases especially cholera disease in Asia and South America [129]. It was shown that the increased UV radiation due to ozone layer depletion provokes the breaking down of bioavailable organic compounds, minerals and micronutrients, stimulating the bacterial activity in aquatic ecosystems [126].

Fishes, green algae and diatoms are often used as water quality indicators of climate changes in waters. Daufresne and Böet [131] observed a change in fish communities due to temperature changes. WHO had also prepared a document called Vision 2030:"The resilience of water supply and sanitation in the face of climate change" that aims to increase our understanding of how anticipated climate change may affect drinking water:http://www.who.int/water_sanitation_health/ publications/9789241598422_cdrom/en/.

8. CONCLUSIONS

The microbial contamination of drinking water and its control constitutes a major issue worldwide, because it is still a major source of infection and can cause mortality, especially in the children of developed countries, and threatens the health of the population of developed regions, as illustrated by recorded outbreaks. The latter are however, considered to be underestimated because the major symptom (*i.e.*, diarrhoea) auto limits by itself without treatment in most healthy people. However, the older and immunocompromized people are at higher risk. Today we also know that this type of infection may have important sequels (e.g., reactive arthritis, irritable bowel syndrome, cancer predisposition, *etc.*). The microbiological controls applied to drinking water, have relied on the analysis of faecal pollution indicators in the finished dinking water. The classical indicators have served together with the improvements on the treatment and disinfection to control waterborne outbreaks. However, the use of these indicators may be substituted by the direct detection of pathogenic microorganisms, e.g., in the case of pathogenic viruses [53,132,133]. In addition, the lessons learned from outbreaks are key elements that should guide the proper management of drinking water. The shortcoming of bacterial indicators to predict parasites and viruses, which can be more resistant to disinfection, and the fact that information derived from the microbiological analysis is not immediate (neither is obtained in a continuous manner), have motivated the development of more preventive approaches, like the Water Safety Plans proposed by the WHO. Their application for the management of drinking water, either in big companies, small ones or in undeveloped countries can be foreseen as an important gain for the future. The adoption of this strategy in the EU legislation, which is in process of being modified, is a promising guarantee for the improvement of quality of dinking water in Europe. The recognition in this modification that small water supplies are at the highest risk, and the introduction of measures to control these supplies more efficiently will contribute to the expected improvement. Furthermore, the accumulated knowledge on the impact of climate change allows preparation of strategies to mitigate its impact on the quality of drinking water.

ACKNOWLEDGEMENTS

This work was supported by funds from the European Commission for the EPIBATHE (SSPI-2006-022618) and HEALTHY WATER (FOOD-CT-2006-036306) projects and partially by an AquaFactory International grant. The authors are solely responsible for the content of this publication. It does not represent the opinion of the European Commission. The European Commission is not responsible for any use that might be made of data appearing therein.

REFERENCES

1. Craun, GF; Brunkard, JM; Yoder, JS; Roberts, VA; Carpenter, J; Wade, T; Calderon, RL; Roberts, JM; Beach, MJ; Roy, SL. Causes of outbreaks associated with drinking water in the United States from 1971 to 2006. *Clin. Microbiol. Rev* **2010**, *23*, 507–528.
2. Vigilancia epidemiológica de brotes de transmisión hídrica en España. 1999–2006. *Bol. Epidemiol. Semanal* **2008**, *16*, 25–28.
3. *Guidelines for Drinking-Water Quality*, 3rd ed; World Health Organization: Geneva, Switzerland, 2008; Volume 1, Recommendations.
4. Bartram, J; Corrales, L; Davison, A; Deere, D; Drury, D; Gordon, B; Howard, G; Rinehold, A; Stevens, M. *Water Safety Plan Manual: Step-by-Step Risk Management for Drinking Water Suppliers*; World Health Organization: Geneva, Switzerland, 2009.
5. Dufour, A; Snozzi, M; Koster, W; Bartram, J; Ronchi, E; Fewtrell, L. *Assessing Microbial Safety of Drinking Water Improving Approaches and Methods*; World Health Organization and the Organisation for Economic Co-operation and Development, IWA Publishing Alliance House: London, UK, 2003. Available online: https://www.who.int/water_sanitation_health/dwq/9241546301full.pdf (accessed on 1 December 2010).

6. *Emerging Issues in Water and Infectious Diseases*; World Health Organization: Geneva, Switzerland, 2003. Available online: http://www.who.int/water_sanitation_ health/emerging/emergingissues/en/ (accessed on 1 December 2010).
7. European Union. *Council Directive 98/83/EC of 3 November 1998 on the quality of water intended for human consumption (98/83/EC)*, 1998. Available online: http://ec.europa.eu/environment/water/water-drink/index_en.html (accessed on 1 December 2010).
8. Borrego, JJ; Figueras, MJ. Microbiological quality of natural waters. *Microbiologia SEM (Int. Microbiol.)* **1997**, *13*, 413–426.
9. Ashbolt, NJ; Grabow, WOK; Snozzi, M. Fewtrell, L, Barthram, J, Eds.; Indicators of microbial water quality. In *Water Quality: Guidelines, Standards and Health*; IWA Publishing: London, UK, 2001; pp. 289–316.
10. Borrego, JJ; Castro, D; Figueras, MJ. Bitton, G, Ed.; Fecal streptococci/enterococci in aquatic environments. In *Encyclopedia of Environmental Microbiology*; John Wiley & Sons: New York, NY, USA, 2002; pp. 1264–1278.
11. Figueras, MJ; Robertson, W; Pike, E; Borrego, JJ. Bartram, J, Rees, G, Eds.; Sanitary inspection and microbiological water quality. In *Monitoring Bathing Waters A Practical Guide to the Design and Implementation of Assessments and Monitoring Programmes*; E & FN Spon: London, UK, 2000; pp. 113–167.
12. Köster, W; Egli, T; Ashbolt, N; Botzenhart, K; Burlion, N; Endo, T; Grimont, P; Guillot, E; Mabilat, C; Newport, L; Niemi, M; Payment, P; Prescott, A; Renaud, P; Rust, A. Analytical methods for microbial water quality testing. In *Assessing Microbial Safety of Drinking Water: Improving Approaches and Methods*; OECD, World Health Organization: Geneva, Switzerland, 2003; pp. 237–292.
13. Borrego, JJ; Castro, D; Figueras, MJ. Bitton, G, Ed.; Salmonella *in aquatic environments*. In *Encyclopedia of Environmental Microbiology*; John Wiley & Sons: New York, NY, USA, 2002; pp. 2789–2802.
14. National Research Council, Evaluation of risk: a phased approach to monitoring microbial water quality. In *Indicators for Waterborne Pathogens*; The National Academic Press: Washington, DC, USA, 2004; pp. 13–16.
15. Toranzos, GA; McFeters, GA; Borrego, JJ; Savill, M. Hurst, CJ, Crawford, RL, Garland, JL, Lipson, DA, Mills, AL, Stetzenbach, LD, Eds.; Detection of microorganisms in environmental freshwaters and drinking waters. In *Manual of Environmental Microbiology*, 3rd ed; American Society for Microbiology Press: Washington, DC, USA, 2007; pp. 249–264.
16. Leclerc, H; Schwartzbrod, L; Dei-Cas, E. Microbial agents associated with waterborne diseases. *Crit. Rev. Microbiol* **2002**, *28*, 371–409.
17. Asbolt, NJ. Microbial contamination of drinking water and disease outcomes in developing regions. *Toxicology* **2004**, *198*, 229–238.
18. Postel, SL. Water and world population growth. *J. Am. Water Works Assoc* **2000**, *92*, 131–138.
19. Morris, J. Filtered or waivered, you need source protection. *N. Engl. Water Works Assoc* **1994**, *9*, 182–191.
20. *Preventing Waterborne Disease: Is Your System at Risk*; American Water Works Association: Denver, CO, USA, 1994.
21. Brettar, I; Höfle, MG. Molecular assessment of bacterial pathogens—A contribution to drinking water safety. *J. Curr. Opin. Biotech* **2008**, *19*, 274–280.
22. Cangelosi, GA. Prospects for applying virulence factor-activity relationships (VFAR) to emerging waterborne pathogens. *J. Water Health* **2009**, *7*, S64–74.
23. Moriñigo, MA; Cornax, R; Castro, D; Martinez-Manzanares, E; Borrego, JJ. Viability of *Salmonella* spp. and indicator microorganisms in seawater using membrane diffusion chambers. *Antonie van Leeuwenhoek* **1990**, *57*, 109–117.
24. Figueras, MJ; Polo, FL; Inza, I; Guarro, J. Past, present and future perspective of the EU bathing water directive. *Marine Pollut. Bull* **1997**, *34*, 148–156.
25. LeChevallier, MW; Welch, WJ; Smith, DB. Full-scale of factors related to coliform regrowth in drinking water. *Appl. Environ. Microbiol* **1996**, *62*, 2201–2211.

26. World Health Organization, Ainsworth, R, Ed.; *Safe Piped Water: Managing Microbial Water Quality in Piped Distribution Systems*; IWA Publishing: London, UK, 2004.
27. LeChevallier, MW. The case for maintaining a disinfectant residual. *J. Am. Water Works Assoc* **1999**, *91*, 86–94.
28. Hazen, TC; Toranzos, GA. McFeters, GA, Ed.; Tropical source water. In *Drinking Water Microbiology*; Springer-Verlag: New York, NY, USA, 1990; pp. 32–54.
29. Lasalde, C; Rodriguez, R; Smith, HH; Toranzos, GA. Heterogeneity of *uid*A gene in environmental *Escherichia coli* populations. *J. Water Health* **2005**, *3*, 297–304.
30. Borrego, JJ; Arrabal, F; De Vicente, A; Gomez, LF; Romero, P. Study of microbial inactivation in the marine environment. *J. Water Pollut. Control Fed* **1983**, *55*, 297–302.
31. Schierack, P; Walk, N; Reiter, K; Weyranch, KD; Wieler, LH. Composition of intestinal *Enterobacteriaceae* populations of healthy domestic pigs. *Microbiology* **2007**, *153*, 3830–3837.
32. Hunter, PR. Drinking water and diarrhoeal disease due to *Escherichia coli*. *J. Water Health* **2003**, *1*, 65–72.
33. Schets, FM; During, M; Italiaander, R; Heijnen, L; Rutjes, SA; van der Zwaluw, WK; de Roda Husman, AM. *Escherichia coli* O157:H7 in drinking water from private water supplies in the Netherlands. *Water Res* **2005**, *39*, 4485–4493.
34. Kay, D; Fleischer, JM; Salmon, RL; Jones, F; Wyer, MD; Goodfree, AF; Zelenauch-Jacquotte, Z; Shore, R. Predicting likelihood of gastroenteritis from sea bathing: results from randomized exposure. *Lancet* **1994**, *344*, 905–909.
35. Audicana, A; Perales, I; Borrego, JJ. Modification of kanamycin-esculin-azide agar to improve selectivity in the enumeration of fecal streptococci from water samples. *Appl. Environ. Microbiol* **1995**, *61*, 4178–4183.
36. Figueras, MJ; Inza, I; Polo, FL; Feliu, MT; Guarro, J. A fast method for the confirmation of fecal streptococci from M-Enterococcus medium. *Appl. Environ. Microbiol* **1996**, *62*, 2177–2178.
37. Figueras, MJ; Inza, I; Polo, FL; Guarro, J. Evaluation of the oxolinic acid-esculin-azide medium for the isolation and enumeration of faecal streptococci in a routine monitoring programme for bathing waters. *Can. J. Microbiol* **1998**, *44*, 998–1002.
38. Kühn, I; Iversen, A; Burman, LG; Olsson-Liljequist, B; Franklin, A; Finn, M; Aarestrup, F; Seyfarth, AM; Blanch, AR; Vilanova, X; Taylor, H; Caplin, J; Moreno, MA; Dominguez, L; Herrero, IA; Möllby, R. Comparison of enterococcal populations in animals, humans, and the environment—a European study. *Int. J. Food. Microbiol* **2003**, *88*, 133–145.
39. Layton, BA; Walters, SP; Lam, LH; Boehm, AB. *Enterococcus* species distribution among human and animal hosts using multiplex PCR. *J. Appl. Microbiol* **2010**, *109*, 539–547.
40. Fricker, CR; DeSarno, M; Warden, PS; Eldred, BJ. False-negative β-D-glucuronidase reactions in membrane lactose glucuronide agar medium used for the simultaneous detection of coliforms and *Escherichia coli* from water. *Lett. Appl. Microbiol* **2008**, *47*, 539–542.
41. Fricker, CR; Warden, PS; Eldred, BJ. Understanding the cause of false negative β-D-glucuronidase reactions in culture media containing fermentable carbohydrate. *Lett. Appl. Microbiol* **2010**, *50*, 547–551.
42. Schets, FM; Nobel, PJ; Strating, S; Mooijman, KA; Engels, GB; Brouwer, A. EU Drinking Water Directive reference methods for enumeration of total coliforms and *Escherichia coli* compared with alternative methods. *Lett. Appl. Microbiol* **2002**, *34*, 227–231.
43. Field, KG; Bernhard, AE; Brodeur, TJ. Molecular approaches to microbiological monitoring: fecal source detection. *Environ. Monit. Assess* **2003**, *81*, 313–326.
44. Moriñigo, MA; Wheeler, D; Berry, C; Jones, C; Muñoz, MA; Cornax, R; Borrego, JJ. Evaluation of different bacteriophage groups as faecal indicators in contaminated natural waters in Southern England. *Water Res* **1992**, *26*, 267–271.
45. Leclerc, H; Edberg, S; Pierzo, V; Delattre, JM. Bacteriophages as indicators of enteric viruses and public health risk in groundwaters. *J. Appl. Microbiol* **2000**, *88*, 5–21.
46. Borrego, JJ. Rana, BC, Ed.; Coliphages as faecal pollution indicators. In *Pollution and Biomonitoring*; McGraw-Hill: New York, NY, USA, 1995; pp. 117–159.

47. Havelaar, AH; Pot-Hogeboom, WH. F-specific RNA-bacteriophages as model viruses in water hygiene: ecological aspects. *Water Sci. Technol* **1988**, *20*, 399–407.
48. Jofre, J; Bosch, A; Lucena, F; Girones, R; Tartera, C. Evaluation of *Bacteroides fragilis* bacteriophages as indicators of the virological quality of water. *Water Sci. Technol* **1986**, *18*, 167–177.
49. Duran, AE; Muniesa, M; Moce-Llivina, L; Campos, C; Jofre, J; Lucena, F. Usefulness of different groups of bacteriophages as model microorganisms for evaluating chlorination. *J. Appl. Microbiol* **2003**, *95*, 29–37.
50. Borrego, JJ; Moriñigo, MA; de Vicente, A; Cornax, R; Romero, P. Coliphages as an indicator of faecal pollution in water. Its relationship with indicator and pathogenic microorganisms. *Water Res* **1987**, *21*, 1473–1480.
51. Paul, JH; Rose, JB; Brown, J; Shinn, EA; Miller, S; Farrah, SR. Viral tracer studies indicate contamination of marine waters by sewage disposal practices in Key Largo, Florida. *Appl. Environ. Microbiol* **1995**, *61*, 2230–2234.
52. Harvey, RW. Microorganisms as tracers in ground water injection and recovery experiments: A review. *FEMS Microbiol. Rev* **1997**, *20*, 461–472.
53. Hot, D; Legeay, O; Jacques, J; Gantzer, C; Caudrelier, Y; Guyard, K; Lange, M; Andreoletti, L. Detection of somatic phages, infectious enteroviruses and enterovirus genomes as indicators of human enteric viral pollution in surface waters. *Water Res* **2003**, *37*, 4703–4710.
54. Havelaar, AH; Furuse, K; Hogeboom, WH. Bacteriophages and indicator bacteria in human and animal faeces. *J. Appl. Bacteriol* **1986**, *60*, 252–262.
55. Cornax, R; Moriñigo, MA; Gonzalez-Jaen, F; Alonso, MC; Borrego, JJ. Bacteriophages presence in human faeces of healthy subjects and patients with gastrointestinal disturbances. *Zentbl. Bakteriol* **1994**, *281*, 214–224.
56. DeBartolomeis, J; Cabelli, VP. Evaluation of an *Escherichia coli* host strain for enumeration of F male-specific bacteriophages. *Appl. Environ. Microbiol* **1991**, *57*, 1301–1305.
57. Lucena, F; Lasofras, J; McIntosh, D; Forcadell, M; Jofre, J. Effect of distance from the pollution focus on relative concentrations of *Bacteroides fragilis* phages and coliphages in mussels. *Appl. Environ. Microbiol* **1994**, *60*, 2272–2277.
58. Cornax, R; Moriñigo, MA; Paez, IG; Muñoz, MA; Borrego, JJ. Application of direct plaque assay for detection and enumeration of bacteriophages of *Bacteroides fragilis* from contaminated-water samples. *Appl. Environ. Microbiol* **1990**, *56*, 3170–3173.
59. Cornax, R; Moriñigo, MA; Balebona, MC; Castro, D; Borrego, JJ. Significance of several bacteriophage groups as indicators of sewage pollution in marine waters. *Water Res* **1991**, *25*, 673–678.
60. Reynolds, KA; Mena, KD; Gerba, CP. Risk of waterborne illness via drinking water in the United States. *Rev. Environ. Contam. Toxicol* **2008**, *192*, 117–58.
61. Risebro, HL; Hunter, PR. Surveillance of waterborne disease in European member states: a qualitative study. *J. Water Health* **2007**, *5*, 19–38.
62. Karanis, P; Kourenti, C; Smith, H. Waterborne transmission of protozoan parasites: A worldwide review of outbreaks and lessons learnt. *J. Water Health* **2007**, *5*, 1–18.
63. Swerdlow, DL; Woodruff, BA; Brady, RC; Griffin, PM; Tippen, S; Donnell, HD; Geldreich, E; Payne, BJ; Meyer, A; Wells, JG; *et al.* A waterborne outbreak in Missouri of *Escherichia coli* O157:H7 associated with bloody diarrhoea and death. *Ann. Intern. Med* **1992**, *117*, 812–819.
64. Fong, TT; Mansfield, LS; Wilson, DL; Schwab, DJ; Molloy, SL; Rose, JB. Massive microbiological groundwater contamination associated with a waterborne outbreak in Lake Erie, South Bass Island, Ohio. *Environ. Health Perspect* **2007**, *115*, 856–864.
65. O'Reilly, CE; Bowen, AB; Perez, NE; Sarisky, JP; Shepherd, CA; Miller, MD; Hubbard, BC; Herring, M; Buchanan, SD; Fitzgerald, CC; Hill, V; Arrowood, MJ; Xiao, LX; Hoekstra, RM; Mintz, ED; Lynch, MF. A waterborne outbreak of gastroenteritis with multiple etiologies among resort island visitors and residents: Ohio, 2004. *Clin. Infect. Dis* **2007**, *44*, 506–512.
66. Figueras, MJ; Collado, L; Guarro, J. A new 16S rDNA-RFLP method for the discrimination of the accepted species of *Arcobacter*. *Diagn. Microbiol. Infect. Dis* **2008**, *62*, 11–15.

67. Collado, L; Inza, I; Guarro, J; Figueras, MJ. Presence of *Arcobacter* spp. in environmental waters correlates with high levels of fecal pollution. *Environ. Microbiol* **2008**, *10*, 1635–1640.
68. Collado, L; Kasimir, G; Perez, U; Bosch, A; Pinto, R; Saucedo, G; Huguet, JM; Figueras, MJ. Occurrence and diversity of *Arcobacter* spp. along the Llobregat River catchment, at sewage effluents and in a drinking water treatment plant. *Water Res* **2010**, *44*, 3696–3702.
69. European Food Safety Authority (EFSA), European Centre for Disease Prevention and Control (ECDC), The Community Summary Report. In *Food-borne Outbreaks in the European Union in 2007*; The EFSA Journal: Brussels, Belgium, 2009; pp. 107–108.
70. Kopilovic, B; Ucakar, V; Koren, N; Krek, M; Kraigher, A. Waterborne outbreak of acute gastroenteritis in a costal area in Slovenia in June and July 2008. *Eurosurveillance* **2008**, *13*, 1–3.
71. Risebro, HL; Doria, MF; Andersson, Y; Medema, G; Osborn, K; Schlosser, O; Hunter, PR. Fault tree analysis of the causes of waterborne outbreaks. *J. Water Health* **2007**, *5*, 1–18.
72. Auld, H; MacIver, D; Klaassen, J. Heavy rainfall and waterborne disease outbreaks: the Walkerton example. *J. Toxicol. Environ. Health. A* **2004**, *67*, 1879–1887.
73. Nichols, G; Lane, C; Asgari, N; Verlander, NQ; Charlett, A. Rainfall and outbreaks of drinking water related disease and in England and Wales. *J. Water Health* **2009**, *7*, 1–8.
74. Kistemann, T; Classen, T; Koch, C; Dangendorf, F; Fischeder, R; Gebel, J; Vacata, V; Exner, M. Microbial load of drinking water reservoir tributaries during extreme rainfall and runoff. *Appl. Environ. Microbiol* **2002**, *68*, 2188–2197.
75. Thurston-Enriquez, JA; Gilley, JE; Eghball, B. Microbial quality of runoff following land application of cattle manure and swine slurry. *J. Water Health* **2005**, *3*, 157–171.
76. Mons, C; Dumetre, A; Gosselin, S; Galliot, C; Moulin, L. Monitoring of *Cryptosporidium* and *Giardia* river contamination in Paris area. *Water Res* **2009**, *43*, 211–217.
77. Howe, AD; Forster, S; Morton, S; Marshall, R; Osborn, KS; Wright, P; Hunter, PR. *Cryptosporidium* oocysts in a water supply associated with a cryptosporidiosis outbreak. *Emerg. Infect. Dis* **2002**, *8*, 619–624.
78. Garg, AX; Pope, JE; Thiessen-Philbrook, H; Clark, WF; Ouimet, J. Walkerton health study investigators. Arthritis risk after acute bacterial gastroenteritis. *Rheumatology* **2008**, *47*, 200–204.
79. Thabane, M; Simunovic, M; Akhtar-Danesh, N; Garg, AX; Clark, WF; Collins, SM; Salvadori, M; Marshall, JK. An outbreak of acute bacterial gastroenteritis is associated with an increased incidence of irritable bowel syndrome in children. *Am. J. Gastroenterol* **2010**, *105*, 933–939.
80. Holme, R. Drinking water contamination in Walkerton, Ontario: positive resolutions from a tragic event. *Water Sci. Technol* **2003**, *47*, 1–6.
81. Lake, IR; Nichols, G; Bentham, G; Harrison, FC; Hunter, PR; Kovats, SR. Cryptosporidiosis decline after regulation, England and Wales, 1989-2005. *Emerg. Infect. Dis* **2007**, *13*, 623–625.
82. Payment, P; Franco, E. *Clostridium perfringens* and somatic coliphages as indicators of the efficiency of drinking water treatment for viruses and protozoan cysts. *Appl. Environ. Microbiol* **1993**, *59*, 2418–2424.
83. Chauret, CP; Radzimiski, CZ; Lepull, M; Creason, R; Andrews, RC. Chlorine dioxide inactivation of *Cryptosporidium parvum* oocyts and bacterial spore indicators. *Appl. Environ. Microbiol* **2001**, *67*, 2993–3001.
84. World Health Organization, Assessing Microbial Safety of Drinking Water. In *Improving Approaches and Methods*; IWA Publishing: London, UK, 2003.
85. Bartram, J; Cotruvo, J; Exnerd, M; Fricker, C; Glasmacher, A. Heterotrophic plate count measurement in drinking water safety management. Report of an Expert Meeting Geneva, 24–25 April 2002. *Int. J. Food Microbiol* **2004**, *92*, 241–247.
86. Ribas, F; Perramon, J; Terradillos, A; Frias, J; Lucena, F. The *Pseudomonas* group as indicator of potential regrowth in water distribution system. *J. Appl. Microbiol* **2000**, *88*, 704–710.

87. Toranzos, GA. Bitton, G, Ed.; Pseudomonas. In *Encyclopedia of Environmental Microbiology*; John Wiley & Sons: New York, NY, USA, 2002; pp. 2632–2639.
88. Mena, KD; Gerba, CP. Risk assessment of *Pseudomonas aeruginosa* in water. *Rev. Environ. Contam. Toxicol* **2009**, *201*, 71–115.
89. Dailloux, M; Laurain, C; Weber, M; Hartemann, PH. Water and nontuberculous mycobacteria. *Water Res* **1999**, *33*, 2219–2228.
90. Figueras, MJ. Clinical relevance of *Aeromonas*. *Rev. Med. Microbiol* **2005**, *16*, 145–153.
91. Martin-Carnahan, A; Joseph, SW. Brenner, DJ, Krieg, NR, Staley, JT, Garrity, GM, Eds.; Order XII. Aeromonadales ord. nov. In *Bergey's Manual of systematic Bacteriology*; Springer: NewYork, NY, USA, 2005; Volume 2, Part B, pp. 556–578.
92. Janda, JM; Abbott, SL. The genus *Aeromonas*. Taxonomy, pathogenicity and infection. *Clin. Microbiol. Rev* **2010**, *23*, 35–73.
93. Demarta, A; Küpfer, M; Riegel, P; Harf-Monteil, C; Tonolla, M; Peduzzi, R; Monera, A; Saavedra, MJ; Martínez-Murcia, A. *Aeromonas tecta* sp. nov., isolated from clinical and environmental sources. *Syst. Appl. Microbiol* **2008**, *31*, 278–286.
94. Figueras, MJ; Alperi, A; Saavedra, MJ; Ko, WC; Gonzalo, N; Navarro, M; Martínez-Murcia, AJ. Clinical relevance of the recently described species *Aeromonas aquariorum*. *J. Clin. Microbiol* **2009**, *47*, 3742–3746.
95. Alperi, A; Martínez-Murcia, AJ; Ko, WC; Monera, A; Saavedra, MJ; Figueras, MJ. *Aeromonas taiwanensis* sp. nov. and *Aeromonas sanarellii* sp. nov., clinical species from Taiwan. *Int. J. Syst. Evol. Microbiol* **2010**, *60*, 2048–2055.
96. Figueras, MJ; Alperi, A; Beaz-Hidalgo, R; Stackebrandt, E; Brambilla, E; Monera, A; Martínez-Murcia, AJ. *Aeromonas rivuli* sp. nov. isolated from the upstream region of a karst water rivulet in Germany. *Int J Syst Evol Microbiol* **2010**.
97. Bomo, AM; Storey, MV; Ashbolt, NJ. Detection, integration and persistence of aeromonads in water distribution pipe biofilms. *J. Water Health* **2004**, *2*, 83–96.
98. Figueras, MJ; Horneman, AJ; Martinez-Murcia, A; Guarro, J. Controversial data on the association of *Aeromonas* with diarrhoea in a recent Hong Kong study. *J. Med. Microbiol* **2007**, *56*, 996–998.
99. Figueras, MJ; Aldea, MJ; Fernández, N; Aspíroz, C; Alperi, A; Guarro, J. *Aeromonas* hemolytic uremic syndrome. A case and a review of the literature. *Diagn. Microbiol. Infect. Dis* **2007**, *58*, 231–234.
100. Chacon, MR; Soler, L; Groisman, EA; Guarro, J; Figueras, MJ. Type III secretion system genes in clinical *Aeromonas* isolates. *J. Clin. Microbiol* **2004**, *42*, 1285–12877.
101. Figueras, MJ; Suarez-Franquet, A; Chacon, MR; Soler, L; Navarro, M; Alejandre, G; Grasa, B; Martinez-Murcia, AJ; Guarro, J. First record of the rare species *Aeromonas culicicola* from a drinking water supply. *Appl. Environ. Microbiol* **2005**, *71*, 538–541.
102. Pablos, M; Rodríguez-Calleja, JM; Santos, JA; Otero, A; García-López, ML. Occurrence of motile *Aeromonas* in municipal drinking water and distribution of genes encoding virulence factors. *Int. J. Food. Microbiol* **2009**, *135*, 158–164.
103. Pablos, M; Remacha, MA; Rodríguez-Calleja, JM; Santos, JA; Otero, A; García-López, ML. Identity, virulence genes, and clonal relatedness of *Aeromonas* isolates from patients with diarrhea and drinking water. *Eur. J. Clin. Microbiol. Infect. Dis* **2010**, *29*, 1163–1172.
104. Yu, CP; Chu, KH. Molecular quantification of virulence gene-containing *Aeromonas* in water samples collected from different drinking water treatment processes. *Environ Monit Assess* **2010**.
105. Alperi, A; Figueras, MJ. Human isolates of *Aeromonas* possess Shiga toxin genes (*stx1* and *stx2*) highly similar to the most virulent gene variants of *Escherichia coli*. *Clin. Microbiol. Infect* **2010**, *16*, 1563–1567.
106. Figueras, MJ; Beaz-Hidalgo, R; Senderovich, Y; Laviad, S; Halpern, M. Re-identification of *Aeromonas* isolates from chironomid egg masses as the potential pathogenic bacteria. *Aeromonas aquariorum Environ Microbiol Rep* **2010**.
107. Pina, S; Puig, M; Lucena, F; Jofre, J; Girones, R. Viral pollution in the environment and in shellfish: human adenovirus detection by PCR as an index of human viruses. *Appl. Environ. Microbiol* **1998**, *64*, 3376–3382.

108. Albinana-Gimenez, N; Miagostovich, MP; Calgua, B; Huguet, JM; Matia, L; Girones, R. Analysis of adenoviruses and polyomaviruses quantified by qPCR as indicators of water quality in source and drinking-water treatment plants. *Water Res* **2009**, *43*, 2011–2019.
109. Kopecka, H; Dubrou, S; Prevot, J; Marechal, J; Lopez-Pila, JM. Detection of naturally occurring enteroviruses in waters by reverse transcription, polymerase chain reaction, and hybridization. *Appl. Environ. Microbiol* **1993**, *59*, 1213–1219.
110. Metcalf, TG; Melnick, JL; Estes, MK. Environmental virology: from detection of virus in sewage and water by isolation to identification by molecular biology—a trip of over 50 years. *Annu. Rev. Microbiol* **1995**, *49*, 461–487.
111. Haramoto, E; Katayama, H; Oguma, K; Yamashita, H; Tajima, A; Nakajima, H; Ohgaki, S. Seasonal profiles of human noroviruses and indicator bacteria in a wastewater treatment plant in Tokyo, Japan. *Water Sci. Technol* **2006**, *54*, 301–308.
112. Griffin, JS; Plummer, JD; Long, SC. Torque teno virus: an improved indicator for viral pathogens in drinking waters. *Virol. J* **2008**, *5*, 112.
113. Huq, A; Rivera, ING; Colwell, RR. Colwell, RR, Grimes, DJ, Eds.; Epidemiological significance of viable but nonculturable microorganisms. In *Nonculturable Microorganisms in the Environment*; ASM Press: Washington, DC, USA, 2000; pp. 301–323.
114. Kahlisch, L; Henne, K; Groebe, L; Draheim, J; Höfle, MG; Brettar, I. Molecular analysis of the bacterial drinking water community with respect to live/dead status. *Water Sci. Technol* **2010**, *61*, 9–14.
115. Cordoba, MA; Del Coco, VF; Minvielle, MC; Basualdo, JA. Influencing factors in the occurrence of injured coliforms in the drinking water distribution system in the city of La Plata, Argentina. *J. Water Health* **2010**, *8*, 205–211.
116. McFeters, GA. McFeters, GA, Ed.; Enumeration, occurrence and significance of injured indicator bacteria in drinking water. In *Drinking Water Microbiology*; Springer-Verlag: New York, NY, USA, 1990; pp. 478–492.
117. Bucklin, KE; McFeters, GA; Amirtharaja, A. Penetration of coliforms through municipal drinking water filters. *Water Res* **1991**, *25*, 1013–1017.
118. Gunnarsdóttir, MJ; Gissurarson, LR. HACCP and water safety plans in Icelandic water supply: preliminary evaluation of experience. *J. Water Health* **2008**, *6*, 377–382.
119. Ferguson, CM; Croke, BF; Beatson, PJ; Ashbolt, NJ; Deere, DA. Development of a process-based model to predict pathogen budgets for the Sydney drinking water catchment. *J. Water Health* **2007**, *5*, 187–208.
120. Delpla, I; Jung, AV; Baures, E; Clement, M; Thomas, O. Impacts of climate change on surface water quality in relation to drinking water production. *Environ. Int* **2009**, *35*, 1225–1233.
121. Van Vliet, MTH; Zwolsman, JJG. Impact of summer droughts on the water quality of the Meuse river. *J. Hydrol* **2008**, *353*, 1–17.
122. Prathumratana, L; Sthiannopkao, S; Kim, KW. The relationship of climatic and hydrological parameters to surface water quality in the lower Mekong River. *Environ. Int* **2008**, *34*, 860–866.
123. Brunetti, M; Maugeri, M; Nanni, T. Changes in total precipitation rainy days and extreme events in northeastern Italy. *Int. J. Climatol* **2001**, *21*, 861–871.
124. Evans, CD; Monteith, DT; Cooper, DM. Long-term increases in surface water dissolved organic carbon: observations, possible causes and environmental impacts. *Environ. Pollut* **2005**, *137*, 55–71.
125. Nichols, G; Lane, C; Asgari, N; Verlander, NQ; Charlett, A. Rainfall and outbreaks of drinking water related disease and in England and Wales. *J. Water Health* **2009**, *7*, 1–8.
126. Soh, YC; Roddick, F; van Leeuwen, J. The future of water in Australia: the potential effects of climate change and ozone depletion on Australian water quality, quantity and treatability. *Environmentalist* **2008**, *28*, 158–165.
127. Charron, DF; Thomas, MK; Waltner-Toews, D; Aramini, JJ; Edge, T; Kent, RA. Vulnerability of waterborne diseases to climate change in Canada: a review. *J. Toxicol. Environ. Health Part A* **2004**, *67*, 1667–1677.
128. Pednekar, AM; Grant, SB; Jeong, Y; Poon, Y; Oancea, C. Influence of climate change, tidal mixing, and watershed urbanization on historical water quality in Newport Bay, a

saltwater wetland and tidal embayment in southern California. *Environ. Sci. Technol* **2005**, *39*, 9071–9082.
129. Hunter, PR. Climate change and waterborne and vector-borne disease. *J. Appl. Microbiol* **2003**, *94*, 37S–46S.
130. Curriero, FC; Patz, JA; Rose, JB; Lele, S. The association between extreme precipitation and waterborne disease outbreaks in the United States, 1948–1994. *Am. J. Public Health* **2001**, *91*, 1194–1199.
131. Daufresne, M; Boët, P. Climate change impacts on structure and diversity of fish communities in rivers. *Global Change Biol* **2007**, *13*, 2467–2478.
132. Verheyen, J; Timmen-Wego, M; Laudien, R; Bousaad, I; Sen, S; Koc, A; Uesbeck, A; Mazou, F; Pfister, H. Detection of adenoviruses and rotaviruses in drinking water sources used in rural areas of Benin, West Africa. *Appl. Environ. Microbiol* **2009**, *75*, 2798–2801.
133. Bosch, A; Guix, S; Sano, D; Pinto, RM. New tools for the study and direct surveillance of viral pathogens in water. *Curr. Opin. Biotechnol* **2008**, *19*, 295–301.

CHAPTER 2

Review on Optimization of Drinking Water Treatment Process

Mohamed Farhaoui[1,2], Mustapha Derraz[2]

[1]*National Office of Electricity and Drinking Water, Meknes, Morocco*

[2]*Biology Department, University of Moulay Ismail, Meknes, Morocco*

ABSTRACT

In the drinking water treatment processes, the optimization of the treatment is an issue of particular concern. In general, the process consists of many units as settling, coagulation, flocculation, sedimentation, filtration and disinfection. The optimization of the process consists of some measures to decrease the managing and monitoring expenses and improve the quality of the produced water. The objective of this study is to provide water treatment operators with methods and practices that enable to attain the most effective use of the facility and in consequence optimize the cubic meter price of the treated water. This paper proposes a review on optimization of drinking water treatment process by analyzing all of the water treatment units and gives some solutions in order to maximize the water treatment performances without compromising the water quality standards. Some practical solutions and methods are performed in the water treatment plant located in the middle of Morocco (Meknes).

Keywords: Optimization, Coagulation Process, Turbidity Removal, Water Treatment

1. INTRODUCTION

The demand on water supply is increasing over the last century due to improved lifestyle, industrial development and population growth. This increased demand is facing a paradox to produce treated water with high quality at lower cost. In order to reduce the water cost, it is very important to optimize the operating expenses in the water treatment plant (power, chemicals, operator's expenses...) and many measures should be performed in this vision.

The optimization of water treatment plant is not a disciplinary to maximize the treatment objective and minimize the cost of the water produced. But it consists of understanding the functioning of the treatment plant and treasure the experience of the operators in dealing with all treatment process related to different aspects.

This paper addresses the problem of optimizing of the conventional drinking water plant and proposes some practical to reach this goal. This paper is organized as fellows. After an introduction of the objective of this study, the water treatment operation is described in Section 2. Review on optimization of conventional drinking water treatment plant is discussed in Section 3. In Section 4, some practical methods are explained.

2. WATER TREATMENT OPERATION

Meknes region is located in the middle of Moroccan Kingdom, has a Mediterranean climate with continental influences. The temperature shifts from cool and cold in winter to hot days in the summer months of June - September. The agriculture and industry are the mean activities in the region.

This study was developed in a water treatment plant located in Meknes, whose source is two big springs Bittit (630 l/s) and Ribaa (400 l/s). The quality of water, produced by the spring, changes according to the rainfall in the region. Sometimes, it can be affected by the snow in the Atlas Mountains. The treatment water plant, as part of other water resources, water to more than 700,000 inhabitants of Meknes city, and it has a nominal capacity of 600 l/s of treated water. This water treatment plant is chosen regarding to the variation of the raw water turbidity and because of the availability of data. Figure 1 presents a schematic overview of the various operations necessary to treat the water.

Many measurements of variables recorded by streaming current detectors such as: turbidity level, PH, conductivity, temperature is needed to carry out the jars test in order to determine the optimal dose of the aluminum sulfate. Table 1 shows the raw water variables used in this study, and they present the following variation intervals.

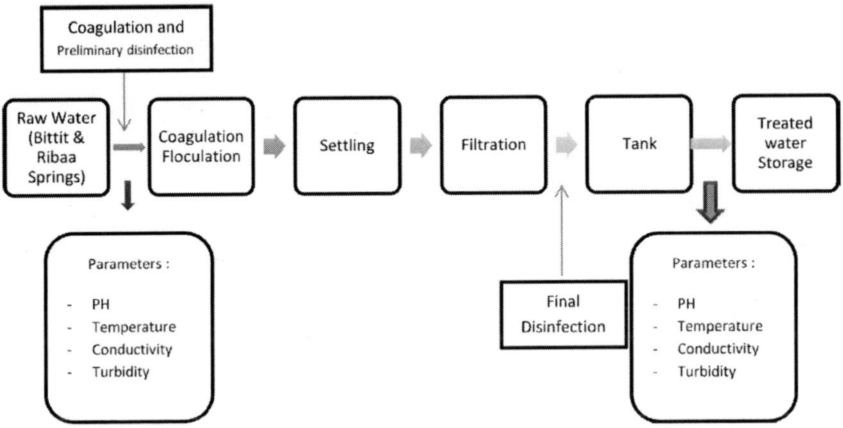

Figure 1. Simplified synopsis of the water treatment plant.

Table 1. Statistical summary of raw water conditions from 01/01/2013 to 31/12/2015 (ONEE, 2015).

Variables	Min	Max
Turbidity: Bittit (NTU)	1.7	850
Turbidity: Ribaa (NTU)	1.62	960
PH	6.80	7.74
Temperature: (°C)	14	24.70
Conductivity micro s/cm	509	624

The variation of PH, conductivity and temperature from 01/01/2013 to 31/12/2015 is given by Figures 2(a)-(c).

In the rainfall period, the turbidity of raw water changes from time to time as shown in the graph below, the turbidity of the raw water can increase to reach levels more than 500 NTU as shown in Figure 3.

However, the turbidity level is less than 10 NTU this three last years (2013, 2014 and 2015) for more than 88% of the year and more than 64% of the year (1637/2556 days); the turbidity is less than 10 NTU for the six last years as shown by Table 2.

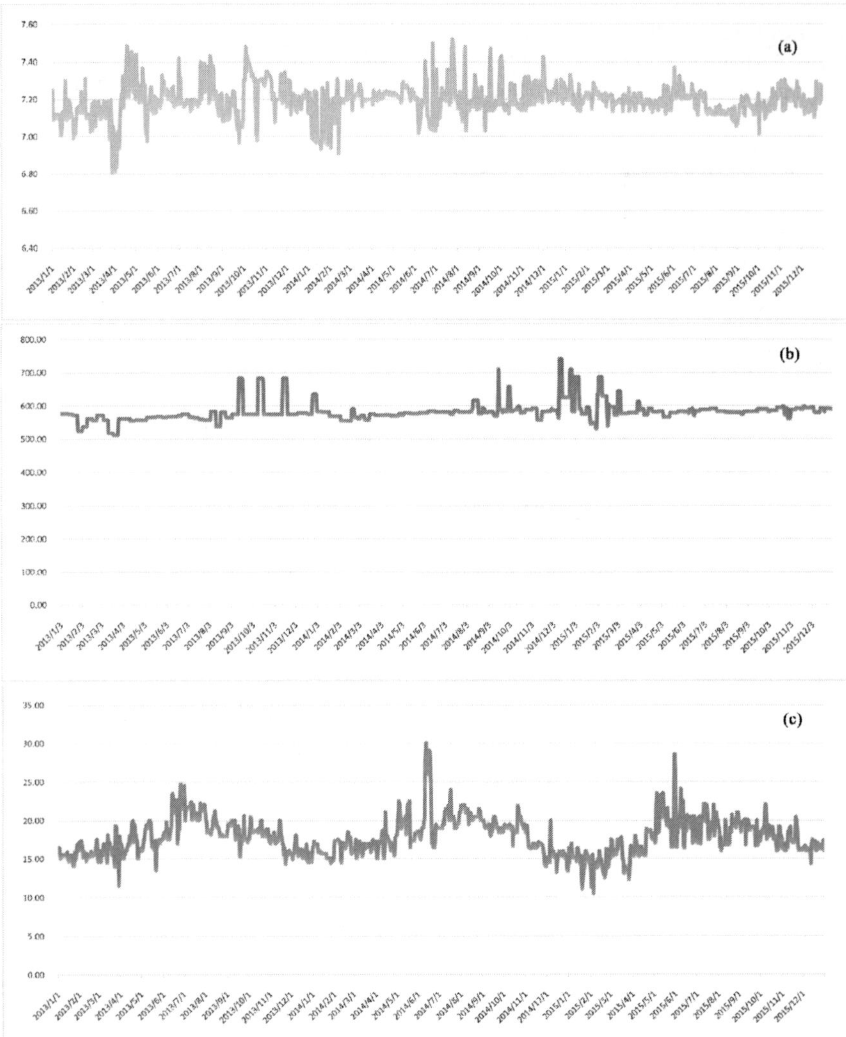

Figure 2. Evolution of PH. (a) conductivity in micro S/cm; (b) and temperature in °C; (c) of raw water from 01/01/2013 to 31/12/2015.

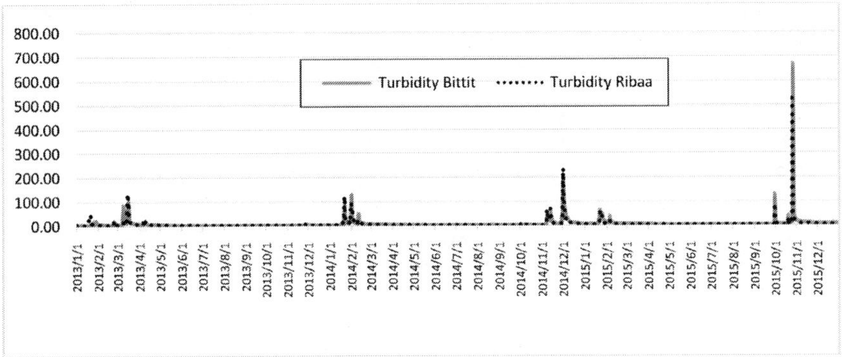

Figure 3. Statistical data of turbidity level of the spring's water from 01/01/2013 to 31/12/2015 (ONEE, 2015).

Table 2. Turbidity levels distribution from 2009 to 2015 (number of days per turbidity level).

Year	Number of days					Total
	Turbidity less or equal than 5 NTU	Turbidity more than 5and less or equal than 10 NTU	Turbidity more than 10 and less or equal than 20 NTU	Turbidity more than 20 and less or equal than 40 NTU	Turbidity more than 40 NTU	
2009	147	60	101	34	23	365
2010	0	0	113	148	104	365
2011	0	132	144	59	30	365
2012	301	38	17	5	5	366
2013	260	74	23	8	0	365
2014	247	62	32	20	4	365
2015	184	132	34	10	5	365
Total	1139	498	464	284	171	2556
Aggregated data	1139	1637	2101	2385	2556	
Percentage	**45%**	**19%**	**18%**	**11%**	**7%**	
Percentage of Aggregated data	**45%**	**64%**	**82%**	**93%**	**100%**	

Table 3 gives the max and min of raw water turbidity by month from 2013 to 2015.

The chemicals used in the water treatment process consume about 50% the total operating expenses of the water treatment. The energy cost is between 10% and 15% related to the total cost in 2013, 2014 and 2015. Figure 4 gives the operations expenses of the water treatment plant in the three last years.

In addition, used as coagulant, the aluminum sulfate (SA) consumption is more than 65% of the total chemicals consumption in the water treatment plant. Le polyelectrolyte (Poly) consumption is less than 10% and the chlorine is between 16% and 26% of the total chemicals used in the water treatment plant according to the water quality as shown by Figure 5.

Table 3. Statistical data of turbidity min and max (NTU) measured in 2013, 2014 and 2015 per month (ONEE, 2015).

Year: 2013				
	Bittit Spring		Ribaa Spring	
Month	min	max	min	max
January	3.70	18.95	10.00	50.00
February	4.50	19.00	3.95	20.65
March	4.95	98.40	4.50	136.66
April	6.40	21.95	4.90	32.77
May	3.93	6.30	3.80	5.30
June	3.72	4.40	3.34	4.89
July	3.30	4.33	2.92	3.62
August	3.00	3.80	2.70	3.46
September	3.10	4.64	2.68	4.15
October	2.90	3.99	2.46	3.62
November	2.70	7.44	2.90	10.00
December	3.07	4.23	2.60	3.95
Year: 2014				
	Bittit Spring		Ribaa Spring	
	min	max	min	max
January	3.06	81.69	3.00	120.66
February	6.48	128.00	5.93	78.90
March	4.86	6.91	3.90	5.65
April	4.30	4.90	3.30	4.19
May	3.17	4.77	2.48	3.37
June	3.00	3.86	2.60	2.87
July	2.36	4.50	2.16	2.49
August	1.74	4.20	1.88	2.28
September	1.80	3.50	1.64	1.90
October	1.89	4.20	1.67	11.64
November	1.80	55.50	1.68	69.29
December	6.57	197.50	5.06	235.89
Year: 2015				
	Bittit Spring		Ribaa Spring	
	min	max	min	max
January	5.09	63.10	3.96	61.59
February	6.33	38.52	3.98	19.45
March	4.97	6.27	3.26	4.64
April	3.65	5.60	2.80	3.76
May	3.50	5.23	2.68	3.06
June	3.40	4.52	2.61	3.19
July	2.80	4.75	2.40	3.60
August	2.60	3.26	2.45	2.74
September	2.70	128.45	2.35	86.63
October	4.35	669.71	3.22	524.03
November	6.76	17.18	3.99	13.19
December	7.18	9.54	3.42	4.02

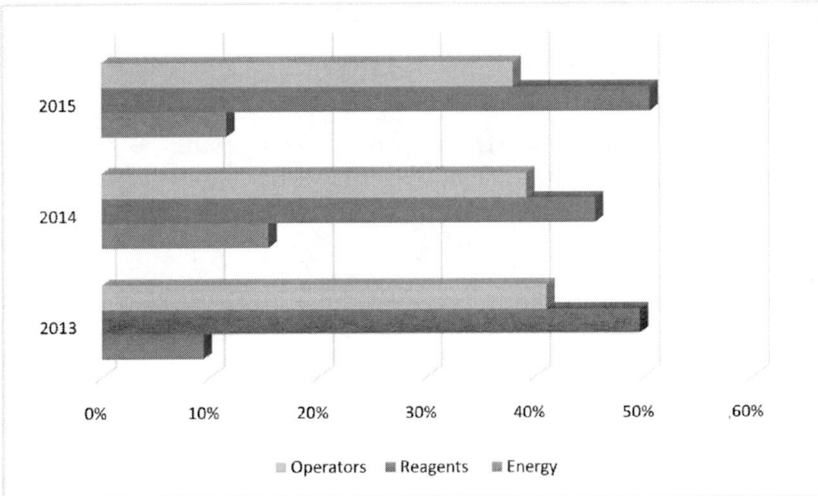

Figure 4. Operations expenses of the water treatment plant in 2013, 2014 and 2015 (ONEE, 2015).

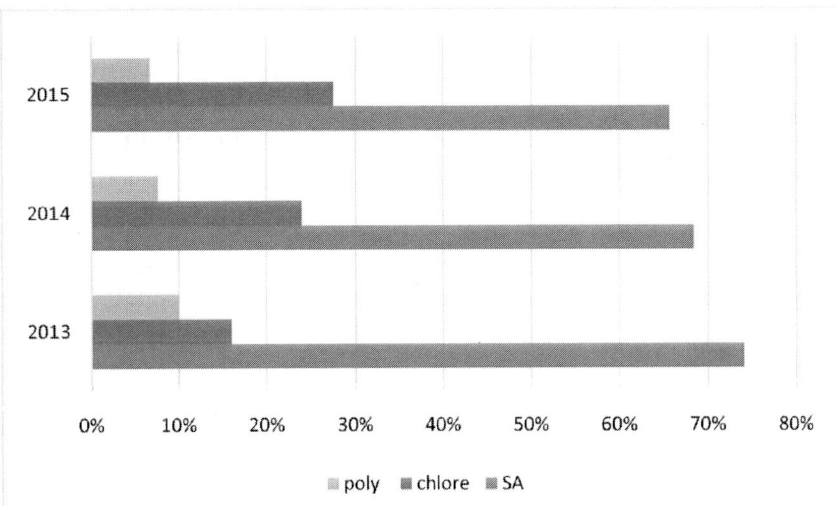

Figure 5. Percentage of chemicals expenses consumed by the water treatment plant in 2013, 2014 and 2015 (ONEE, 2015).

3. REVIEW ON OPTIMIZATION OF DRINKING WATER TREATMENT PLANT

In order to remove turbidity and contaminants from raw water, the treatment comprises many steps as the settling, aeration, coagulation, sedimentation, filtration and disinfection. The optimization of water treatment plant

concerns the different steps of the process. The main goals are to:

- Achieve the chemical cost performance.
- Manage and control the energy consumption.
- Reduce the water wastes.

Many research and studies are performed in order to increase the water treatment plant performances. In this section, one aspect related to the optimization is treated which is the coagulant consumption.

3.1. Coagulant and Coagulant Aid

In the water treatment plant, colloidal particles are separated by means of a chemical coagulation process: consisting in the charge destabilization of the suspended particles by adding coagulant. The coagulant used is aluminum sulfate; it is the most widely used coagulant in Morocco as well as many other countries in the drinking water industry. It is mainly used because of its effectiveness, accessibility and low price. As a common practice, aluminum sulfate is applied according to the jars test results. The main difficulty is how to optimize the aluminum sulfate dosage related to raw water characteristics by using other cheaper products. Some attempts have been made to improve the effectiveness of the aluminum sulfate or to substitute this coagulant by another natural, available and cheaper. Finding of various coagulation processes have been reported in literature. Some of these include; studying the effect of using the bentonite on the coagulation in the treatment of low turbidity [1]. Also, Mukheled (2012) [2]used Date seeds and Pollen Sheath as coagulant to treat different levels of turbidity (75, 150 and 300 NTU). Eman et al. (2010) [3]tried Moringa Oleifera seed in the coagulation process to treat low turbid water in Malaysia. However, Aho et al. (2012) [4]highly recommended the use of this natural coagulant in the domestic turbid water purification in Nigeria. Other natural coagulants are proposed as an important alternative in the water treatment plant. The coagulants are from plant origin such as nirmali seed and maize [5], cassia angustifolia seed [6], mesquite bean and cactus latifaria [7], chestnut and acorn [8], Cocciniaindica fruit mucilage [9]and from different leguminous species [10]. Also, the Bhindi seed, Methi, Beheda, Guar seeds and Drum stick can be used as coagulant in 100 to 1200 NTU turbid water range with remarkable percentage removal from 70% to 93% [11]. In the other hand, Cocciniaindica, Strychnospotatorum and Cactus opuntia were used as natural coagulant and improved the quality of the filtered water [12]. Those natural products have coagulating activity in the treatment of turbid water and can be used as coagulant or as coagulant aid with other synthetic and industrial coagulants (aluminum sulfate...) in order to reduce the coagulant consumption in the water treatment plant [13].

3.2. Coagulant Control and Monitoring

As a common practice, aluminum sulfate is applied according to the jars test results. The problem is to determine the optimal dose of aluminum sulfate related to raw water characteristics. Both manual and automatic methods are used to predict optimum coagulant dose [14][15]. Automatic method is ensured by streaming current detectors [16]. However, manual method is consisting to determine the quantity of the coagulant to apply experimentally and based on the jar test results. Jar test involves taking a raw water samples and applying different quantities of coagulant to each sample [15]. After a short period of time, each sample is assessed for water quality and the dosage that produces the optimal result used a set point. This operation should be repeated by the operators each time when the quality of raw water changes. The aluminum sulfate is the compound likely to be mathematically modeled and therefore its value can be estimated according to the data available in the treatment plant. The optimization of using the coagulant is very interesting approach because under dosing of coagulant can lead to poor quality drinking water while too much coagulant leads to many operating problems (less efficient filtration and sedimentation, PH), healthy problems and can increase the cost of treated water.

Some attempts have been made to model the relationships between raw water quality characteristics and the optima coagulant dosage rate.

4. PRACTICAL METHODS OF WATER TREATMENT PLANT OPTIMIZATION

4.1. Understanding of the Water Treatment Process

The understanding of the water treatment functioning is a very important step on the performances optimization program.

A systematic gathering of information about plant performances are need such as:

- Data trending and analysis.
- Check plant design criteria against actual and the controlling operations to improve the plant performances.
- Track chemical dosing and use indicators.
- Field measurements and visual observation.
- Collaboration of operators (quality water operators, maintenance team and managing operator).
- Plant specific operating guidelines and procedures.
- Accuracy and control of the on-line measurement devices.

4.2. The Aluminum Sulfate Consumption

The sludge produced in the settling step of the treatment process is used not only to reduce the aluminum sulfate dose and the sludge volume but also in order to improve the water quality.

A standard jar test apparatus equipped with six paddles rotating in a set of six beakers is used to simulate coagulation, flocculation and sedimentation processes. At the first, Control experiments for coagulation tests are performed in order to determine the optimal dose of the aluminum sulfate in normal conditions. The selected level of turbid water (1 L) is filled into the beakers and various doses in the range from 10 to 100 mg/l of sludge and the aluminum sulfate according to the results of the first jar test determining the optimal dose of the inorganic coagulant in normal conditions are separately added in the beakers and mixed rapidly (300 rpm) for one minute. The mixing speed was then reduced to 40 rpm for 20 minutes. Then the stirrer is turned off and the suspensions are allowed to settle for different periods of time ranging from 30 to 120 minutes under quiescent conditions. After each period of settling time, supernatant samples of each beaker in the jar test is withdrawn from located 10 cm below the water level and residual turbidity is measured.

Standard Jar test experiments were performed to determine the optimal dose of aluminum sulfate. A set of experiments were performed using jar test to investigate the optimal dose of sludge to add in order to improve the quality of water.

The sludge produced by the treatment plant and used with the aluminum sulfate improves the turbidity removal percentage and achieve good removal efficiencies. The maximum value of turbidity removal at turbidity of 10 NTU is 98.41 corresponding to the optimum dose of sludge of 35 mg/l.

A set of experiments is performed using the optimal sludge dose obtained from the experiments below (35 mg/l) while the aluminum sulfate was changed from 0 to 10 mg/l. Figure 6 shows the effect of aluminum sulfate and sludge dose under various settling time on turbidity removal.

Table 4 gives the results of turbidity removal percentage using aluminum sulfate only, optimal dose of AS and sludge and the optimal dose of sludge and the proposed dose of AS.

The results show that the sludge used as coagulant aid with the aluminum sulfate improve not only the water quality produced by the treatment plant (turbidity removal percentage from 96.71% to 97.56% but also it can be used to reduce the aluminum sulfate dose in the coagulation process by 40%.

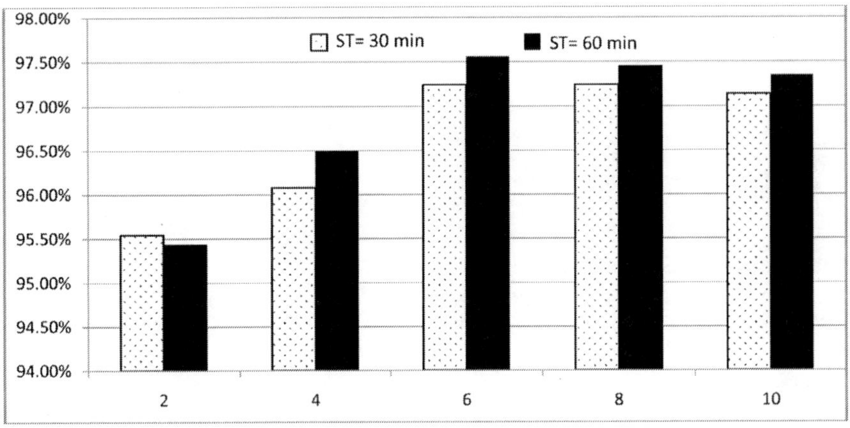

Figure 6. Effect of Aluminum sulfate doses (mg/l) and sludge dose (35 mg/l) under various settling time (ST) on turbidity removal percent (%).

The chlorine is the second chemical which impacts the cost of the treated water. The optimization of the consumption of this chemical can reduce the cost of the water produced by the water treatment plant. Some measures should be taken such as:

- On-line and laboratory instruments verification.
- Disinfection data handling and reporting.
- Verification of on-line chlorine residual analyzes.

- Verification of the performances of the on-line instruments with a grab sample measurement at least once every week.
- Respect of the manufacturer's instructions for instrument calibration.
- Completion of grab sample verification after changing reagents or doses or doing any other maintenance operations on the instrument.
- Incorporation of all manufacturer recommended instrument maintenance.
- Using a portable kit to verify performances of on-line instruments.

Table 4. Turbidity removal percentage (%) using AS and sludge at settling time of 60 minutes.

	Turbidity = 10 NTU
Initial turbidity	10
Optimal dose of Aluminum sulfate (mg/l) (1)	10
Optimal dose of sludge (mg/l) used within AS (2)	35
Dose of AS (mg/l) proposed to be used with optimal dose of sludge (3)	6
Turbidity removal percentage using only AS (1)	96.71%
Turbidity removal percentage using AS and sludge as coagulant aid (1) + (2)	97.14%
Turbidity removal percentage using optimal dose of sludge and AS (2) + (3)	97.56%

4.4. Energy

The energy consumed by the water treatment plant can be optimized by adopting many measures such as:

- Using of variables speed drives on pump motors.
- Lighting controls upgrade.
- The preventive maintenance of different equipment and motors.
- The replacement of existing pumps and motors by more efficient pumps and motors.

The implementation of such measures results in annual electric savings of more than 6% of the energy consumed by the water treatment plant.

5. CONCLUSION

This paper has presented some results concerning many practical methods used in the management of the water treatment plant, the use of sludge as coagulant aid with the aluminum sulfate in the water treatment plant. The sludge improves the quality of the produced water by increasing the coagulation efficiency and the aluminum sulfate dosage is decreased. The sludge volume is reduced and subsequently sludge management costs. The coagulation process is optimizing and this approach offers several significant economics and operational benefits such as it minimizes the aluminum dose required to 40%, reduces the sludge volume produced by the treatment plant and decreases the residual turbidity of the water. These advantages are very important not only for process economy through reducing the cost of treatment but also in the management of the sludge volume. Therefore, it is reasonable to consider this approach to be applied in the treatment plant for water with similar turbidity level. Also, the chlorine and energy consumption can be reduced in order to optimize the water treatment plant performances.

REFERENCES

1. Shen, Y.H. (2005) Treatment of Low Turbidity Water by Sweep Coagulation Using Bentonite. Journal of Chemical Technology and Biotechnology, 80, 581-586. http://dx.doi.org/10.1002/jctb.1244
2. Mukheled, A. (2012) A Novel Water Pretreatment Approach for Turbidity Removal Using Date Seeds and Pollen Sheath. Journal of Water Resource and Protection, 4, 79-92. http://dx.doi.org/10.4236/jwarp.2012.42010
3. Eman, N.A., Suleyman, A.M., Hamzah, M.S., Zahangir, Md.A. and Salleh, M.R.M. (2010) Production of Natural Coagulant from Moringa oleifera Seed for Application in Treatment of Low Turbidity Water. Journal of Water Resource and Protection, 2, 259-266. http://dx.doi.org/10.4236/jwarp.2010.23030
4. Aho, I.M. and Lagasi, J.E. (2012) A New Water Treatment System Using Moringa oleifera Seed. American Journal of Scietific and Industrial Research, 3, 487-492.
5. Raghuwanshi, P.K., Mandloi, M., Sharma, A.J., Malviya, H.S. and Chaudhari, S. (2002) Improving Filtrate Quality Using Agrobased Materials as Coagulant Aids. Water Quality Research, 37, 745-756.
6. Sanghi, R., Bhatttacharya, B. and Singh, V. (2002) Cassia An-Gustifolia Seed Gum as an Effective Natural Coagulant for Decolourisation of Dye Solutions. Green Chemistry, 4, 252-254.
7. Diaz, A., Rincon, N., Escorihuela, A., Fernandez, N., Chacin, E. and Forster, C.F. (1999) A Preliminary Evaluation of Turbidity Removal by Natural Coagulants Indigenous to Venezuela. Process Biochemistry, 35, 391-395. http://dx.doi.org/10.1016/S0032-9592(99)00085-0
8. Sciban, M., Klasnja, M., Antov, M. and Skrbic, B. (2009) Removal of Water Turbidity by Natural Coagulants Obtained from Chestnut and Acorn. Bioresource Technology, 100, 6639-6643.
9. Patale, V. and Parikh, P. (2010) A Preliminary Study on Coccinia indica Fruit Mucilage Extract as Coagulant-Flocculent for Turbid Water Treatment. Journal of Pure and Applied Sciences, 18, 27-30.
10. Sciban, M., Klasnja, M. and Stojimirovic, J. (2005) Investigation of Coagulation Activity of Natural Coagulants from Seeds of Different Leguminose Species. Acta Periodica Technologica, 36, 81-87.
11. Amitkumar, A.G. and Farooq, I.C. (2015) Performance Evaluation of Some Natural Coagulants. International Journal of Civil and Structural Engineering, 3, 368-375.

12. Nirmala Rani, C. and Jadhav, M.V. (2012) Enhancing Filtrate Quality of Turbid Water Incorporating Seeds of Strychnospotatorum, Pads of Cactus Opuntia and Mucilage Extracted from the Fruits of Coccinia indica as Coagulants. Journal of Environmental Research and Development, 7, 668-674.
13. Farhaoui, M. and Derraz, M. (2016) Optimizing Coagulation Process by Using Sludge Produced in the Water Treatment Plant. Journal of Chemical and Pharmaceutical Research, 8, 749-756.
14. Lind, C. (1994) Coagulation Control and Optimization: Part One. Public Works for October, 56-57.
15. Lind, C. (1994) Coagulation Control and Optimization: Part Two. Public Works for October, 32-33.
16. Liu, H.Y., Rong, Q.W., Xu, L.H. and Zhang, Y. (2013) Application of Automatic Control System in Water Plant in China. Journal of Convergence Information Technology (JCIT), 8, 243-250.

CHAPTER 3

Global Access to Safe Water: Accounting for Water Quality and the Resulting Impact on MDG Progress

Kyle Onda, Joe LoBuglio and Jamie Bartram *

The Water Institute, Gillings School of Global Public Health, University of North Carolina at Chapel Hill, 135 Dauer Drive, CB #7431, Chapel Hill, NC 27599, USA

ABSTRACT

Monitoring of progress towards the Millennium Development Goal (MDG) drinking water target relies on classification of water sources as "improved" or "unimproved" as an indicator for water safety. We adjust the current Joint Monitoring Programme (JMP) estimate by accounting for microbial water quality and sanitary risk using the only-nationally representative water quality data currently available, that from the WHO and UNICEF "Rapid Assessment of Drinking Water Quality". A principal components analysis (PCA) of national environmental and development indicators was used to create models that predicted, for most countries, the proportions of piped and of other-improved water supplies that are faecally contaminated; and of these sources, the proportions that lack basic sanitary protection against contamination. We estimate that 1.8 billion people (28% of the global population) used unsafe water in 2010. The 2010 JMP estimate is that 783 million people (11%) use unimproved sources. Our estimates revise the 1990 baseline from 23% to 37%, and the target from 12% to 18%, resulting in a shortfall of 10% of the global population towards the MDG target in 2010. In contrast, using the indicator "use of an improved source" suggests that the MDG target for drinking-water has already been achieved. We estimate that an additional 1.2 billion (18%) use water from sources or systems with significant sanitary risks. While our estimate is imprecise, the magnitude of the estimate and the health and development implications suggest that greater attention is needed to better understand and manage drinking water safety.

Keywords: water quality; drinking water; water safety; Millennium Development Goals (MDGs); Joint Monitoring Programme (JMP)

1. INTRODUCTION

The United Nations Millennium Development Goals (MDGs) include Target 7c, to halve the "proportion of the population without sustainable access to safe drinking-water" between 1990 and 2015 [1]. The Joint Monitoring Programme for Water Supply and Sanitation (JMP) of the World Health Organization (WHO) and the United Nations Children's Fund (UNICEF) reports progress towards meeting this goal [2].

The corresponding MDG indicator is the "proportion of households using water from an improved source," and is reported on a country-by-country basis [1]. Sources are classified as improved or unimproved as shown in Table 1, according to whether they are "protected from outside contamination" [2].

The MDG indicator thereby conflates access to certain water sources with use of safe water. However, data and monitoring mechanisms regarding the safety of water sources at a national scale when the MDG targets were cast were, and remain, scant. As such, at the time the MDGs were developed, there was no credible alternative approach to an indicator that allows for the calculation of a percentage figure easily aggregated to the country and global scales and amenable for use as a target given the types of data available.

Using this approach, WHO and UNICEF estimate that 5.8 billion people used improved sources in 2010, with 783 million using unimproved water sources [3]. Treating use of an improved source as an indicator for use of safe water is likely to overestimate the population using safe water, since some improved sources may provide water that is microbiologically or chemically contaminated whether at source or by the time it reaches the home and is consumed [4,5]. On the other hand, most unimproved sources do not provide safe drinking water, so under-accounting of safe water coverage due to unimproved sources providing safe water is likely to be small [6,7].

Table 1. JMP Classification of drinking-water source types as improved or unimproved [2].

Source class	Type of source
Unimproved drinking-water source	Unprotected dug well, unprotected spring, cart with small tank or drum, surface water (e.g., river, dam, lake, pond, stream, canal or irrigation channel) and bottled water
Improved drinking-water source (piped to dwelling, plot or yard)	Piped water connection located inside the user's dwelling, plot or yard
Improved drinking-water source (other sources)	Public taps or standpipes, tube wells or boreholes, protected dug wells, protected springs and rainwater collection

In 2010, WHO and UNICEF released data on water quality and sanitary risk (*i.e.*, risk of contamination) associated with improved sources from five countries as part of the Rapid Assessment of Drinking-Water Quality (RADWQ) study which had been undertaken between October 2004 and April 2005 [8,9,10,11,12].

The RADWQ study is the only source of nationally-representative drinking water quality data amenable to analysis for microbial contamination disaggregated by water source type. A recent analysis of these data found that accounting for water source compliance with WHO water quality guidelines significantly reduced the estimates of safe water access relative to the JMP figures, as well as national-level progress towards the MDG target, in four of the five countries [13]. However, this study did not account for sanitary risk, and did not extrapolate beyond the five countries for which data were available.

We use both the water quality and sanitary risk data from the RADWQ studies to extrapolate to other countries, and estimate global figures. We believe our study is the first to use the available nationally representative water safety data to estimate the proportion of the global population with access to safe drinking water and the concomitant impact on estimates of global progress towards MDG Target 7c with a method that accounts for differences between countries in factors affecting water source safety.

2. METHODS

2.1. Drinking Water Source Type Classification

Here we considered "piped" sources as the subset of MDG-classified improved sources that are piped connections to a user's dwelling, plot, or yard, and "other-improved" sources as all other improved drinking water sources as described in Table 1.

2.2. Drinking Water Quality Data

The RADWQ data was collected following the methodology described in Howard *et al.* [14]. The RADWQ data includes a nationally-representative sample of approximately 1,600 improved water sources in each of Ethiopia, Jordan, Nicaragua, Nigeria, and Tajikistan [8,9,10,11,12]. A population-weighted sample was taken of each source type that provided water to at least 5% of the national population in four of the five countries [8,9,10,11,12]. In Nicaragua, no single improved source type other than utility piped supplies covered more than 5% of the population, so several other-improved source types, each covering less than 5% of the population, were sampled [11]. Each sampled source was tested for thermotolerant coliform bacteria (TTC), fluoride, arsenic, and nitrate, and subjected to a standardized sanitary inspection [14]. Sanitary inspections in the RADWQ surveys identified risk factors for faecal contamination for a water source from a standardized list for each water source type of 10 common risks [8,9,10,11,12]. Examples of sanitary risks on these lists included pipe breaks, supply discontinuities, poor drainage, and proximity to latrines and animal waste. The RADWQ studies reported sanitary risk as a categorical variable with four levels: 0–, 3–5, 6–8, and 9–10 sanitary risks [8,9,10,11,12]. The RADWQ studies also provided cross-tabulations of sanitary risk levels and TTC contamination levels, except in the case of Nicaragua, which only reported the percentage of contaminated water sources in each sanitary risk level [8,9,10,11,12].

For each of the five countries, we used the RADWQ data to compute the percentages of piped water sources and other-improved water sources that were "safe" by virtue of not testing positive for TCC [15]. We computed the percentages of safe sampled piped and other-improved sources that also had greater than two sanitary risks on the sanitary risk inspection that accompanied the water quality tests for each sampled water source. We used this threshold to identify systems with all but the lowest aggregated level of sanitary risk reported in the RADWQ studies.

We considered any one sanitary risk to represent an elevated risk of water contamination or re-contamination that is not captured by water quality testing at a single point in space and time. Where multiple piped or other-improved water source types were tested within a country, the percentages were computed as an average weighted by the population using each source type. The population receiving drinking water from each source in 2004 to 2005 was estimated from RADWQ project reports.

The RADWQ study made no attempt to assess the safety of water from unimproved sources, other than the case of tanker trucks in Nigeria [9]. In the absence of such data, we assumed that 100% of unimproved sources do not provide safe water.

2.3. Synthetic Covariates from Principal Components Analysis

In order to estimate global access to safe water, empirical statistical models capturing the relationship between faecal contamination or sanitary risk proportions and country-level economic, governance, health, social and environmental characteristics were built. Covariates were chosen based on the availability of data for countries with 2010 JMP estimates and on their relation to drinking water quality. The following country-level indicators were explored: gross domestic product per capita (GDP) [16], the World Bank's Government Effectiveness (GE) score [17], the Human Development Index [18], the Water Quality Index (WQI) from Yale's Environmental Performance Index [19], annual aggregate precipitation [16], percent of population attaining tertiary education [20], and under-5 diarrheal morbidity rates [16]. Due to many of these variables being highly correlated, and the small size of our dataset, the models were limited to one or two covariates. Principal Components Analysis (PCA) was used to create uncorrelated synthetic variables that captured the most variance in those national characteristics. This analysis was performed with Stata 10.1.

2.4. Predicting Drinking Water Safety Proportions for Piped and Other-Improved Sources

Four fractional logit models were built based on the data points listed in Table 2, one for each of the proportions of: safe piped sources, safe other-improved sources, safe piped sources with elevated sanitary risk, and safe other-improved sources with elevated sanitary risk. We modeled the sanitary risk proportions of safe sources, since faecally-contaminated sources were assumed to be unsafe regardless of their level of sanitary risk.

Table 2. Cross-tabulation of TTC contamination and sanitary risk for piped and other-improved water sources [8,9,10,11,12].

Piped *					
	% Sanitary risk category safe	TTC Count (cfu/100 mL)			
		<1	1–10	10–100	>100
Ethiopia					
0–2 (Very Low San. Risk)	88%	78	1	8	2
3–5 (Low San. Risk)	89%	220	9	14	5
6–8 (Med. San. Risk)	70%	40	5	11	1
9–10 (High San. Risk)	0%	0	1	0	0
Jordan					
0–2 (Very Low San. Risk)	100%	1233	1	0	0
3–5 (Low San. Risk)	100%	404	0	0	0
6–8 (Med. San. Risk)	NA	0	0	0	0
9–10 (High San. Risk)	NA	0	0	0	0
Nigeria					
0–2 (Very Low San. Risk)	89%	108	10	3	1
3–5 (Low San. Risk)	79%	263	48	23	1
6–8 (Med. San. Risk)	72%	115	14	26	5
9–10 (High San. Risk)	23%	3	3	5	2
Tajikistan					
0–2 (Very Low San. Risk)	91%	1038	100	0	0
3–5 (Low San. Risk)	72%	91	35	1	0
6–8 (Med. San. Risk)	48%	10	11	0	0
9–10 (High San. Risk)	NA	0	0	0	0
Other-Improved **					
	% Sanitary risk category safe	TTC Count (cfu/100 mL)			
		<1	1–10	10–100	>100
Ethiopia					
0–2 (Very Low San. Risk)	94%	192	2	3	8
3–5 (Low San. Risk)	72%	567	76	94	46
6–8 (Med. San. Risk)	41%	122	46	76	53
9–10 (High San. Risk)	0%	0	2	0	3
Nigeria					
0–2 (Very Low San. Risk)	78%	364	71	31	0
3–5 (Low San. Risk)	80%	256	31	33	1
6–8 (Med. San. Risk)	81%	105	9	15	1
9–10 (High San. Risk)	38%	12	9	10	1
Tajikistan					
0–2 (Very Low San. Risk)	77%	121	28	6	2
3–5 (Low San. Risk)	89%	146	15	3	0
6–8 (Med. San. Risk)	54%	7	4	1	1
9–10 (High San. Risk)	NA	0	0	0	0

* All crosstabulations from RADWQ project reports; The Nicaragua report did not include them.

** No other-improved sources tested in Jordan.

The fractional logit model as developed by Papke and Wooldridge [21], rather than minimizing the sum-of-squares error, is a quasi-maximum likelihood method. The fractional logit is superior to an ordinary least squares (OLS) regression because the predicted values of such models are not guaranteed to be restricted to values between 0 and 1. This method is also superior to the common alternative of performing an OLS regression on the logit transformation of the proportion (log-odds regression), in that the predicted values of the proportions from such a model are not recoverable without making significant assumptions.

Candidate covariates used in the models were the first three PCA components from the analysis described in section 2.3. For each of the four dependent variables, models using combinations of these three components were run and the models having the greatest log-likelihoods while maintaining a significant difference from the null model were chosen.

The resulting models were used to extrapolate the proportions calculated from the RADWQ countries to the remaining countries with JMP data for 2010. The predicted proportion of piped or other-improved sources uncontaminated with TTC for each country was multiplied by the corresponding JMP estimate of the population using piped or other-improved water, respectively, to estimate the population with access to microbiologically safe piped or other-improved water in each country. Similarly, the predicted proportion of uncontaminated piped or other-improved sources with elevated sanitary risk was multiplied by the estimated population with access to uncontaminated piped and other-improved water sources. 95%-confidence interval upper and lower bounds for each figure were calculated by using the standard errors of the model coefficients. This method of calculating confidence intervals assumes the RADWQ studies are perfectly accurate in their reported percentages. This assumption is necessary as the complete RADWQ data sets were not available and we were therefore unable to calculate standard errors for the RADWQ country percentages.

2.5. Adjusting MDG Target 7c Progress Estimates

We re-estimated MDG progress towards global safe water access by accounting for faecal contamination of piped or other-improved water sources, as well as sanitary risks of piped or other-improved sources for which faecal contamination was not detected. The 1990 baseline estimates obtained from JMP were multiplied by the estimated proportions of people with piped and other improved water sources for which such sources were estimated to be unsafe by virtue of having tested positive for TTC. This was repeated to account for piped and other-improved sources with sanitary risks. In the absence of any trend data, proportions estimated based on the 2004–2005 RADWQ data were used to estimate both 2010 and 1990 water source safety. Projections of the global proportion of people without access to safe water in 2015 according to JMP data, and accounting for water quality, were based on linear extrapolations of the 1990–2010 JMP and water quality-adjusted data, respectively. This is the same method used by JMP to make its projections [22]. The 2015 MDG targets were calculated by halving the original and recalculated 1990 baselines.

3. RESULTS

The results of the RADWQ studies in the form of cross-tabulations of detected TTC contamination levels and sanitary risk scores, aggregated into piped and other-improved sources, are shown in Table 2. Within each country, TTC contamination rates for each water source type generally increase with increasing sanitary risk, indicating the expected association between water contamination and the presence of sanitary risk factors.

The aggregated results of the RADWQ studies are presented in Table 3. The relationship between sanitary risk and TTC contamination differs between countries. For example, even though all Jordan's piped water supplies are uncontaminated, almost a quarter have more than two sanitary risks; while less than 9% of Tajikistan's uncontaminated piped supplies have more than two sanitary risks.

Data were available such that the PCA could be performed for 150 countries, accounting for 92.5% of the global population (Table 4). The first three components explained 84% of the variance among these countries. The first component shows that GDP, GE, HDI, and Tertiary Education rates all covary and contribute similar information. The second component is dominated by annual aggregate precipitation, and the third component by the WQI. The spread of the five RADWQ countries with respect to the dominant variables of the first three components are shown in Table 5.

We used the fractional logit models shown in Table 6 to predict proportions of piped and other-improved sources that provide safe water in every country for which 2010 JMP estimates were available. A comparison of the RADWQ countries' measured and predicted proportions is shown in Table 7. We estimate that 1 billion (confidence interval 0.75 to 1.6 billion) of the 5.8 billion using piped or other-improved sources receive faecally-contaminated water. This lowers the number of people estimated to use safe water from 5.8 billion (the 2010 JMP figure) to 4.8 billion, and increases the number or people with unsafe water from 0.78 billion to 1.8 billion as of 2010.

Table 3. Contamination of piped and other-improved sampled sources with TTC, and sanitary risk of safe sources [8,9,10,11,12].

Source type, by country	Thermotolerant coliform (TTC) contamination *		Sanitary risk of safe sources
	TTC-Uncontaminated (Safe) (%)	Sources sampled (n)	>2 Sanitary Risks (%)
Ethiopia			
Piped	87.6	838	60.7
Other-Improved	55.0	764	67.1
Jordan			
Piped	99.9	1639	24.7
Other-Improved	NA	0	NA
Nicaragua			
Piped	85.5	600	47.3
Other-Improved	33.5	888	56.5
Nigeria			
Piped	77.0	630	77.9
Other-Improved	76.0	949	50.6
Tajikistan			
Piped	88.6	1286	8.9
Other-Improved **	82.0	334	55.8

* Data aggregated from RADWQ reports. TTC contamination judged against guideline value of <1 TTC per 100 mL as indicated in the WHO Guidelines for Drinking Water Quality [15]; ** Only 44% of the protected springs in Tajikistan were found to be adequately protected for designation as an improved source.

Table 4. Factor loadings, Eigenvalues, and Variances accounted for from PCA.

Variable	Comp 1	Comp 2	Comp 3	Comp 4	Comp 5	Comp 6	Comp 7
Eigenvalue	4.19	1.06	0.623	0.476	0.326	0.180	0.138
pc GDP	0.422	0.121	0.353	0.204	−0.429	0.647	0.199
GE	0.422	0.151	0.336	0.348	−0.0785	−0.715	0.219
HDI	0.441	0.0920	0.258	−0.101	0.399	0.0799	−0.745
Precipitation	−0.102	0.913	0.0051	−0.380	0.0029	−0.0178	0.106
WQI	0.331	0.237	−0.793	0.433	−0.0363	0.0497	−0.123
U5 Diarrhea	−0.384	0.214	0.227	0.642	0.537	0.210	0.100
Tert. Educ.	0.427	−0.137	−0.127	−0.280	0.601	0.131	0.567
Proportion	0.599	0.152	0.0885	0.0679	0.0466	0.0258	0.0197
Cumulative	0.599	0.751	0.839	0.908	0.955	0.980	1.00

Table 5. Spread of countries on dominant PCA Covariates.

	GE	Precipitation (mm)	WQI
Ethiopia	−0.35	848	55.3
Jordan	0.08	111	11.9
Nicaragua	−0.96	2391	57.1
Nigeria	−1.2	1150	20.1
Tajikistan	−0.91	691	42.8

Table 6. Fractional Logit Models of Water Source Safety Proportions.

	Model			
Variable	% Piped Sources Safe	% Other-Improved Sources Safe	% Safe Piped Sources w/Elevated Sanitary Risk	% Safe Oth. Imp. Sources w/Elevated Sanitary Risk
Comp 1	0.879 **		−1.131 ***	
	(0.286)		(0.285)	
Comp 2	−0.436 *	−0.792 ***		
	(0.209)	(0.159)		
Comp 3				−0.350 ***
				(0.099)
Constant	3.188	0.370	−1.819	0.345
	(0.566)	(0.236)	(0.707)	(0.075)
N	5	4	5	4
Residual df	2	2	3	2
Log pseudo-likelihood	−1.2305	−1.65	−2.119	−1.773

* $p < 0.05$; ** $p < 0.01$; *** $p < 0.001$; one-tailed tests for hypothesized effects, robust standard errors in parentheses.

Of these 4.8 billion using safe water, approximately 1.2 billion people (confidence interval 0.75 to 2.1 billion) receive water from sources that are at risk of faecal contamination by virtue of having greater than two of the common sanitary risks for that source type as defined by RADWQ [14]. If a more stringent definition of safety (requiring both no faecal contamination and low sanitary risk) is used, then the estimate of the number of people with unsafe water is 3 billion, (confidence interval 1.5 billion to 3.9 billion). 302 million people reside in countries for which JMP estimates do not exist. An additional 370 million people reside in countries for which JMP estimates exist, but for which the data necessary for principal components analysis was unavailable. The safety of the water received by these 370 million is not estimated.

Table 7. Comparison of Measured and Predicted Proportions.

Country	*RADWQ*	*Predicted*	*RADWQ*	*Predicted*	*RADWQ*	*Predicted*	*RADWQ*	*Predicted*
Ethiopia	88%	85%	55%	55%	61%	59%	67%	57%
Jordan	100%	98%	NA	87%	25%	18%	NA	48%
Nicaragua	86%	85%	34%	33%	47%	33%	57%	56%
Nigeria	77%	77%	76%	68%	78%	73%	51%	51%
Tajikistan	88%	93%	82%	76%	9%	18%	56%	55%

The full categorization of the world population of piped and other-improved water source users into categories based on faecal contamination and sanitary risk using our estimation methods is shown in Figure 1, where upper and lower bounds are shown in parentheses where available. The world's population is divided into those using improved sources, unimproved sources, and sources of unknown type. The improved source population is divided into those using piped and other-improved sources. Each of these two populations is divided into those using sources that are safe, unsafe, or unestimated safety. The populations using safe piped and safe other-improved sources are divided into those using sources with low and elevated levels of sanitary risks.

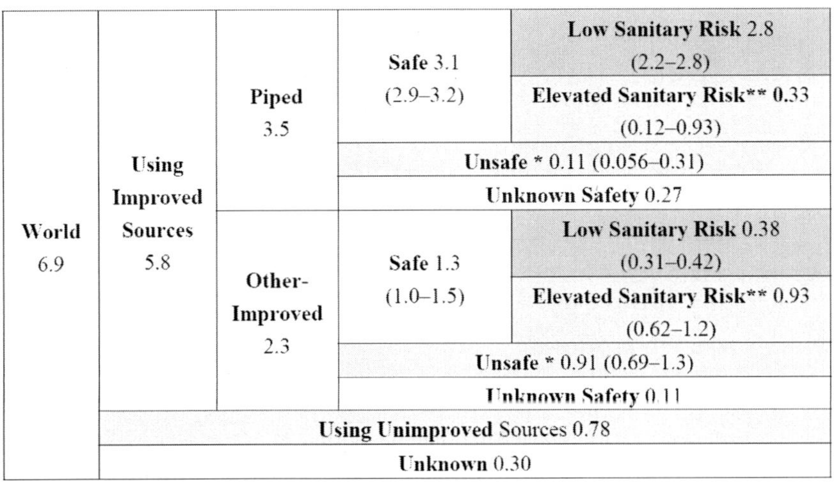

* One billion people with improved sources that may be unsafe; ** 1.2 billion people with otherwise safe water with high sanitary risk

Figure 1. World population by water contamination status and sanitary risk for 2010 (billions).

Percentages of the world population for the purposes of adjusting the MDG Target 7c progress are calculated by assuming the populations with unknown water source types and water source safety are distributed similarly to the rest of the world. Accounting for faecal contamination based on our models increases the 1990 baseline estimate of the population without access to a safe water source from 23% to 37% (see Figure 2). In order to meet MDG Target 7c, the proportion without access to safe water would need to be reduced to 18% by 2015, while trends indicate this figure will be 26%. For 2010, the shortfall is 10 percentage points (680 million people), and we project an 8-percentage point shortfall in 2015, while JMP estimates, based on the indicator of use of an improved source, indicate that the MDG target has already been met.

Using a more stringent definition of water safety that requires water sources be both uncontaminated and of low sanitary risk would result in a 1990 baseline of 53% of the population, a target of 26%, and a 2015 projection of 46%. This would result in a 21-percentage point shortfall in 2010 and a 20-percentage point shortfall in 2015.

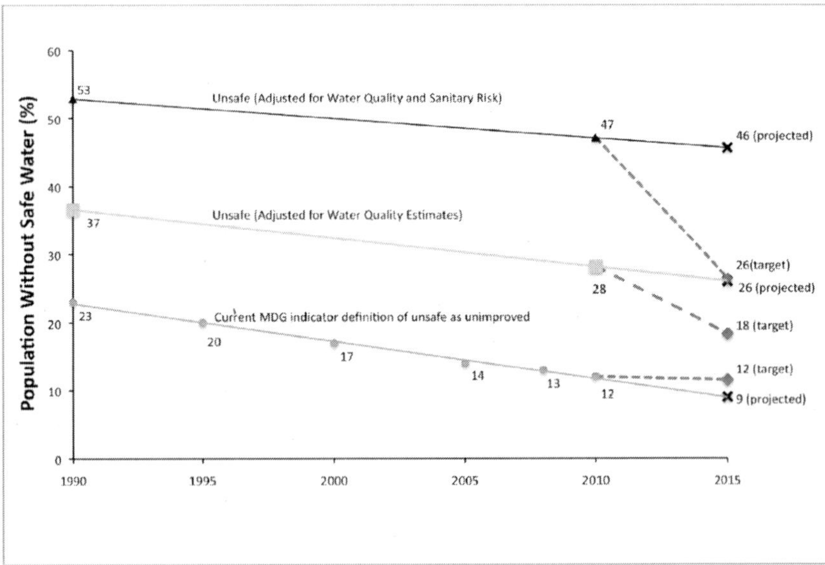

Figure 2. Comparison of MDG Target 7c baseline and target when including and excluding faecal contamination and sanitary risk in water safety.

4. DISCUSSION

Our estimates of the population with access to safe water were based on several assumptions that could contribute to over- or under-estimation.

We used the >1 cfu/100 mL TTC level as a threshold for considering water microbiologically safe, as did the RADWQ studies. *E. coli* and *Enterrococci* have been shown to be better indicators of waterborne disease risks than total TTC [23]. This could result in an overestimate of the population actually at risk of waterborne disease, as some sources testing positive for TTC may not be positive for *E. coli*.

We assumed that the relationships between water quality and national characteristics described by the models in Table 3 hold for countries other than the five countries assessed by RADWQ. The causal factors determining water quality for a given source type may differ from country to country and the causal relationships determining water quality for a given source type may differ from country to country. In addition, countries differ substantially regarding the mixture of predominant source types. For instance, protected springs were the only non-piped improved water source assessed by RADWQ in Tajikistan, while boreholes and protected wells were assessed in Ethiopia, Nicaragua, and Nigeria [8,9,10,11,12]. All non-piped "improved" sources were aggregated and treated as "other-improved". Given these country-to-country differences, the direction of error will differ from country to country although the overall effect on the global estimates may be relatively small as the errors are aggregated.

The sanitary risk estimates are based on the proportion of safe sources assessed as having greater than two of the sanitary risks listed as possible on the RADWQ survey forms. Moreover, the RADWQ project reports stated that the lists of sanitary risks used in the survey could have been better tailored to country conditions [8,9,10,11,12]. Each of the possible risks on these forms probably contributes to a different level of actual risk for water contamination, but are weighted equally in the RADWQ methodology [8,9,10,11,12,14]. The relationship between the number of sanitary risks and the prevalence and degree of water contamination differs between countries (Table 2). Since any single sanitary risk is significant, use of this threshold (three or more risk factors) is likely to underestimate the population using unsafe water.

We assumed that no contamination occurs between the water point and the point and time of use. However, such contamination is known to occur [5]. Water that is safe in a distribution system at one point may become contaminated at another point before it is received by the user, due to deficiencies in the distribution system [24]. Additionally, water that must be transported manually from the source to the home, and any water stored in the home, as is common with other-improved sources, can become contaminated due to unsanitary storage conditions [25]. We did not account for this due to lack of representative data. The effect may be greater for the estimates for "other-improved" sources than for piped sources. However, the effect is likely to be significant in the case of piped sources with discontinuous service that encourage users to store water in the home. These assumptions would tend to lead to an underestimate of the population using unsafe water.

We assumed that the proportion of piped and other-improved sources that provide safe water found in the RADWQ reports represent the safety of those sources over an entire year. This is likely incorrect, due to the transient nature of many contamination events [26]. One round of water quality testing is unlikely to capture the true extent of microbial contamination that might occur over a long period of time at a given source. As such, the RADWQ data likely substantively underestimate the proportion of water sources that are contaminated over an extended time period. The effect would be to underestimate the population using unsafe water.

We did not consider contamination from nitrates, arsenic, fluoride, or any of the chemical contaminants not tested for by the RADWQ project. This is because waterborne pathogens from faecal sources cause more disease than any other waterborne contaminant, and thermotolerant coliform bacteria measurement serves as an indicator for the presence of faecal contamination [15]. There is likely not complete overlap in terms of the types of contamination a given water source might face, so more water sources are probably unsafe than reliance on microbiological indicators alone would indicate. The effect of these assumptions is to underestimate the population using unsafe water.

We assumed that the proportion of piped and other-improved sources that are faecally contaminated, and the proportion of these with elevated sanitary risk, remained constant across time, at least backwards to 1990 for the purposes of recalculating the MDG baseline, and forwards to 2015 for projected proportions. This assumption was necessary due to a lack of equivalent data for the relevant time periods. The effect of this assumption will vary from country to country, and the overall effect on the global estimates may be relatively small as the errors are aggregated.

Since the RADWQ data was not disaggregated by rural and urban settings, we assumed that the proportions of sampled water sources complying with WHO guidelines and having significant levels of sanitary risk were the same in urban and rural settings. The effect of this assumption will vary from country to country.

We assumed that 100% of unimproved sources do not provide safe water. This is based on a lack of water quality data and a consequent reliance on the same methodology as JMP, treating all such sources as unsafe due to the lack of sustainability of access and lack of protection against contamination characteristic of such sources. This assumption could result in an overestimate of the number of people receiving faecally contaminated water. However, this assumption does not overestimate those receiving water from sources with elevated sanitary risks, since unimproved sources by definition lack sanitary protection. In addition, the MDG indicator accounts for both access and water quality [1]. Since unimproved sources may not meet the access criterion of the indicator, the potential error of the assumption is further reduced.

We did not account for household water treatment (HWT). Use of adequate HWT strategies in households can reduce the health risks of water from contaminated sources (both improved and unimproved sources) and also the risk arising from water contamination during transport and household storage [25,27]. Among a sample of 67 low and middle-income countries, using data from the nationally representative Demographic and Health Surveys (DHS), 1.1 billion people use HWT [27]. However, the DHS surveys used did not confirm responses indicating HWT use with physical indicators in the home of such use. Additionally, it is unknown how these people might be distributed over the actual quality of their source water, how many of these people only use water that they have treated, or to what degree these people sustain the HWT strategy for long periods of time. There also exists the risk of recontamination of water post-treatment [28]. The number of people relying on HWT receiving safe water is likely much smaller than 1.1 billion. As such, the effect of the assumption is likely to be a relatively small overestimate of the population using unsafe water.

The effects of the above assumptions will vary between countries. As such, the country-level estimates for water source compliance and sanitary risk are not presented, as these are likely to be imprecise. However, the overall effect on the global estimates may be relatively small as the errors are aggregated.

The confidence intervals were calculated using the standard errors of the model parameter estimates, and did not account for the standard error of the dependent variable. The standard errors of the reported RADWQ figures were unavailable. This results in an underestimate of the upper bounds and an overestimate of the lower bounds.

5. CONCLUSIONS

We show that the MDG indicator (proportion of the population using an improved water source) used in assessing progress towards MDG Target 7c results in a substantive underestimate of the proportion of the population using unsafe water. The 2010 JMP data shows that 780 million people (11% of the population) use unimproved sources. We estimate that 1 billion (between a lower bound of 750 million and an upper bound of 1.6 billion) people using piped or other-improved water sources receive unsafe water, meaning 1.8 billion people did not have access to safe water in 2010.

Using a definition for safe water that includes the absence of faecal contamination in a one-off sample from a piped or other-improved source shows that the current indicator, based on the definition of an improved source alone, underestimated the progress required to meet the drinking-water component of MDG Target 7c by 10% of the global population, whereas the 2010 JMP progress estimate suggests that the MDG drinking-water target has already been achieved.

We estimate that an additional 1.2 billion (between 750 million and 2.1 billion) people using safe piped or other-improved sources are using sources that are at elevated risk of contamination unlikely to be detected by one-off, or perhaps even routine monitoring.

While these estimates are imprecise, their magnitude and health and development implications suggest that greater attention is needed to better understand and manage the problem of contamination of improved water sources.

This study highlights the substantive differences between population estimates using water from an improved source and estimates of populations using water free of faecal contamination and water with low risk of faecal contamination. As increasing populations use piped and other-improved water sources it is likely that increasing attention will be paid to water safety. The demand for evidence to inform effective policy making will increase in parallel. Here we use data from large scale dedicated surveys to gain preliminary insights. While the RADWQ methodology has provided valuable information regarding national water safety in terms of contamination and sanitary risk factors, further work is required to identify the most appropriate means to secure greater and more targeted evidence to inform decision-making.

ACKNOWLEDGMENTS

We thank Rolf Luyendijk at UNICEF and Robert Bain for valuable feedback. This work was supported by the International Association of Plumbing and Mechanical Officials (IAPMO). The University of North Carolina at Chapel Hill's Libraries provided support for open access funding.

REFERENCES

1. United Nations Development Group. *Indicators for Monitoring the Millenium Development Goals*; United Nations: New York, NY, USA, 2003.
2. WHO/UNICEF Joint Monitoring Programme for Water Supply and Sanitation. *Progress on Sanitation and Drinking-Water: 2010 Update*; World Health Organization: Geneva, Switzerland, 2010.
3. Data Resources and Estimates of the WHO/UNICEF Joint Monitoring Programme for Water Supply and Sanitation for Water. Available online: http://www.wssinfo.org/data-estimates/introduction/ (accessed on 6 March 2012).
4. Godfrey, S.; Labhasetwar, P.; Wate, S.; Pimpalkar, S. How safe are the global water coverage figures? Case study from Madhya Pradesh, India. *Environ. Monit. Assess.* **2011**, *176*, 561–574.
5. Wright, J.; Gundry, S.; Conroy, R. Household drinking water in developing countries: A systematic review of microbiological contamination between source and point-of-use. *Trop. Med. Int. Health* **2004**, *9*, 106–117.
6. Kravitz, J.D.; Nyaphisi, M.; Mandel, R.; Petersen, E. Quantitative bacterial examination of domestic water supplies in the Lesotho Highlands: Water quality, sanitation, and village health. *Bull. World Health Organ.* **1999**, *77*, 829–836.
7. Guillemin, F.; Henry, P.; Uwechue, N.; Monjour, L. Faecal contamination of rural water supply in the Sahelian area. *Water Res.* **1991**, *25*, 923–927.
8. Tadesse, D.; Desta, A.; Geyid, A.; Girma, W.; Fisseha, S.; Schmoll, O. *Rapid Assessment of Drinking-Water Quality in the Federal Democratic Republic of Ethiopia: Country Report of the Pilot Project Implementation in 2004–2005*; World Health Organization: Geneva, Switzerland, 2010.
9. Ince, M.; Bashire, D.; Oni, O.; Awe, E.O.; Ogbechie, V.; Korve, K. *Rapid Assessment of Drinking-Water Quality in the Federal Republic of Nigeria: Country Report of the Pilot Project Implementation in 2004–2005*; World Health Organization: Geneva, Switzerland, 2010.
10. Properzi, F. *Rapid Assessment of Drinking-Water Quality in the Hashemite Kingdom of Jordan: Country Report of the Pilot Project Implementation in 2004–2005*; World Health Organization: Geneva, Switzerland, 2010.
11. Aldana, J.M. *Rapid Assessment of Drinking-Water Quality in the Republic of Nicaragua: Country Report of the Pilot Project Implementation in 2004–2005*; World Health Organization: Geneva, Switzerland, 2010.
12. Aliev, S.; Shodmonov, P.; Babakhanova, N.; Schmoll, O. *Rapid Assessment of Drinking-Water Quality in the Republic of Tajikistan: Country Report Project Implementation in 2004–2005*; World Health Organization: Geneva, Switzerland, 2010.
13. Bain, R.; Gundry, S.; Wright, J.; Yang, H.; Pedley, J.; Bartram, J. Accounting for water quality in MDG monitoring: Lessons from five countries. *Bull. World Health Organ.* Available online: http://www.who.int/bulletin/online_first/11-094284.pdf (accessed on 9 February 2012). in press..

14. Howard, D.; Ince, M.; Smith, M. *Rapid Assessment of Drinking Water Quality: A Handbook for Implementation*; World Health Organization & United Nations Children's Fund: Geneva, Switzerland, 2003.
15. World Health Organization. *Guidelines for Drinking-Water Quality*, 4th; World Health Organization: Geneva, Switzerland, 2011.
16. World Development Indicators. Available online: http://data.worldbank.org/data-catalog/world-development-indicators (accessed on 30 November 2011).
17. Worldwide Governance Indicators. Available online: http://info.worldbank.org/governance/wgi/sc_country.asp (accessed on 30 November 2011).
18. Country Profiles and International Human Development Indicators. Available online: http://info.worldbank.org/governance/wgi/sc_country.asp (accessed on 30 November 2011).
19. Water Quality Index. Available online: http://www.epi2010.yale.edu/Metrics/WaterQualityIndex (accessed on 30 November 2011).
20. EdStats-Education Statistics. Available online: http://web.worldbank.org/ WBSITE/EXTERNAL/TOPICS/EXTEDUCATION/EXTDATASTATISTICS/EXTEDSTATS/0,,menuPK:3232818~pagePK:64168427~piPK:64168435~theSitePK:3232764,00.html (accessed on 30 November 2011).
21. Papke, L.E.; Wooldridge, J.M. Econometric methods for fractional response variables with an application to 401(k) plan participation rates. *J. Appl. Econ.* **1996**, *11*, 619–632.
22. JMP's method-deriving progress estimates. Available online: http://www.wssinfo.org/definitions-methods/method/ (accessed on 3 February 2012).
23. Moe, C.L.; Sobsey, M.D.; Samsa, G.P.; Mesolo, V. Bacterial indicators of risk of diarrhoeal disease from drinking-water in the Philippines. *Bull. World Health Organ.* **1991**, *69*, 305–317.
24. Lee, E.J.; Schwab, K.J. Deficiencies in drinking water distribution systems in developing countries. *J. Water Health* **2005**, 109–127.
25. Sobsey, M. *Managing Water in the Home: Accelerated Health Gains from Improved Water Supply*; World Health Organization: Geneva, Switzerland, 2002.
26. Nwachuku, N.; Craun, G.F.; Calderon, R.L. How effective is the TCR in assessing outbreak vulnerability? *J. Am. Water Works Assoc.* **2002**, *94*, 88–96.
27. Rosa, G.; Clasen, T. Estimating the scope of household water treatment in low- and medium-income countries. *Am. J. Trop. Med. Hyg.* **2010**, *82*, 289–300.
28. McLaughlin, L.A.; Levy, K.; Beck, N.K.; Shin, G.-A.; Meschke, J.S.; Eisenberg, J.N. An observational study on the effectiveness of point-of-use chlorination. *J. Environ. Health* **2009**, *71*, 48–53.

CHAPTER 4

Health Risk Assessment after Exposure to Aluminium in Drinking Water between Two Different Villages

M. S. Qaiyum[1], M. S. Shaharudin[1], A. I. Syazwan[1*], A. Muhaimin[2]

[1]*Environmental and Occupational Health Science Unit, Department of Community Health, Faculty of Medicine and Health Sciences, Universiti Putra Malaysia, Selangor, Malaysia*

[2]*Department of Environmental Management, Faculty of Environmental Studies, Universiti Putra Malaysia, Selangor, Malaysia*

ABSTRACT

A cross-sectional study was conducted at Mukim Parit Lubok (MPL) and Parit Raja (PR), Batu Pahat, Malaysia. The main objective of this study was to determine the aluminium concentration in drinking water and to perform health risk assessment prediction among respondents from these two residential areas. A total of 100 respondents were selected from the study areas based on inclusive and exclusive criteria. Two duplicates of treated water samples were taken from each respondent's house using 200mL high-density polyethylene (HDPE) bottles and 0.4 mL (69%) pure concentrated nitric acid were added as a preservative. Aluminium concentrations were analyzed using a Lambda 25 UV/V spectrophotometer. The result showed that aluminium concentration in drinking water from MPL was 0.18 ± 0.022 mg/L and 0.22 ± 0.044 mg/L for PR. Statistical analysis showed that 14 (28%) water samples collected from MPL and 35

(70%) from PR recorded concentration of aluminium above the standard limit set by the Ministry of Health, Malaysia for drinking water guideline (0.2 mg/L). The mean value of Chronic Daily Intake (CDI) of aluminium in drinking water from PR (0.00707 mg/kg/day) was significantly higher compared to MPL (0.00164 mg/kg/day). Hazard Index (HI) calculation showed that all respondents had "HI" of less than 1. In conclusion, there was an unlikely potential for adverse health effects from aluminium intake in drinking water from both study areas. However, it was necessary for some actions to be taken in order to reduce aluminium levels found in drinking water for both locations.

Keywords: Aluminium, Drinking Water, Health Risk Assessment, Chronic Daily Intake (CDI), Hazard Index (HI)

1. INTRODUCTION

Water is of major importance to all living things, in some organisms, up to 90% of their body weight comes from water and up to 60% of the human body is water [1]. Malaysia has an abundant water resources and adequate rainfall and yet this nation has water shortages and water quality problem [2,27]. A national monitoring network was established in 1978 by the Department of Environment to monitor river quality and detect changes in water quality as a result of development [2]. A study by Calderon in year 2000 has identified a number of chemical contaminants in drinking water and one of those contaminants is aluminium [3]. According to Malaysia Drinking Water Quality 2009, the permissible level of aluminium in drinking water must not exceed 0.2 mg/L [4]. Based on the Report of Study on Water Supply System 2005 that was done on two water treatment plants (WTPs) in Batu Pahat (Parit Raja IV WTP and Sri Gading WTP), there have been violations of water quality standard for aluminium levels [5]. There were 8 sampling stations from the Parit Raja IV WTP that showed violation of aluminium standard along the reticulation system with the highest value of 8.00 mg/L [5].

Aluminium can be either beneficial or harmful, depending on a number of factors, including the amounts to which someone being exposed. Aluminium is the most abundant metal and the third most abundant element, after oxygen and silicon, in the earth's crust [6]. It is widely distributed and constitutes approximately 8.8% of the earth's crust or approximately 88 000 parts of aluminium per million parts of the earth's crust [7]. Dissolved aluminium concentrations in waters with near-neutral pH values usually range from 0.001 to 0.05 mg/L but rise to 0.5-1 mg/L in more acidic waters or water rich in organic matter [8]. Although acute exposure to high doses of aluminium are well tolerated, aluminium may be a contributing factor in certain neurodegenerative diseases such as Alzheimer's disease (AD), amyotrophic lateral sclerosis (ALS) and Parkinson's Dementia [9]. Aluminium levels in drinking water vary according to the levels found in the source water and whether aluminium coagulants are used during water treatment [8]. Aluminium salts such as aluminium sulphate (Al_2SO_4) are widely used in water treatment plant as coagulants to reduce organic matter, colour, turbidity and microorganism levels in untreated water [8].

There is considerable concern throughout the world over the levels of aluminium found in drinking water sources (raw water) and treated drinking water [10]. Then, in a study, it was found out that high aluminium levels in drinking water (≥ 0.1 mg/L) were associated with an elevated risk of dementia and Alzheimer's disease [11]. Although there has been debate regarding relation of aluminium and Alzheimer Disease (AD), there were eighteen drinking water studies that have linked aluminium level to elevated risks of AD and elderly cognitive impairment [12].

Total removal of harmful pollutants from the environment is impossible or infeasible and many natural occurring substances such as aluminium itself also pose health risks [13]. Besides that, it has been hypothesised that aluminium exposure is a risk factor for the development or acceleration of onset of AD in humans [8]. Therefore, health risk assessment is used to estimate whether current exposure will pose health risks to the community studied [13,26].

In this study, water samples will be taken from respondents' houses to determine their daily intake of aluminium from drinking water. In addition, water sources for both areas are different because it involves two villages with different water treatment plants, therefore comparison of aluminium levels between two different populations can be done and then compared to permissible level from Malaysia National Standard for Drinking Water Quality (2009) [4].

2. METHODS AND MATERIALS

2.1. Subject Recruiting and Selection of Study Locations

A cross sectional study was conducted involving 100 respondents from two villages in state of Johor in Peninsular Malaysia, namely Mukim Parit Lubok (MPL) and Parit Raja (PR). These two specific locations were selected due to non-compliance in aluminium levels in drinking water. This study were aimed at determining health risk from intake of aluminium in drinking water and the risk associate with the aluminium exposure using chronic daily intake estimation. The respondents were purposively selected from each village based on identified inclusive criteria, such as exclusive use of tap water and has lived in the village for at least 6 months. On the other hand, the exclusive criteria include consumption of other sources of drinking water such as bottled water and the usage of a water filter.

2.2. Questionnaires and Body Weight

A set of structured questionnaires were used in this study which comprised of two sections. The first section contained questions regarding respondents' background data such as age, household income, duration of residence and education level. The second section contained questions regarding exposure duration, chronic daily intake and water consumption during living period in the selected villages. The second section of the questionnaire was modified from the Baseline, Descriptive and Time-Activity Questionnaires used in the National Human Exposure Assessment Survey (NHEXAS)-Arizona study [14]. During the data collection period, the respondents were interviewed face to face with the researcher at their respective houses. In determining the chronic daily intake, body weight measurement was taken using a Tanita Digital Weight Scale.

2.3. Analysis of Aluminium in Drinking Water

Drinking water samples were collected at the respondents' kitchen tap. The tap was turned on and water was left to run for 3-5 minutes. A 200 milliliter (mL) nonacidified high-density polyethylene (HDPE) bottle was used for water sample collection. Two replicates of water samples were taken from each respondent's house. The samples were then preserved using 0.4 mL 69% pure concentrated nitric acid before being analyzed at the laboratory to ensure bacterial removal from the samples and to lengthen the storage time of the samples [15].

2.4. Estimation of Health Risk Associated with Aluminium Drinking Water

In order to estimate health risk associated with aluminium in drinking water, chronic daily intake (CDI) where calculated first using the following equation [17]:

$$CDI(I)=(C_1 R_1 F_E D_t)/W_B T_{AVG})$$

where CDI (I) wasthe chronic daily intake (mg/kg/d), C_1 was level of aluminium concentration in drinking water (mg/L), R_1 was ingestion rate (2 L/day), F_E was exposure frequency (day/year), $D_t$was exposure duration (year), W_B was body weight (kg) and T_{AVG} was average of exposure duration (D × 365 days/year).

2.5. Calculation of Significant Exposure for Non-carcinogen Health Effect

To conclude the significant exposure and overall potential for non-carcinogenic health effect posed by aluminium in drinking water, the Health Index (HI) was calculated using the following equation [17]:

$$\text{Hazard Index}=(\text{CDI})/\text{RfD})$$

CDI was chronic daily intake and RfD was reference dose. RfD for Aluminium is 7 mg/kg/day. [18] In cases where the non-cancer HI does not exceed unity (HI < 1), it is assumed that no chronic risks are likely to occur at the site [17,19].

2.6. Statistical Analysis

Data from questionnaire, aluminium concentration in drinking water, respondent body weight, CDI and HI of both study areas were analyzed using Statistical Package for Social Science (SPSS) version 18.0 with the significant value ($p < 0.05$). Kolmogorov -Smirnov, Shapiro Wilk and Skewness tests were used to determine the normality of distribution of the interval/ratio scale variables. The data were analyzed in two stages. The first stage was univariate analysis. Bivariate analysis was then calculated to test the hypothesis, which was made to test for testing differences between two villages' study variables.

2.7. Quality Control

Pre-test questionnaires were performed on 10% of sample size before the start of the study on selected adults to ensure the understanding of the questions (face validity). In order to control the selection bias, matching procedure was conducted controlling for: i) source of water and ii) the usage of water filter.

2.8. Ethical

Approval from Medical Researcher Ethics Committee, Faculty of Medicine and Health Sciences, Universiti Putra Malaysia (UPM) was obtained. Written consent was signed before data collection commenced.

Table 1. Respondents' age, household income and duration of residence.

Variable	MPL (n=50)	PR (n=50)	Statistical Test	
	Mean ± (S.D.)	Mean ± (S.D.)	Z/t Value	p Value
Age (years)	43.2 ± (18.68)	44.2 ± (16.53)	− 0.31	0.756
Household Income (RM)	805.6 ± (482.9)	1065 ± (662.19)	− 2.26	0.024*
Duration of Residence (years)	27.4 ± (19.05)	31.86 ± (20.96)	− 1.12	0.264

N = 100; *Significant at $p < 0.05$

Table 2. Respondents' levels of education.

Levels of Education	MPL (n = 50)	PR (n = 50)
	Frequency (%)	Frequency (%)
No Formal Education	4 (8)	4 (8)
Primary School	20 (40)	13 (26)
Penilaian Menengah Rendah (PMR)	6 (12)	5 (10)
Sijil Pelajaran Malaysia (SPM)	9 (18)	21 (42)
Sijil Tinggi Pelajaran Malaysia (STPM)	3 (6)	3 (6)
Higher Education	8 (16)	4 (8)

N = 100

3. RESULTS

3.1. Socio-demographic Information

This study was conducted in M. Pt. Lubok (MPL) and Pt. Raja (PR) where respondents use treated water as their main source of drinking water. Respondents were Malay, aged between 19 to 88 years old and from 100 respondents, 69 (69%) were males and the 31 (31%) were females. Table 1 shows data regarding age, monthly household income and duration of residence while Table 2 shows respondents' education levels.

3.2. Aluminium Levels in Drinking Water

It was found that aluminium levels in both locations at MPL and PR ranged from 0.12 mg/L and 0.36 mg/L with a mean and standard deviation of 0.200 mg/L and 0.039 mg/L, respectively. Table 3 shows information regarding aluminium levels in drinking water in both study locations. Parametric test (t-test) was used to determine the difference of aluminium concentration in drinking water between these two areas.

Results showed that there was significant difference of aluminium concentration in drinking water between the two study areas ($p < 0.05$, Z = – 0.4509) (Table 3).

Table 3. Aluminium level in drinking water.

Variable	MPL (n=50)	PR (n=50)	t Value	p Value
	Range (Mean ± S.D.)	Range (Mean ± S.D.)		
Aluminium levels (mg/L)	0.13 - 0.23 (0.200 ± 0.039)	0.14 - 0.36 (0.22 ± 0.044)	– 0.4509	p < 0.001*

N = 100 * Significant at p < 0.001

Table 4. Difference of aluminium levels between MPL and Malaysia drinking water standard.

MPL (n=50)		t Value	p Value
Mean (mg/kg/day)	S.D.		
0.2	± 0.039	4.782	0.007*

N = 50 * Significant at $p < 0.01$

Table 5. Difference of aluminium levels between PR and Malaysia drinking water standard.

Pt. Raja (n = 50)	Observed Proportion Value		Z Value	p Value
Mean ± S.D.	Group 1 (≤ 0.199 mg/L)	Group 2 (> 0.199 mg/L)		
0.22 ± 0.044	0.3	0.7	0.50	0.007*

N=50*Significant at $p<0.01$

3.3. Difference of Aluminium Levels between MPL and Malaysia Drinking Water Standard

The difference in aluminium concentration in drinking water from MPL with Malaysia Drinking Water Standard was analyzed using a parametric test (one sample t-test) with reference value of 0.2 mg/L. Result in Table 4 shows significant difference in aluminium concentration in drinking water from MPL with Malaysia Drinking Water Standard ($p < 0.05$, $t = 4.782$).

3.4. Difference of Aluminium Levels between PR and Malaysia Drinking Water Standard

Non parametric test was used to analyze the difference in aluminium concentration in drinking water from PR with Malaysia Drinking Water

Standard with the reference value of 0.2 mg/L. From binomial test, there was significant difference ($p < 0.05$, $z = 0.50$) between aluminium levels in drinking water at PR with the Malaysia Drinking Water Standard (Table 5).

3.5. Chronic Daily Intake and Hazard Index

Respondents' chronic daily intake (CDI) for both study area was calculated using the CDI equation. The variables that used in calculated CDI included aluminium concentration in water, duration of residence and body weight. Table 6 shows the Chronic Daily Intake (CDI) of respondents from both study areas. From analysis, it was found out that there was significant difference of CDI of aluminium intake between both locations ($p < 0.005$, $t = - 2.268$). Hazard index (HI) was used to determine the heath risk of aluminium by dividing Chronic Daily Intake (CDI) with Reference Dose (RfD) of aluminium RfD (7 mg/kg/day) [18]. From the result shown in Table7, hazard index of both locations were less than 1.

4. DISCUSSION

4.1. Demography Data and Respondents' Background

Respondents were chosen from two different locations: M. Pt. Lubok (MPL) and Pt. Raja (PR). It was found that respondents of these two locations were not much different in terms of age, levels of education and other demographic background. Most respondents were in the range of 40 to 50 years old. With regards to gender, 69 (69%) of them were male and 31 (31%) were female. Males exceeded females due to most of the respondents were self-employed, so chances to meet the male respondents were higher than the female respondents especially when involving village people.

From the statistical analysis, it was found that the only difference between respondents from both locations was household monthly income, where MPL had a lower household monthly income compared to PR. This situation occurred probably because most of the respondents in MPL were self-employed such as farmers and this location was quite far from the main city compared to PR. However, for respondents at PR, they have a better standard of living, for example as government servants.

Variables that were similar for respondents in both locations were age and duration of residence. It was found out that through statistical analysis, respondents' age and duration of residence showed no significant difference between these two locations. Duration of residence is important because it showed how long the respondents were being exposed to aluminium in drinking water at each location.

Table 6. Chronic daily intake.

Variable	MPL (n = 50)		PR (n=50)		t	p
	Range	Mean (S.D.)	Range	Mean (S.D.)		
Chronic Daily Intake (mg/kg/day)	0.00303 - 0.01158	0.00619 (± 0.00164)	0.0027 - 0.01274	0.00707 (± 0.0022)	− 2.268	0.026

N = 100

Table 7. Hazard index.

Hazard Index (HI)	MPL		PR	
	Frequency (%)	(Mean)	Frequency (%)	(Mean)
HI<1	50(100)	(0.00088)	50(100)	(0.00101)
HI>1	0	0	0	0

N = 100

4.2. Difference of Aluminium Levels between Two Locations

The design and operation of processes at water treatment plant influence the aluminium levels in the treated water that will be delivered to consumers [20]. If there is another additional treatment process such as lime softening added during aeration process for acidic raw water, this process may influence aluminium levels in treated water [20]. Consequently, aluminium levels in treated water may differ for different treatment plants. Therefore, this study did find the same conclusion in which there was significant difference of aluminium levels in drinking water between MPL and PR ($P < 0.05$, $Z = -0.4509$).

Nevertheless, this finding was similar to a previous study done in Romania, where there was a significant difference of aluminium levels in drinking water from different water treatment plants [21]. Another study in Malaysia, showed significant different of aluminium levels between two locations in the southern region of Malaysia [22]. Therefore, as stated in the result, the 'p' value was less than alpha set in this study (0.05), thus the null hypothesis is rejected where there is significant difference of aluminium levels in two different villages.

4.3. Difference of Aluminium Levels between MPL and Malaysia Drinking Water Standard

From the analysis of difference test, it showed that there was significant difference between aluminium levels at the study location and standard of aluminium for drinking water ($p < 0.05$, $t = 4.782$). Next, it was found out that 14 of 50 (28%) water samples collected exceeded the standard aluminium levels in Malaysia, whereas in a study done in Galicia, Spain 19% of water samples collected contained aluminium levels exceeded the maximum concentration level of 0.2 mg/L [23].

Aluminium levels in drinking water vary according whether aluminium sulphate (alum) were used during water treatment. The dose of alum used depended on the properties of the raw water, turbidity, temperature and alkalinity [24]. Raw water with low turbidity often needs higher dose of alum to bring about coagulation than more turbid waters since waters of low turbidity and high colour are most difficult to treat [24]. Nevertheless, under optimal conditions, conventional method is capable of achieving a minimum aluminium levels in the treated water of around 0.03mg/L [20].

4.4. Difference of Aluminium Levels between PR and Malaysia Drinking Water Standard

Aluminium levels in PR ranged between 0.14 mg/L to 0.36 mg/L. From 50 samples collected, (70%) samples exceeded the standard level (0.2 mg/L) with the highest value of 0.36 mg/L. In addition, it was found that there was significant difference in aluminium levels between PR and Malaysia Drinking Water Standard. Besides that, from the year 2001 to 2005, there occurred violations of aluminium standard where 154 water samples collected from 211 (73%) samples had exceeded the Malaysian Drinking Water Standard [5].

Aluminium levels in treated water that above of 0.3 mg/L reflect a lack of optimization in the coagulation, sedimentation and filtration stages of the conventional method [20]. According to WHO, in the coagulation process, what is important is that a correct dose be applied to the raw water as effectively as possible. This means that certain information must be known accurately, such as the dose required, amount of coagulant per unit of volume in each batch, amount of dilution water and also the dosing equipment applies the desired dose all the time [8].

4.5. Difference of Respondents' Chronic Daily Intake (CDI) of Aluminium Intake between the Two Locations

From the statistical analysis, it was found out that there was significant difference of CDI of aluminium intake between both locations ($p < 0.005$, $t = -2.268$). CDI of aluminium intake for MPL was in ranged of 0.00303 mg/kg/day to 0.01158 mg/kg/day, whereas for PR, CDI ranged from 0.0027 mg/kg/day to 0.01274 mg/kg/day. This is mostly due to low aluminium levels measured in drinking water for both locations since the most contributable factor in CDI equation is aluminium levels rather than other factors such as respondents' body weight and exposure duration to aluminium in drinking water. These findings were far lower if compared to a previous study in Selangor, Malaysia, where mean of first location, KSM was 0.0516 mg/kg/day and second location, was 0.0391 mg/kg/day [25].

4.6. Hazard Index of Aluminium Intake between the Two Locations

The HI was used to determine the hazard risk, where, if the HI value was less than 1, it indicates an unlikely potential for adverse health effects, while if the HI value was greater than 1, it indicates a concern for adverse health effects or the need for further study [19]. The result from this study was 100% for both locations showing HI lower than 1, which means that the risk was acceptable. When comparing CDI of aluminium intake between study locations with Reference Dose (RfD) which is 7 mg/kg/day, CDI of aluminium intake for both locations were far lower than RfD.

As a conclusion, this study showed that mean of aluminium levels in both locations had exceeded the Malaysian Drinking Water Standard which is 0.2 mg/L. Even though the degree of difference was not significantly higher, action must be taken to reduce the aluminium levels in drinking water to similar or below the national standard. This is important to ensure the quality of water received by the consumers follow the standards put up by the health authority. Optimization of raw water process at the water treatment plant must be implemented since the conventional method is capable of achieving a very minimum aluminium levels in treated water.

ACKNOWLEDGEMENTS

M. S. Qaiyum, M. S. Shaharudin, A. I. Syazwan, A. Muhaimin The researchers wish to thank for the cooperation of the laboratory staff, academic staff for the sampling and analysis of water at the Environmental and Occupational Health Science Unit, Department of Community Health, UPM. Give special token of appreciation to Ministry of Health, especially all management staff and health inspectors in Batu Pahat District Office Malaysia for their assistance during data collection and sampling. Lastly, not forget all respondents at Mukim Parit Lubok and Parit Raja, who were very helpful in making the study a success.

REFERENCES

1. United States Geological Surve, USA Geological Survey, National Field Manual for the Collection of Water-Quality Data, 2011. http://pubs.water.usgs.gov/twri9A/
2. M. S. Aini, A. Fakhrul-Razi, O. Mumtazah and J. C. Meow Chen, "Malaysian Households' Drinking Water Practices: A Case Study," International Journal of Sustainable Development & World Ecology, Vol. 14, pp. 503-505, 2007.
3. R. L. Calderon, "The Epidemiology of Chemical Contaminants of Drinking Water," Food and Chemical Toxicology, Vol. 38, pp. S13-S20, 2000. doi:10.1016/S0278-6915(99)00133-7
4. MOH, "Drinking Water Quality Standard Drinking Water Quality Surveillance Programme," Engineering Services Division Ministry of Health, Putrajaya, Malaysia, 2009.
5. MOH, "Sistem Bekalan Air Loji Rawatan Air Parit Raja IV dan Sri Gading, Batu Pahat," Batu Pahat, Johor, Malaysia, 2005.
6. D. Krewski, R. A. Yokel, E. Nieboer, D. Borchelt, J. Cohen, J. Harry, S. Kacew, J. Lindsay, A. M. Mahfouz and V. Rondeau, "Human Health Risk Assessment for Aluminium, Aluminium Oxide, and Aluminium Hydroxide," Journal of Toxicology and Environmental Health-Part B-Critical Reviews, Vol. 10, pp. 1-269, 2007. doi:10.1080/10937400701597766
7. ATSDR, "Public Health Statement: Aluminium," Department of Health and Human Services, Public Health Service: Agency for Toxic Substances and Disease Registry (ATSDR), Atlanta, 2006.
8. WHO, "Upgrading Water Treatment Plants," World Health Organization Publication, 2001.
9. M. Gidding, "Aluminum," Environmental and Workplace Health Canadian Drinking Water Quality, Canada, 1998.
10. P. T. Srinivasan, T. Viraraghavan and K. S Subramanyam, "Aluminium in Drinking Water: An Overview," Water SA, Vol. 25, pp. 47-55, 1999.
11. V. Rondeau, D. Commenges, H. J. Acqmin-Gadda, J. F. Dartigues, "Relation Between Aluminum Concentrations in Drinking Water and Alzheimer's Disease: An 8-year Follow-up Study," American Journal of Epidemiology, Vol. 152, pp. 59-66, 2000. doi:10. 1093/aje/152.1.59
12. C. Janson, et al., "Insomnia is More Common among Subjects Living in Damp Buildings," Occupational and Environmental Medicine, Vol. 62, pp. 113-118, 2005.
13. EPA, "A Guide to Health Risk Assessment," California Office of Environmental Health Hazard Assessment, California, 2001.

14. P. Kavcar, et al., "A Health Risk Assessment for Exposure to Trace Metals via Drinking Water Ingestion Pathway," International Journal of Hygiene and Environmental Health, Vol. 212, pp. 216-227, 2009. doi:10.1016/j.ijheh.2008.05.002
15. MOH, "National Standard for Drinking Water Quality," Engineering Sevices Division, Ministry of Health Malaysia, Putrajaya, 2004.
16. PerkinElmer, "Water and Environmental Analysis According to US EPA Regulations Using the Lambda 25/35/45 UV/Vis Spectrophotometers," 2006.
17. USEPA, "Risk Assessment Guidance for Superfund. Vol. I. Human Health Evaluation Manual. Part A. Interim Final," Office of Emergency and Remedial Response, U. S. Environmental Protection Agency, Washington, DC, 1989.
18. FAO and WHO, "Aluminum: Evaluation of Certain Food Additives and Contaminants, 32," Report of the Joint FAO/WHO Expert Committee on Food Additives, WHO, Geneva, 1989, pp. 28-31.
19. USGS., U. S. Geological Survey, National Field Manual for the Collection of Water-Quality Data. http://pubs.water.usgs.gov/twri9A/
20. Health Canada, "Aluminum [Technical docment–Chemical/Physical Parameters]," 2008, p. 3. http://www.hc-sc.gc.ca/ewh-semt/pubs/water-eau/aluminum/aluminum-aluminium-eng.php
21. D. Diaconu, et al., "Aluminum Concentration in Drinking Water from Moldova Territory, Romania," Ovidius University Annals of Chemistry, Vol. 20, pp. 115-118, 2009.
22. Nora, "Kepekatan Aluminium Dalam Air Minum dan Penilaian Risiko Kesihatan di dua petempatan di Johor," Unievrsiti Putra Malaysia, Selangor, 2003.
23. D. A. Rubinos, et al., "Aluminium Contents Indrinking Water from Public Water Supplies of Galicia (Northwest Spain)," The Fourth Inter-Celtic Colloquium on Hydrology and Management of Water Resources, Portugal, 2005.
24. NHRMC, "Australian Drinking Water Guidelines," National Water Quality Management Strategy, National Resource Management Ministerial Council (NHMRC), Australia, 2004
25. Azmir, "Kepekatan Aluminium di dalam air minum dan Penilaian Risiko Kesihatan di dua buah petempatan di Selangor," BSc. Environmental and Occupational Health, Department of Community Health, Universiti Putra Malaysia, Selangor, 2003.
26. M. S. Shaharuddin, Y. M. Kamil, Y. M. Ismail, R. M. Firuz, I. S. Aizat and A. M. Yunus, "Fluoride Concentration in Malaysian Drinking Water," American-Eurasian Journal of Agricultural & Environmental Sciences, Vol. 6, No. 4, pp. 417-420, 2009.
27. M. S. Shaharuddin, M. N. Kidahus, S. Sumarlan, Y. M. Kamil and Y. M. Ismail et al. "Dental Fluorosis (DF) and Its Relationship With Fluoride Levels in Drinking Water in Three States in Malaysia," Research Journal of Medical Sciences, Vol. 4, No. 1, pp. 20-24, 2010. doi: 10.3923/rjmsci.2010.20.24

CHAPTER 5

Water Microbiology. Bacterial Pathogens and Water

João P. S. Cabral

Center for Interdisciplinary Marine and Environmental Research (C. I. I. M. A. R.), Faculty of Sciences, Oporto University, Rua do Campo Alegre, 4169-007 Oporto, Portugal

ABSTRACT

Water is essential to life, but many people do not have access to clean and safe drinking water and many die of waterborne bacterial infections. In this review a general characterization of the most important bacterial diseases transmitted through water—cholera, typhoid fever and bacillary dysentery—is presented, focusing on the biology and ecology of the causal agents and on the diseases' characteristics and their life cycles in the environment. The importance of pathogenic *Escherichia coli* strains and emerging pathogens in drinking water-transmitted diseases is also briefly discussed. Microbiological water analysis is mainly based on the concept of fecal indicator bacteria. The main bacteria present in human and animal feces (focusing on their behavior in their hosts and in the environment) and the most important fecal indicator bacteria are presented and discussed (focusing on the advantages and limitations of their use as markers). Important sources of bacterial fecal pollution of environmental waters are also briefly indicated. In the last topic it is discussed which indicators of fecal pollution should be used in current drinking water microbiological analysis. It was concluded that safe drinking water for all is one of the major challenges of the 21st century and that microbiological control of drinking water should be the norm everywhere. Routine basic microbiological analysis of drinking water should be carried out by assaying the presence of *Escherichia coli* by culture methods. Whenever financial resources are available, fecal coliform determinations should be complemented with the quantification of enterococci. More studies are needed in order to check if ammonia is reliable for a preliminary screening for emergency fecal pollution outbreaks. Financial resources should be devoted to a better understanding of the ecology and behavior of human and animal fecal bacteria in environmental waters.

Keywords: drinking water; cholera; typhoid fever; bacillary dysentery; fecal indicator bacteria; coliforms; ammonia

1. DRINKING WATER AS A VEHICLE OF DISEASES

Water is essential to life. An adequate, safe and accessible supply must be available to all. Improving access to safe drinking-water can result in significant benefits to health. Every effort should be made to achieve a drinking water quality as safe as possible [1].

Many people struggle to obtain access to safe water. A clean and treated water supply to each house may be the norm in Europe and North America, but in developing countries, access to both clean water and sanitation are not the rule, and waterborne infections are common. Two and a half billion people have no access to improved sanitation, and more than 1.5 million children die each year from diarrheal diseases [2]. According to the WHO, the mortality of water associated diseases exceeds 5 million people per year. From these, more that 50% are microbial intestinal infections, with cholera standing out in the first place.

In general terms, the greatest microbial risks are associated with ingestion of water that is contaminated with human or animal feces. Wastewater discharges in fresh waters and costal seawaters are the major source of fecal microorganisms, including pathogens [1–4].

Acute microbial diarrheal diseases are a major public health problem in developing countries. People affected by diarrheal diseases are those with the lowest financial resources and poorest hygienic facilities. Children under five, primarily in Asian and African countries, are the most affected by microbial diseases transmitted through water [5].

Microbial waterborne diseases also affect developed countries. In the USA, it has been estimated that each year 560,000 people suffer from severe waterborne diseases, and 7.1 million suffer from a mild to moderate infections, resulting in estimated 12,000 deaths a year [6]. The most important bacterial diseases transmitted through water are listed in Table 1.

Table 1. The main bacterial diseases transmitted through drinking water.

Disease	Causal bacterial agent
Cholera	*Vibrio cholerae*, serovarieties O1 and O139
Gastroenteritis caused by vibrios	Mainly *Vibrio parahaemolyticus*
Typhoid fever and other serious salmonellosis	*Salmonella enterica* subsp. *enterica* serovar Paratyphi *Salmonella enterica* subsp. *enterica* serovar Typhi *Salmonella enterica* subsp. *enterica* serovar Typhimurium
Bacillary dysentery or shigellosis	*Shigella dysenteriae* *Shigella flexneri* *Shigella boydii* *Shigella sonnei*
Acute diarrheas and gastroenteritis	*Escherichia coli*, particularly serotypes such as O148, O157 and O124

2. CHOLERA

2.1. The Genus Vibrio

Vibrio are small, curved-shaped Gram-negative rods, with a single polar flagellum. Vibrios are facultative anaerobes capable of both fermentative and respiratory metabolism. Sodium stimulates growth of all species and is an absolute requirement for most. Most species are oxidase-positive and reduce nitrate to nitrite. Cells of certain species (*V. cholerae*, *V. parahaemolyticus* and *V. vulnificus*) have pili (fimbriae), structures composed of protein TcpA. TcpA formation is co-regulated with cholera toxin expression and is a key determinant of *in vivo* colonization (see below) [7,8].

Several *Vibrio* species can infect humans (Table 2). *V. cholerae* is, by far, the most important of these species. *V. alginolyticus* has been isolated from several types of soft tissue infections.

V. fluvialis, *Grimontia hollisae* (*V. hollisae*), and *V. mimicus* can cause diarrhea or infections of the gastrointestinal tract. *V. furnissii* has been isolated from a few individuals with diarrhea, but there is no evidence that it can actually cause this pathology. *V. parahaemolyticus* is a well-documented causal agent of acute food-borne gastroenteritis, particularly in Japan and South East Asia. Cases are associated with the consumption of raw or undercooked shellfish such as oysters, shrimp, crabs, and lobster. *V. vulnificus* is an important cause of (often fatal) septicemia and wound infections. Other vibrios, namely *Allivibrio fischeri* (*Vibrio fischeri*) and *V. natriegens*, have no relation with humans [7,8].

Table 2. Main species of *Vibrio* and their occurrence in human clinical specimensa.

Main species	Occurrence in human clinical specimens	
	Intestinal	Extra-intestinal
Vibrio alginolyticus	+	++
Vibrio cholerae O1 and O139	+++++	+
Vibrio cholerae non O1 or O139	++	++
Aliivibrio fischeri (*Vibrio fischeri*)	-	-
Vibrio fluvialis	++	-
Vibrio furnissii	++	-
Vibrio harveyi	-	+
Grimontia hollisae (*Vibrio hollisae*)	++	-
Vibrio mimicus	++	+
Vibrio natriegens	-	-
Vibrio parahaemolyticus	++++	+
Vibrio vulnificus	+	+++

[a] Adapted from [7,8]. Nomenclature according to [9]. The symbols give the relative frequency of each organism in human clinical specimens, and apply to the whole World, rather than to a particular country.

Vibrios are primarily aquatic bacteria. Species distribution depends on sodium concentration and water temperature. Vibrios are very common in marine and estuarine environments, living free or on the surfaces and in the intestinal contents of marine animals. Species with a low sodium requirement are also found in freshwater habitats [7,8].

2.2. The Species Vibrio Cholerae

Vibrio cholerae cells can grow at 40 °C with pH 9–10. The growth is stimulated by the presence of sodium chloride. *Vibrio cholerae* is a very diverse bacterial species (Table 3). It is divided in ca. 200 serovarieties, characterized by the structure of the lipopolysaccharide (LPS) (O antigens). Only serovarieties O1 and O139 are involved in "true" cholera. Some other serovarieties can cause gastroenteritis, but not cholera. The distinction between Classical and El Tor biotypes is based on biochemical and virological characteristics [1,7,8,10,11].

Table 3. Subdivision of *Vibrio cholerae* below the species levela.

Serovariety	Serotype	Biotype
O1	Inaba	Classical
		El Tor
	Ogawa	Classical
		El Tor
	Hikojima	
O139		
others		

[a] Adapted from [8].

2.3. Cholera

2.3.1. Characterization of the disease

The incubation period for cholera is ca. 1–3 days. The disease is characterized by an acute and very intense diarrhea that can exceed one liter per hour. Cholera patients feel thirsty, have muscular pains and general weakness, and show signs of oliguria, hypovolemia, hemoconcentration, followed by anuria. Potassium in blood drops to very low levels. Patients feel lethargic. Finally, circulatory collapse and dehydration with cyanosis occurs [7].

The severity of the disease depends on several factors: (1) personal immunity: this may be conferred by both previous infections and by vaccines; (2) inoculum: the disease only occurs after ingestion of a minimum amount of cells, ca. 10^8[1,7,8,10,11]; (3) The gastric barrier: *V. cholera* cells likes basic media and therefore the stomach, normally very acidic, is an adverse medium for bacterial survival. Patients consuming anti-acidic medications are more susceptible to infection than healthy people; (4) blood group: for still unknown reasons, people with O-group blood are more susceptible than others [1,7,8,10,11].

In the absence of treatment, the mortality of cholera-patients is ca. 50%. It is mandatory to replace not only lost water but also lost salts, mainly potassium. In light dehydrations, water and salts can be orally-administered, but in severe conditions, rapid and intravenous-administration is obligatory. The most efficient antibiotic is currently doxicyclin. If no antibiotic is available for treatment, the administration of water with salts and sugar can, in many cases, save the patient and help in the recovery [1,7,8,10,11].

There are two main determinants of infection: (1) the adhesion of the bacterial cells to the intestinal mucous membrane. This depends on the presence of pili and adesins at the cell's surface; (2) the production of cholera toxin [1,7,8,10,11].

2.3.2. Cholera toxin

Cholera toxin is an exotoxin with a very precise action on target cells. The toxin attaches to a specific receptor (ganglioside Gl) on the cell membrane of intestinal cells and activates the enzyme adenylate cyclase. This results in a non-stop degradation of internal ATP, with release of cAMP and inorganic phosphate. The rise in the internal concentration of cAMP causes an efflux of water, sodium, potassium, chloride and carbonate ions from the cells of the mucous membrane, and this is the main cause of diarrhea [7].

2.3.3. Cholera pandemics and the emergence of El Tor biotype and O139 serovariety. New facts about cholera epidemiology

Cholera has been a well known disease since the 19th century. In the 19th and 20th centuries, seven major pandemics are recognized. The first six pandemics occurred during the following periods: 1st: 1816–1826, 2nd: 1829–1851, 3rd: 1852–1860, 4th: 1863–1875, 5th: 1881–1896, 6th: 1899–1923. These pandemics all started in Asia, passed through Europe and then reached South America. The Classical biotype was involved. The seventh pandemic, still in course, started in 1961 in the Celebes Isles, in Asia. In the 1960s, the disease spread through Asia, in the 1970s reached the Middle East and Africa, and in 1991 streaked violently across South America. Now El Tor has replaced the Classical biotype. El Tor biotype had been detected before, in 1905, but only in the development of the seventh pandemic did this biotype replace the Classical one and become dominant [1,7,8,10,11].

In 1992, a new serovariety (O139), which was coined the Bengal serovariety, was detected for the first time in Bangladesh. This new serovariety quickly spread to India and to southeastern Asia, displacing O1. Although serovariety O1 El Tor has reappeared in 1994 and 1995, the Bengal serovariety still remains the dominant one. The illness caused by serovarieties O139 and O1 are indistinguishable [8,12,13].

In 1991, the seventh pandemic entered South America through the coastal area of Peru. On 23 January, in Chancay, north Peru, *Vibrio cholerae* O1 El Tor was isolated from patients with cholera symptoms, confirming the disease. In this region, between 24 January and 9 February, 1,859 people were hospitalized and 66 died. From Peru, the disease spread rapidly to other countries in South America. Two routes have been proposed for the entrance of the bacterium in Peru: (1) ballast water from a boat coming from Asia; (2) the El Niño current may have transported zooplankton harboring *V. cholerae* cells. Shellfish and fish nourishing on this zooplankton became contaminated and the bacterium was transmitted to humans who ate these marine foods [14–17].

The misfortune of people who died in the first months of this disastrous South American cholera epidemic appeared to have unleashed scientists to study the disease harder and, indeed, important epidemiological studies were carried out during this outbreak. These studies confirmed that contaminated uncooked food and beverages can also be a vehicle for transmission of cholera [18].

2.3.4. Genes for toxin and pili protein production

The genes responsible for toxin production are harbored in the CTXΦ segment (7–9.7 kb) of the chromosome (only in toxigenic strains). The CTXΦ segment carries at least six genes. In addition to the gene encoding cholera toxin production, this segment (virulence cassette) include an accessory cholera toxin (*ace*), a zonula occludens toxin (*zot*), core encoded pilin (*cep*), and an open reading frame of unknown function. During the replication of the chromosome, the CTXΦ fragment can form an autonomous copy and this can constitute an independent plasmid. The plasmid can give rise to virus-like particles—CTXΦ bacteriophages, which can infect non-toxigenic strains. The CTXΦ segment incorporates into the chromosome of the infected cells which became toxigenic. This process was demonstrated *in vitro* in cell suspensions and *in vivo* in the gut of the rat [8,13,19,20].

Epidemic and pandemic strains of *V. cholerae* contain another chromosomal segment designated as VPI. VPI is 39.5 kb in size and contains two ToxR-regulated genes: a regulator of virulence genes (ToxT) and a gene cluster containing colonization factors, including the toxin co-regulated pili (TCP). The tcp gene encodes for the 20.5-kDa TcpA pili protein. This VPI segment appears to be transferable from *V. cholerae* 01 to non-01 strains. *V. cholerae* 0139 strains, like 01, carry the structural genes encoded by the CTX operon and TCP. *V. cholerae* strains non-01 or 0139 normally lack cholera toxin genes and have never been found to carry TCP [8].

2.3.5. Ecology of the bacterium and the cycle of the disease

V. cholerae non-01 or 0139 strains are common in the environment, especially in estuaries. They have been isolated from many estuarine animals such as birds, frogs, fishes and shellfish, and survive and multiply on the surface of phytoplankton and zooplankton cells [8,21].

V. cholerae O1 and O139 strains are isolable from the environment only in epidemic areas. They survive in the cultivable state in water and aquatic and marine organisms for a considerable period of time [8,12,22–24]. When *V. cholerae* cells face adverse environmental conditions, they reduce cell size, became coccoid and enter a dormant stage inside exopolysaccharide biofilms. Cells display a certain metabolism, but are not able to growth and multiply on the surface of agarized media and give rise to colonies. Cells in this viable but non-culturable state retain viability as well as the potential for pathogenicity for significant periods of time [25–27].

Viable but non-culturable cells can leave their dormant stage and multiply again, resulting in an explosion of their concentration in the environment. Since the presence of non-toxigenic strains is common in aquatic milieu, especially in estuaries, if a horizontal transfer of cholera exotoxin producing genes occurs between toxigenic and non-toxigenic strains, the number of toxigenic cells in the environment can rise rapidly and pronouncedly. The episodic nature and the sudden appearance of violent cholera outbreaks, followed by a rapid slowing down, are probably related with these phenomena.

3. SALMONELLOSIS

3.1. The Genus Salmonella. Pathogenicity of Main Serovars

The genus *Salmonella* was designated by Lignières in 1900 [28,29]. Antigenic analysis began when Castellani described, in 1902, a method for absorbing antisera. The first antigenic scheme for *Salmonella* was published by White in 1926, and subsequently developed extensively by Kauffmann, in two classical works published in 1966 and 1978 [28,29]. The Kauffmann-White antigenic scheme contained, by 1988, about 2,250 different serovars [28,29].

The genus *Salmonella*, a member of the family *Enterobacteriaceae*, include Gram-negative motile straight rods. Cells are oxidase-negative and catalase-positive, produce gas from D-glucose and utilize citrate as a sole carbon source. Salmonellae have several endotoxins: antigens O, H and Vi [28,29].

The concept "one serovar-one species", in use for many years, is no longer acceptable. The taxonomy and nomenclature of the genus

Salmonella has been subject of debate since Le Minor and Popoff proposed changes in a paper published in 1987. The issue was settled by a decision of the International Committee on the Systematics of Prokaryotes and published in 2005. The current taxonomy of the genus is presented in Table 4. According to the rules of bacterial nomenclature, the names of the serovars are not italicized and the first letter must be a capital [28–30].

Table 4. Current taxonomy and nomenclature of the genus *Salmonella*. Habitat and pathogenicity of main serovars[a].

Species	Sub-species	Main serovars (from a total of ca. 1,443)	Habitat and pathogenicity
Salmonella enterica	*Salmonella enterica* subsp. *enterica*	Abortusovis	Pathogenic to sheeps.
		Choleraesuis	Pathogenic to humans and animals.
		Enteritidis	Ubiquitous and frequently the cause of infections in humans and animals. Very frequent agent of gastroenteritis in humans.
		Gallinarum	Isolated chiefly from chickens and other birds. Causal agent of fowl thyphoid.
		Paratyphi A	Pathogenic only to humans. Causes paratyphoid fever.
		Paratyphi B	Causes paratyphoid fever in humans and very rarely infects animals.
		Paratyphi C	Causes paratyphoid fever in humans.
		Typhi	Pathogenic only to humans, causing typhoid fever. Transmitted by water and food contaminated with feces.
		Typhimurium	Ubiquitous and frequently the cause of infections in humans and animals. Very frequently, the causal agent of gastroenteritis in humans.
		Typhisuis	Pathogenic to swines.
	Salmonella enterica subsp. *arizonae*	At least 94 serovars.	
	Salmonella enterica subsp. *diarizonae*	At least 323 serovars.	
	Salmonella enterica subsp. *houtenae*	At least 70 serovars.	Isolated mainly from cold-blooded animals and from the environment. Not pathogenic to humans.
	Salmonella enterica subsp. *indica*	At least 11 serovars.	
	Salmonella enterica subsp. *salamae*	At least 488 serovars.	
Salmonella bongori		At least 20 serovars.	

[a] Adapted from [29]. Nomenclature according to [9].

Table 5. Current taxonomy and nomenclature of the genus *Shigella*. Habitat and pathogenicity of speciesa.

Species	Main serotypes	Habitat and pathogenicity
Shigella dysenteriae	15 serotypes.	Intestinal pathogens of humans and primates, causing bacillary dysentery.
Shigella flexneri	8 serotypes 9 subserotypes	Humans are the primary reservoir. A long-term carrier state occurs in few cases.
Shigella boydii	19 serotypes	*Shigella dysenteriae* serotype 1 causes more severe disease then other serotypes and produces a potent exotoxin (Shiga toxin). Large epidemics in developing countries are commonly caused by serotype 1. Diseases caused by other serotypes may be mild or severe.
Shigella sonnei	1 serotype	*Shigella sonnei* illness is usually milder than that caused by other *Shigella* species.

[a] Adapted from [34]. Nomenclature according to [9].

S. enterica subsp. *enterica* serovar Enteritidis is the most frequently isolated serovar from humans all over the world. However, locally, other serovars can be predominant. In the period 1994–2004, Tunisia was exposed to salmonellosis outbreaks in 1997, 1999, 2002 and 2004. In 1997, salmonellosis outbreak was caused by serovar Mbandaka. In 1999, three salmonellosis outbreaks were reported from hospitals located in three different regions. Each outbreak was associated with a different serotype: Mbandaka, Livingstone and Typhi Vi+. In 2002, a *S. enterica* subsp. *enterica* serovar Livingstone infection occurred in the same hospital that reported an outbreak caused by serovar Typhi Vi+ in 1999, but in a different unit. In that year, the Livingstone serovar jumped to the first position in human infection in Tunisia. In 2004, a second outbreak by serovar Typhi Vi+ was reported. The source of isolation was a fermented juice traditionally extracted from palm-tree [31].

3.2. Characterization of the Diseases

Salmonellae pathogenic to humans can cause two types of salmonellosis: (1) typhoid and paratyphoid fever (do not confuse with typhus, a disease caused by a rickettsia); (2) gastroenteritis [28]. Low infective doses (less than 1,000 cells) are sufficient to cause clinical symptoms. Salmonellosis of newborns and infants presents diverse clinical symptoms, from a grave typhoid-like illness with septicemia to a mild or asymptomatic infection. In pediatric wards, the infection is usually transmitted by the hands of staff [29].

Food-borne *Salmonella* gastroenteritis are frequently caused by ubiquitous *Salmonella* serovars such as Typhimurium. About 12 h following ingestion of contaminated food, symptoms (diarrhea, vomiting and fever) appear and last 2–5 days. Spontaneous cure usually occurs. *Salmonella* may be associated with all kinds of food. Prevention of *Salmonella* food-borne infection relies on avoiding contamination (improvement of hygiene), preventing multiplication of *Salmonella* in food (constant storage of food at 4 °C), and use of pasteurization (milk) or sterilization when possible (other foods). Vegetables and fruits may carry *Salmonella* when contaminated with fertilizers of fecal origin, or when washed with polluted water [28].

The incidence of typhoid fever decreases when the level of development of a country increases (*i.e.*, controlled water sewage systems, pasteurization of milk and dairy products). Where these hygienic conditions are missing, the probability of fecal contamination of water and food remains high and so is the incidence of typhoid fever [29].

3.3. Ecology of Salmonellae and the Cycle of Salmonellosis

The principal habitat of *Salmonella* is the intestinal tract of humans and animals [28]. Salmonellae are constantly found in environmental samples, because they are excreted by humans, pets, farm animals, and wild life. Municipal sewage, agriculture pollution, and storm water runoff are the main sources of these pathogens in natural waters [1,32]. Salmonellae do not seem to multiply significantly in the natural environment, but they can survive several weeks in water and in soil if conditions of temperature, humidity, and pH are favorable [28].

Salmonellae isolated from environmental sources are predominantly non-Typhi or Paratyphi serovars. In a study carried out in Tunisia during 1994–2004, *S. enterica* subsp. *enterica* serovars Anatum, Enteritidis and Corvallis were the most common serotypes isolated from food. The great majority of the strains were isolated from poultry, red meat, milk and dairy products, vegetables and fruits. From environmental sources, 73% of the isolates were from tap water. Serovars Corvallis, Enteritidis, and Anatum were the commonest [31]. Arvanitidou *et al.* [32] reported a comparative study carried out in Rivers Aliakmon and Axios, in northern Greece, during a 1-year period, from May 2002 to April 2003. A total of 29 *Salmonella* species were recovered from the water samples. Many of the isolated *Salmonella* serovars were of non-human animal origin such as Mbantaka, Virchow, Hadar, Infantis and Senftenberg, commonly isolated from poultry farm.

Unlike cholera, humans infected with salmonellae can carry the bacteria in the gut without signs of disease. Infected humans can harbor the bacteria for considerable periods of time. About 5% of patients clinically cured from typhoid fever remain carriers for months or even years.

These people can be chronic holders of the bacterium in the gut, and constitute the main reservoir of the bacteria in the environment [29].

The salmonellosis cycle in the environment can involve shellfish. Salmonellae survive sewage treatments if suitable germicides are not used in sewage processing. If effluent from the sewage plant passes into a coastal area, edible shellfish (mussels, oysters) can become contaminated. Shellfish concentrate bacteria as they filter several liters of water per hour. Ingestion by humans of these seafoods (uncooked or superficially cooked) may cause typhoid fever or other salmonellosis. Evidence of such a cycle has been obtained by the use of strain markers, including phage typing [29].

4. SHIGELLOSIS OR BACILLARY DYSENTERY

4.1. The Genus Shigella

Shigella are Gram-negative, non-sporeforming, non-motile, straight rod-like members of the family *Enterobacteriaceae*. Cells ferment sugars without gas production. Salicin, adonitol and myo-inositol are not fermented. Cells do not utilize citrate, malonate and acetate as sole carbon source and do not produce H_2S. Lysine is not decarboxylated. Cells are oxidase-negative and catalase-positive. Members of the genus have a complex antigenic pattern, and taxonomy is based on their somatic O antigens [1,33,34].

4.2. Characterization of the Disease

The incubation period is 1–4 days. The disease usually begins with fever, anorexia, fatigue and malaise. Patients display frequent bloody stools of small volume (sometimes grossly purulent) and abdominal cramps. Twelve to 36 hours later, diarrhea progresses to dysentery, blood, mucus and pus appearing in feces that decreases in volume (no more than 30 mL of fluid per kg per day) [34–36].

Although the molecular basis of shigellosis is complex, the initial step in pathogenesis is penetration of the colonic mucosa. The resulting focus of *Shigella* infection is characterized by degeneration of the epithelium and by an acute inflammatory colitis in the lamina propria. Ultimately, desquamation and ulceration of the mucosa cause leakage of blood, inflammatory elements, and mucus into the intestinal lumen. Under these conditions the absorption of water by the colon is inhibited and the volume of stool is dependent upon the ileocecal flow. As a result, the patient will pass frequent, scanty, dysenteric stools [37,38].

In order for *Shigella* to enter an epithelial cell, the bacterium must first adhere to its target cell. Generally, the bacterium is internalized via an endosome, which it subsequently lyses to gain access to the cytoplasm where multiplication occurs [37,38].

4.3. Virulence Factors

S. dysenteriae serotype 1 produces high levels of a cytotoxic Shiga toxin. *S. sonnei* and *S. flexneri* produce much lower amounts of this toxin. Shiga toxin binds to Galotl-4Galp (galabiose) glycolipid receptors and inhibits mammalian protein synthesis by cleaving the N-glycosidic bond at adenine 4324 in 28S rRNA. The toxic mechanism is identical to that of the plant toxin ricin, produced by *Ricinus communis*. Shigella also release a LPS endotoxin (O antigens), that cause an inflammatory response [37,38].

Shigella 180- to 230-kb plasmids encode genes essential for virulence, namely for: production of adhesins involved in the adherence of bacteria onto the surface of target epithelial cells; production of invasion plasmid antigens (Ipa) that have a direct role in the *Shigella* invasion process; transport or processing functions that ensure the correct surface expression of the Ipa proteins; induction of endocytic uptake of bacteria and disruption of endocytic vacuoles; regulation of plasmid-encoded virulence genes [37,38].

Shigella emerged from *E. coli* during evolution. The acquisition and evolution of the pathogenicity island which encodes all of the genes required for cell invasion and phagolysosomal lysis, permitted a major alteration in pathogenesis [37,38].

4.4. Risk Factors

Many studies have identified risk factors and protective effects for shigellosis incidence and fatality. Despite gradual improvements in water supply, shigellosis continues to be endemic among the disadvantaged populations living in the tropics, often among displaced populations following natural disasters and political crises. In Guatemala, young children, the elderly, and 15–44-year-old males were found to be most susceptible to *S. dysenteriae* serotype 1. In Sierra Leone, the attack rate was higher among children younger than 5 years of age than in the rest of the population. In rural Bangladesh, shigellosis was most common in children aged 1–2 years and in people 60 years or older. In Dhaka,

Bangladesh, it was found that shigellosis mortality was most common in severely malnourished people of all ages, in children under 2 who were not being breastfed, and in all children under 1. In a 3-year study carried out in Matlab, Bangladesh, during 1992 to 1994, it was found that the incidence of *S. dysenteriae* serotype 1 and *S. flexneri* was highest in children under 2 followed by children from 2 to 5. The location of *S. dysenteriae* serotype 1 risk varies in time but *S. flexneri* risk areas were persistent in time. Neighborhoods near bazaars with many non-septic latrines were at highest risk for *S. dysenteriae* serotype 1. *S. flexneri* was most common in flood-controlled areas. It was concluded that *S. dysenteriae* serotype 1 risk was more related to hygiene and sanitation whereas *S. flexneri* was more related to the environment [35].

4.5. Shigellosis through the World

The total number of *Shigella* episodes that occur each year throughout the World is estimated to be 164.7 million, including 163.2 million cases in developing countries, 1.1 million of which result in death. Children under 5 account for 61% of all deaths attributable to shigellosis [35,36].

Shigella species are not uniformly distributed in the world. *S. dysenteriae* is usually found in densely populated areas of South America, Africa and Asia. Infections usually result in significant epidemic outbreaks. Serotype 1 has been distinguished by both its virulence and its ability to produce ravaging epidemics. It predominates in India, Malaysia and Guatemala. Serotype 2 predominates in Yemen and Nigeria. *S. flexneri* is usually found in areas where endemic shigellosis occurs. *S. boydii* occurs sporadically, except in the Indian subcontinent where it was first identified. *S. sonnei* usually occurs in Western developed countries, such as France and USA [35,36].

Important epidemics were reported in the last decades: (1) in 1970 in Central America where 112,000 people were affected and 13,000 died; (2) in 1985, in Texas (USA), 5,000 people became infected after ingestion of contaminated lettuce; (3) in May–June 1994, domestic cases of *S. sonnei* infection were detected in several European countries, including Norway, Sweden, and the United Kingdom. Epidemiological evidence incriminated imported iceberg lettuce as the vehicle of transmission; (4) in 1996, in Paris, with 153 reported patients [33].

4.6. Ecology of Shigellae and the Cycle of Shigellosis

Shigella is typically an inhabitant of the intestinal tract of humans and other primates [1,33,34,36,39]. It is typically spread by fecal-contaminated drinking water or food, or by direct contact with an infected person. In water, shigellae can survive for at least six months at room temperature, and this high survival favors transmission through water. Flies have been implicated on the transmission of *Shigella* cells from human feces to foods. The hand is an important vehicle for transmission of shigellosis, since *S. dysenteriae* serotype 1 cells survives for up to one hour on a human's skin and a very small inoculum is required to unchain infection and disease. Indeed, studies on American volunteers experimentally infected with *Shigella* have shown that as few as one hundred *Shigella* cells given orally cause the disease in 25–50% of the cases. Resistance of *Shigella* to gastric juice certainly accounts, although not exclusively, for this high infectivity [36,40]. Asymptomatic and inappropriately-treated patients with shigellosis can harbor the bacteria in the gut and these appear to be the main reservoirs of the bacteria in the environment [41].

Recent reports on the ecology of shigellae have brought new elements for the understanding of the cycle of the disease. In environmental waters of regions with high numbers of shigellosis' cases, it has been found that, although numbers of cultivable cells were low, genetic elements such as plasmids and genetic fragments with bacteriophage origin, could be detected. Many of the genes that code for exotoxin production are precisely found in these genetic elements. These results suggest that the sudden rise of the number of virulent strains in the environment can result from the incorporation, by cells with reduced virulence, of this type of genetic elements present in the waters. If this hypothesis is confirmed, there is a certain similarity between the cholera and the shigellosis cycles in the environment. It remains to be elucidated if shigellae can also exist in environmental waters in a viable but non-culturable state, as vibrios [41].

5. PATHOGENIC *ESCHERICHIA COLI* STRAINS

E. coli strains isolated from intestinal diseases have been grouped into at least six different main groups, based on epidemiological evidence, phenotypic traits, clinical features of the disease and specific virulence factors. From these, enterotoxigenic (ETEC, namely O148), enterohemorrhagic (EHEC, namely O157) and enteroinvasive serotypes (EIEC, namely O124) are of outstanding importance and can be transmitted through contaminated water [42,43].

5.1. Enterotoxigenic E. coli (ETEC) Strains

Enterotoxigenic *E. coli* (ETEC) serotypes can cause infantile gastroenteritis. The number of reports of their occurrence in developed countries is comparatively small, but it is an extremely important cause of diarrhea in the developing world, where there is no adequate clean water and poor sanitation. In developing countries, these strains are the most commonly isolated bacterial enteropathogen in children below 5 years of age, and account for several hundred million cases of diarrhea and several ten of thousand deaths each year [42–44].

Disease caused by ETEC follows ingestion of contaminated food or water and is characterized by profuse watery diarrhea lasting for several days that often leads to dehydration and malnutrition in young children [42–44]. ETEC also are the most common cause of "travelers' diarrhea" that affects individuals from industrialized countries travelling to developing regions of the World [42–44].

5.2. Enterohemorrhagic E. coli (EHEC) Strains

Reported outbreaks had been associated mainly with the consumption of contaminated foods, such as raw or undercooked ground meat products and raw milk. The primary reservoir of this bacterium has been found to be healthy cattle [42,45,46].

E. coli serotype O157:H7 causes abdominal pain, bloody diarrhea, and hemolytic uremic syndrome. This bacterium produces Shiga-like toxins. The incubation period is 3–4 days, and the symptoms occur for 7–10 days. It is estimated that 2–7% of *E. coli* O157:H7 infections result in acute renal failure [42,45,46].

Although *E. coli* O157:H7 is not usually a concern in treated drinking water, outbreaks involving consumption of drinking water contaminated with human sewage or cattle feces have been documented. An increasing number of outbreaks are associated with the consumption of fruits and vegetables (sprouts, lettuce, coleslaw, salad) contaminated with feces from domestic or wild animals at some stage during cultivation or handling. EHEC has also been isolated from bodies of water (ponds, streams), wells and water troughs, and has been found to survive for months in manure and water-trough sediments [45,46].

Person-to-person contact is an important mode of transmission through the oral-fecal route. An asymptomatic carrier state has been reported, where individuals show no clinical signs of disease but are capable of infecting others [45,46].

5.3. Enteroinvasive E. coli (EIEC) Strains

Enteroinvasive *E. coli* (EIEC) behave in many respects like shigellae. They are capable of invading and multiplying in the intestinal epithelial cells of the distal large bowel in humans. The illness is characterized by abdominal cramps, diarrhea, vomiting, fever, chills, a generalized malaise, and the appearance of blood and mucus in the stools of infected individuals. [42,43,47].

EIEC strains were isolated, for instance, from 28 subjects in the Jesreel district of Israel during a peak period for dysentery. An investigation in Croatia showed that *E. coli* 0124 could frequently be isolated from cases of gastroenteritis, enterocolitis, and dysentery. The dysentery was more common among the older age groups, while the two other types of disease occurred equally in all age groups. A 1985 survey was carried out in Bankok, Thailand in which 410 children with diarrhea and an equal number of control children without diarrhea were examined for the presence of strains of *Shigella*, EIEC, and other pathogens. It was found that 17 of the children with diarrhea and six without yielded EIEC [42,43].

Any food contaminated with human feces from an ill individual, either directly or via contaminated water, could cause disease in others. Outbreaks have been associated with hamburger meat and unpasteurized milk [47].

6. EMERGING WATERBORNE BACTERIAL PATHOGENS

The emerging pathogenic bacteria of concern outlined here have the potential to be spread through drinking water, but they do not correlate with the presence of *E. coli* or with other commonly used drinking water quality indicators, such as coliform bacteria. In most cases, there are no satisfactory microbiological indicators of their presence. More studies are needed in order to understand the real significance and dimension of the diseases caused by water contaminated with these bacteria, and the ecology of these pathogens [45].

6.1. Mycobacterium Avium Complex (Mac)

The *Mycobacterium avium* complex (Mac) consists of 28 serovars of two distinct species: *Mycobacterium avium* and *Mycobacterium intracellulare.*

The importance of Mac organisms was recognized with the discovery of disseminated infection in immunocompromised people, particularly people with HIV and AIDS. Members of MAC are considered opportunistic human pathogens [45,48].

Mac organisms have been identified in a broad range of environmental sources, including marine waters, rivers, lakes, streams, ponds, springs, soil, piped water supplies, plants, and house dust. Mac organisms have been isolated from natural water and drinking water distribution systems in the USA [45,49,50].

The ubiquitous nature of Mac organisms results from their ability to survive and grow under varied conditions. Mac organisms can proliferate in water at temperatures up to 51 °C and can grow in natural waters over a wide pH range [45]. These mycobacteria are highly resistant to chlorine and the other chemical disinfectants used for the treatment of drinking-water. Standard drinking-water treatments will not eliminate Mac organisms but, if operating satisfactorily, will significantly reduce the numbers that may be present in the source water to a level that represents a negligible risk to the general population. The entryway of these mycobacteria in distribution systems is through leaks. Growth of Mac organisms in biofilms is probably important for their continuous presence in distribution systems. Slow growing mycobacteria can be found at densities greater than 4,000 per cm^2 in the surface biofilm, creating a potentially high level of exposure [48].

The symptoms encountered with Mac infections result from colonization of either the respiratory or the gastrointestinal tract, with possible dissemination to other locations in the body. Exposure to Mac organisms may occur through the consumption of contaminated food, the inhalation of air with contaminated soil particles, or contact with or ingestion, aspiration, or aerosolization of potable water containing the organisms [45].

With respect to water supplies, infection with *M. avium* and *M. intracellulare* has been well documented. Unlike gastrointestinal pathogens, where *E. coli* can be used to indicate their potential presence, no suitable indicators have been identified to signal increasing concentrations of Mac organisms in water systems [45].

6.2. Helicobacter Pylori

Helicobacter pylori has been cited as a major etiologic agent for gastritis and has been implicated in the pathogenesis of peptic and duodenal ulcer disease and gastric carcinoma. However, most individuals that are infected by this pathogen remain asymptomatic [45].

Using culture-based methods, *H. pylori* has not been isolated from environmental sources, including water [45,51]. On the contrary, molecular methods have been successful in detecting this pathogen. Fluorescence *in situ* hybridization has been successfully used to detect this pathogen in drinking water distribution systems and other water bodies. Polymerase chain reaction has also been used to detect the presence of *H. pylori* DNA in drinking water, especially associated with biofilms [45,51,52]. In drinking-water biofilms, *H. pylori* cells rapidly lose culturability, entering a viable but non-culturable state. In these biofilms, cells can persist for more than one month, with densities exceeding 10^6 cells per cm^2 [51].

How the organism is transmitted is still not fully understood. However, the fact that it has been recovered from saliva, dental plaques, the stomach, and fecal samples strongly indicates oral-oral or fecal-oral transmission. Water and food appear to be of lesser direct importance, but they can still play a significant role in situations with improper sanitation and hygiene [45].

6.3. Aeromonas Hydrophyla

In recent years, *A. hydrophila* has gained public health recognition as an opportunistic pathogen. It has been implicated as a potential agent of gastroenteritis, septicemia, meningitis, and wound infections. It can play a significant role in intestinal disorders in children under five years old, the elderly, and immunosuppressed people. [45,53,54].

Aeromonas hydrophila are Gram-negative, non-sporeforming, rod-shaped, facultative anaerobic bacilli belonging to the family *Aeromonadaceae*. Although *A. hydrophila* is usually the dominant species, other aeromonads, such as *A. caviae* and *A. sobria*, have also been isolated from human feces and from water sources [45,54].

Aeromonas species, including *A. hydrophila*, are ubiquitous in the environment. It is frequently isolated from food, drinking water, and aquatic environments [45,53,54]. In clean rivers and lakes, concentrations of *Aeromonas* spp. are usually around 10^2 colony-forming units (CFU)/mL. Groundwaters generally contain less than 1 CFU/mL. Drinking water immediately leaving the treatment plant has been found to contain between 0 and 10^2 CFU/mL. Drinking water in distribution systems can display higher *Aeromonas* concentrations, due to the growth in biofilms [45,55]. *Aeromonas* spp. have been found to grow between 5 °C and 45 °C [44,54]. *A. hydrophila* is resistant to standard chlorine treatments, probably surviving inside biofilms [56].

The common routes of infection suggested for *Aeromonas* are the ingestion of contaminated water or food or contact of the organism with a break in the skin. Drinking or natural mineral water can be a possible source of contamination for humans. No person-to-person transmission has been reported [45,54].

7. MICROBIOLOGICAL WATER ANALYSIS

7.1. The Rationale of the Use of Fecal Indicator Bacteria

The most important bacterial gastrointestinal diseases transmitted through water are cholera, salmonellosis and shigellosis. These diseases are mainly transmitted through water (and food) contaminated with feces of patients. Drinking water can be contaminated with these pathogenic bacteria, and this is an issue of great concern. However, the presence of pathogenic bacteria in water is sporadic and erratic, levels are low, and the isolation and culture of these bacteria is not straightforward. For these reasons, routine water microbiological analysis does not include the detection of pathogenic bacteria. However, safe water demands that water is free from pathogenic bacteria [57].

The conciliation of the two needs was met by the discovery and testing of indicator bacteria. Water contaminated with pathogenic species also has the normal inhabitants of the human intestine. A good bacterial indicator of fecal pollution should fulfill the following criteria: (1) exist in high numbers in the human intestine and feces; (2) not be pathogenic to humans; (3) easily, reliably and cheaply detectable in environmental waters. Additionally, the following requisites should be met if possible: (4) does not multiply outside the enteric environment; (5) in environmental waters, the indicator should exist in greater numbers than eventual pathogenic bacteria; (6) the indicators should have a similar die-off behavior as the pathogens; (7) if human fecal pollution is to be separated from animal pollution, the indicator should not be very common in the intestine of farm and domestic animals [1,4,6,57,58]. The usefulness of indicator bacteria in predicting the presence of pathogens was well illustrated in many studies, namely by Wilkes *et al.* [59].

7.2. The Composition of Human and Animal Feces

Microbiological analysis of the human feces was important in order to structure and validate the use of fecal indicator bacteria in environmental waters. Bacteria present in feces are naturally derived from the microbiota of the human gastrointestinal tract.

Although bacteria are distributed throughout the human gastrointestinal tract, the major concentration of microbes and metabolic activity can be found in the large intestine. The upper bowel (stomach, duodenum, and jejunum) has a sparse microbiota with up to 10^5 CFU/ml of contents. From the ileum on, bacterial concentrations gradually increase reaching in the colon 10^{10} to 10^{11} CFU/g [60].

It has been estimated that at least 500–1,000 different microbial species exist in the human gastrointestinal microbiota, although on a quantitative basis 10–20 genera usually predominate (Table 6). The total number of microbial genes in the human gastrointestinal tract has been estimated as 2–4 million.

Table 6. Total viable count in feces of healthy humans (children, adults and elderly)a.

Microbial group	$\log_{10}$ CFU/g feces
Bacteroides	11.3*
Eubacterium	10.7*
Bifidobacterium	10.2*
Ruminococcus	10.2*
Peptostreptococcus	10.1*
Peptococcus	10.0*
Clostridium	9.8*
Lactobacillus	9.6*
Propionobacterium	9.4*
Actinomyces	9.2*
Methanobrevibacter	8.8*
Desulphovibrio	8.4*
Fusobacterium	8.4*
Enterococci	3.5–7.2**
Enterobacteriaceae	5.9–8.0**
Escherichia coli	7.5–7.7**
Citrobacter	3.3**
Klebsiella	2.4**
Yeasts	1.0–2.5**

[a] Adapted from [61-63]** and [64]*. * Values expressed as dry weight. ** Values expressed as wet weight.

This represents an enormous metabolic potential which is far greater than that possessed by the human host [60,64].

The composition of feces from an individual is stable at genus level, but the species composition can vary markedly from day to day. The relative proportion of intestinal bacterial groups can vary between individuals [60,64].

The microflora of the human gastrointestinal tract is dominated by obligate anaerobes, which are ca. 10^3 more abundant than facultative anaerobes. The main anaerobic genera are *Bacteroides, Eubacterium* and Bifidobacteria. These organisms account for ca. 90% of the cultivable human fecal bacteria. *Bacteroides* (mainly *B. thetaiotaomicron* and *B. vulgatus*) are the most abundant organism in the human feces and account for 20–30% of cultivable bacteria. The most abundant facultative anaerobes are Enterococci and *Enterobacteriaceae*. The main *Enterobacteriaceae* genera are *Escherichia, Citrobacter, Klebsiella, Proteus* and *Enterobacter. Citrobacter* and *Klebsiella* are present in most individuals although in low numbers. *Proteus* and *Enterobacter* are only present in a minority of humans [64].

A variety of molecular techniques have been used to study the microbial composition of the human gastrointestinal tract. Results yielded by these studies have shown that many microbes detected by molecular techniques are not isolable by conventional culture-based methods. The presence of high proportions of bifidobacteria detected by culture-based methods is not supported by the results of molecular-based studies. However, the results of molecular-based approaches support many of the findings derived from culture-based methods: the dominance of the obligate anaerobes over facultative anaerobes; the presence of high counts of *Bacteroides, Clostridium* and *Eubacterium* [64].

Anaerobic bacteria such as *Bacteroides* and *Eubacterium* are not easily cultured by conventional techniques since require incubation chambers with nitrogen atmosphere. *Bifidobacterium* and *Lactobacillus* tolerate some oxygen but are fastidious bacteria growing very slowly in culture media. Therefore, these four genera are not adequate to be used as indicators of fecal pollution (the introduction of molecular techniques may improve the situation). *Citrobacter, Klebsiella* and *Enterobacter* are present in low numbers in the human intestine and are widespread in environmental waters, and therefore are also not suitable as indicators of fecal pollution. *Clostridium, Streptococcus* and *Escherichia* do not suffer from these drawbacks. Therefore, their suitability as fecal indicators has been tested since several decades.

7.3. Fecal Bacteria in Their Hosts and in the Environment

7.3.1. Bacteroides

The traditional genus *Bacteroides* included Gram-negative, non-sporeforming, anaerobic pleiomorphic rods. Many species have been transferred to other genera—*Mitsuokella, Porphyromonas, Prevotella, Ruminobacter*. Bacteroides are the most abundant bacteria in human feces. In animal feces, on the contrary, *Bacteroides* are present at low numbers. Although anaerobic, *Bacteroides* are among the most tolerant to oxygen of all anaerobic human gastrointestinal species. *B. thetaiotaomicron* is one of the most abundant species in the lower regions of the human gastrointestinal tract. *Bacteroides* have a high pathogenic potential and account for approximately two-thirds of all anaerobes isolated from clinical specimens. The most frequently isolated species has been *B. fragilis*. The survival of *Bacteroides* in environmental waters is usually much lower than the survival of coliforms [64,65].

7.3.2. Eubacterium

The traditional genus *Eubacterium* included anaerobic non-sporeforming Gram-positive rods. Some species have been transferred to other genera—*Actinobaculum, Atopobium, Collinsella, Dorea, Eggerthella, Mogibacterium, Pseudoramibacter* and *Slackia*. Cells are not very aerotolerant. Species isolated from the human gastrointestinal tract include: *E. barkeri*, E. *biforme, E. contortum, E. cylindrioides, E. hadrum, E. limosum, E. moniliforme, E. rectal* and *E. ventricosum* [64].

7.3.3. Bifidobacterium

Bifidobacteria are Gram-positive, non-sporeforming, pleiomorphic rods. Bifidobacteria are anaerobic (some species tolerate oxygen in the presence of carbon dioxide) or facultative anaerobic. The optimum growth temperature is 35–39 °C. The genus *Bifidobacterium* contains ca. 25 species, most of which have been detected in the human gastrointestinal tract [64–66].

Bifidobacteria are present in high numbers in the feces of humans and some animals. Several *Bifidobacterium* species are specific either for humans or for animals. *B. cuniculi* and *B. magnum* have only been found in rabbit fecal samples, *B. gallinarum* and *B. pullorum* only in the intestine of chickens and *B. suis* only in piglet feces. In human feces, the species composition changes with the age of the individual. In the intestine of infants *B. breve* and *B. longum* generally predominate. In the adult, *B. adolescentis, B. catenulatum, B. pseudocatenulatum* and *B.*

longum are the dominant species. In both human and animal feces, bifibobacteria are always much more abundant than coliforms [64–66].

Bifidobacteria have been found in sewage and polluted environmental waters, but appears to be absent from unpolluted or pristine environments such as springs and unpolluted soil. This results from the fact that upon introduction into the environment, bifidobacteria decrease appreciably in numbers, probably due to their stringent growth requirements. Bifidobacteria grow poorly below 30 °C and have rigorous nutrient requirements. Reports on the survival of bifidobacteria in environmental waters indicate that their survival is lower than that of coliforms [64–66].

The presence of bifidobacteria in the environment is therefore considered an indicator of fecal contamination. Since some species are specific for humans and animals, the identification of *Bifidobacterium* species present in the polluted water could, in principle, provide information on the origin of fecal pollution [64–66].

A study carried out in a highly contaminated stream near Bologna, Italy, revealed that *B. adolescentis*, *B. catenulatum*, *B. longum*, *B. pseudocatenulatum* and *B. thermophilum* were the most representative species, whereas *B. angulatum*, *B. animalis* subsp. *animalis (B. animalis)*, *B. breve*, *B. choerinum*, *B. minimum*, *B. pseudolongum* subsp. *globosum* (*B. globosum*) and *B. subtile* occurred only in low numbers [66].

Bifidobacteria are the less studied of all fecal bacteria, due to the technical difficulties in their isolation and cultivation. Other Gram-positive bacteria, such as *Streptococcus* and *Lactobacillus*, which may occur in higher numbers than bifidobacteria, can inhibit their growth. Although selective media has been designed for the isolation of bifidobacteria from environmental waters, the outcome is still unsatisfactory, with appreciable numbers of false positives and low recovery percentages [64–66].

7.3.4. Clostridia

The genus *Clostridium* is one of the largest genera of the prokaryotes containing 168 validly published species. From these, 77 (including *C. perfringens*) are considered to belong to a united group—*Clostridium sensu stricto* [64,67,68].

Clostridia are Gram-positive rods, forming endospores. Most of the clostridial species are motile with peritrichous flagellation. Cells are catalase-negative and do not carry out a dissimilatory sulphate reduction. Clostridia usually produce mixtures of organic acids and alcohols from carbohydrates and proteins. Many species are saccharolytic and proteolytic. Some species fix atmospheric dinitrogen [64,67,68].

The genus *Clostridium* includes psychrophilic, mesophilic, and thermophilic species. The major role of these organisms in nature is in the degradation of organic material to acids, alcohols, CO_2, H_2, and minerals. Frequently, a butyric acid smell is associated with the proliferation of clostridia. The ability to form spores that resist dryness, heat, and aerobic conditions makes the clostridia ubiquitous [64,67,68].

Most species are obligate anaerobic, although tolerance to oxygen occurs. Oxygen sensitivity restricts the habitat of the clostridia to anaerobic areas or areas with low oxygen tensions. Growing and dividing clostridia will, therefore, not be found in air saturated surface layers of lakes and rivers or on the surface of organic material and soil. Clostridial spores, however, are present with high probability in these environments, and will germinate when oxygen is exhausted and when appropriate nutrients are present [64,67,68].

C. perfringens ferment lactose, sucrose and inositol with the production of gas, produce a stormy clot fermentation with milk, reduce nitrate, hydrolyze gelatin and produce lecithinase and acid phosphatase. The species is divided into five types, A to E, on the basis of production of major lethal toxins [68,69].

C. perfringens appears to be a universal component of the human and animal intestine, since has been isolated from the intestinal contents of every animal that has been studied. Humans carry *C. perfringens* as part of the normal endogenous flora. The main site of carriage is the distal gastrointestinal tract. The principal habitats of type A are the soil and the intestines of humans, animals, and birds. Types B, C, D, and E appears to be obligate parasites of animals and occasionally are found in humans [68,69].

Clostridium perfringens is the most frequently isolated *Clostridium* in clinical microbiology laboratories, although it seldom causes serious infections. *C. perfringens* is isolated from infections in humans and the organism most commonly found in gas gangrene in humans. *C. perfringens* is most commonly isolated from infections derived from the colonic flora, namely peritonitis or abdominal abscess [68,69].

This organism is a common cause of food poisoning due to the formation of the enterotoxin in the intestine. *C. perfringens* food poisoning is seldom fatal, being marked by diarrhea and nausea, with no vomiting and no fever [68,69].

Sources yielding *C. perfringens* include soil and marine sediment samples worldwide, clothing, raw milk, cheese, semi-preserved meat products, and venison. Like *E. coli*, *C. perfringens* does not multiply in most water environments and is a highly specific indicator of fecal pollution. Berzirtzoglou *et al.* [70] reported a comparative study on the occurrence of vegetative cells and spores of *Clostridium perfringens* in a polluted station of the lake Pamvotis, in rural North-West Greece. The numbers of *C. perfringens* varied according to the water depth. Sporulated forms were found in all sampling sites with the exception of the surface sampling.

7.3.5. Lactobacillus

Lactobacilli are non-sporeforming Gram-positive long rods. There are more than thirty species in the genus. Most are microaerophillic, although some are obligate anaerobes. Cells are catalase-negative and obtain their energy by the fermentation of sugars, producing a variety of acids, alcohol and carbon dioxide. Lactobacilli have complex nutritional requirements and in agarized media may need the supplementation with aminoacids, peptides, fatty-acid esters, salts, nucleic acid derivatives and vitamins. Lactobacilli very rarely cause infections in humans [64].

7.3.6. Enterococci

Enterococci are Gram-positive, non-sporeforming, catalase-negative ovoid cells. Cells occur singly, in pairs or short chains. Optimal growth for most species is 35–37 °C. Some will grow at 42–45 °C and at 10 °C. Growth requires complex nutrients but is usually abundant on commonly used bacteriological media. Cells are resistant to 40% bile, 0.4% azide, 6.5% sodium chloride, have β-glucosidase and hydrolyze esculin. The enterococci are facultative anaerobic but prefer anaerobic conditions [64,71].

The genus was separated from *Streptococcus* in the 1980s. Enterococci form relatively distinct groups. Members of such groups exhibit similar phenotypic characteristics and species delimitation can be difficult. The *E. faecalis* group contains, among others, *E. faecalis*. The *E. avium* group contains, among others, *E. avium*. The *E. faecium* group contains, among others, *E. faecium*, *E. durans* and *E. hirae*. The *E. gallinarum* group contains, among others, *E. gallinarum* [64,71].

Most species are part of the intestinal flora of mammals, reptiles, birds, and other animals. In the human digestive tract, *E. faecalis* is the prevailing species, although in particular situations, *E. faecium* may predominate. In poultry, *E. cecorum*, *E. durans*, *E. faecalis*, *E. faecium* and *E. hirae* and dominate the intestinal flora [64,71].

Enterococci have been increasingly isolated from a variety of nosocomial and other infections, mainly from the urinary tract and wound infections, bacteremias, and endocarditis [64,71].

Although enterococci are considered only a temporary part of the microflora of plants, in optimal conditions, cells can proliferate on their surfaces. *E. casseliflavus, E. faecalis, E. faecium, E. hirae, E. mundtii* and *E. sulfureus* have been isolated from plants. They are generally isolated more often from flowers than from buds or leaves [64,71].

Enterococci are naturally present in many kinds of foods, especially those of animal origin such as milk and milk products, meat and fermented sausages. Enterococci are usually considered secondary contaminants of food, although they often play a positive role in ripening and aroma development of some types of cheeses [64,71]. Although soil is not a natural habitat for enterococci, cells can be found in this habitat due to the transport by rain [64,71].

Environmental waters are not a natural habitat for enterococci and their presence in this milieu is considered the result of fecal pollution. The most common species found in environmental waters are *E. durans, E. faecalis, E. faecium* and *E. hirae*, and less commonly, *E. avium, E. cecorum, E. columbae* and *E. gallinarum*. However, pristine waters in Finland have been reported to contain *E. casseliflavus* [64,71].

In environmental samples (compost, sewage effluent, harbor sediments, brackish water and swimming pool water), Pinto *et al.* [72] reported the isolation of *E. casseliflavus, E. durans, E. faecalis, E. faecium, E. gallinarum* and *E. hirae. E. durans, E. faecium* and *E. hirae* were isolated from all sources except from harbor sediments. *E. raffinosus* was only isolated from compost and swimming pool water. *E. faecalis* and *E. faecium* accounted for the vast majority of enterococcal strains.

7.3.7. Escherichia

Escherichia, a member of *Enterobacteriaceae*, are oxidase-negative catalase-positive straight rods that ferment lactose. Cells are positive in the Methyl-Red test, but negative in the Voges-Proskauer assay. Cells do not use citrate, do not produce H_2S or lipase, and do not hydrolyze urea [73]. *E. coli* is a natural and essential part of the bacterial flora in the gut of humans and animals. Most *E. coli* strains are nonpathogenic and reside harmlessly in the colon. However, certain serotypes do play a role in intestinal and extra-intestinal diseases, such as urinary tract infections [43]. In a study of the enteric bacteria present in the feces of Australian mammals, Gordon and FitzGibbon [74] reported that *E. coli* was the commonest species, being isolated from nearly half of the species studied.

7.3.8. Citrobacter

Citrobacter, a member of *Enterobacteriaceae*, are motile straight rods. Cells are oxidase-negative, catalase-positive and positive in the Methyl-Red test. Cells use citrate, are negative in the Voges-Proskauer test and do not decarboxylate lysine [73].

In a study of the enteric bacteria present in the feces of Australian mammals, Gordon and FitzGibbon [74] reported the isolation of *C. amalonaticus, C. freundii* and *C. koseri* (*C. diversus*). *Citrobacter* species can be isolated from different clinical sites. In particular, *C. freundii* is an intestinal inhabitant of humans that may sometimes have—or acquire—the ability to produce an enterotoxin and thus become an intestinal pathogen. *Citrobacter* is reported to occur in environments such as water, sewage, soil and food [75,76].

7.3.9. Klebsiella and Raoultella

Klebsiella and *Raoultella* are *Enterobacteriaceae*, oxidase-negative catalase-positive non-motile straight rods, surrounded by a capsule. Cells decarboxylate lysine, but are ornithine and arginine dihydrolase negative. Cells grow on KCN, do not produce H_2S and ferment most carbohydrates [73].

In humans, *K. pneumoniae* is present as commensal in the nasopharynx and in the intestinal tract. *Klebsiella* spp. can cause human diseases, ranging from asymptomatic colonization of the intestinal, urinary, or respiratory tract to fatal septicemia. *Klebsiella* are mostly considered nosocomial pathogens. *K. pneumoniae* and *Enterobacter aerogenes* (*K. mobilis*) are most frequently involved, although *K. oxytoca* and *R. planticola*, and rarely *R. terrigena*, can be found. In the hospital, the principal reservoir of *K. pneumoniae* is the gastrointestinal tract of patients. The principal vectors are the hands of personnel [77,78]. In a study of the enteric bacteria present in the feces of Australian mammals, Gordon and FitzGibbon [74] reported the isolation of *K. pneumoniae* and *K. oxytoca*.

Klebsiellae are ubitiquous in the environment. They have been found in a variety of environmental situations, such as soil, vegetation, or water, and they influence many biochemical and geochemical processes. They have been recovered from aquatic environments receiving industrial wastewaters, plant products, fresh vegetables, food with a high content of sugars and acids, frozen orange juice concentrate, sugarcane wastes, living trees, and plants and plant byproducts. They are commonly associated with wood, sawdust, and waters receiving industrial effluents from pulp and paper mills and textile finishing plants (see below). *Klebsiella* have been isolated from the root surfaces of various plants. *K.*

pneumoniae, *K. oxytoca*, and *R. planticola* are all capable of fixing dinitrogen [77,78].

7.3.10. Enterobacter

Enterobacter a member of *Enterobacteriaceae*, are motile straight rods. Cells are positive in the Voges-Proskauer test VP and in Simmons citrate agar. Cells do not decarboxylate lysine, but are ornithine positive. Malonate is usually utilized and gelatin is slowly liquefied. Cells do not produce H_2S, deoxyribonuclease and lipase [73].

In a study of the enteric bacteria present in the feces of Australian mammals, Gordon and FitzGibbon [74] reported the isolation of *Enterobacter cloacae* subsp. *cloacae* (*E. cloacae*), *E. cancerogenus* (*E. taylorae*) and *E. aerogenes* (*Klebsiella mobilis*).

Before the widespread use of antibiotics, *Enterobacter* species were rarely found as pathogens, but these organisms are now increasingly encountered, causing nosocomial infections such as urinary tract infections and bacteremia. In addition, they occasionally cause community-acquired infections [79,80].

In the USA, the Surveillance and Control of Pathogens of Epidemiological Importance project analyzed 24,179 nosocomial bloodstream infections, from 1995–2002. *Enterobacter* species were the second most common gram-negative organism, behind *Pseudomonas aeruginosa*. Both bacteria were reported to each represent 4.7% of bloodstream infections in intensive care units. *Enterobacter* species represented 3.1% of bloodstream infections in non-intensive care units. Of nearly 75,000 gram-negative organisms collected from intensive care units' patients in the USA, between 1993 and 2004, *Enterobacter* species comprised 13.5% of the isolates. Multidrug resistance increased over time, especially in infections caused by *E. cloacae* [81].

In the USA, the National Healthcare Safety Network reported a study on healthcare-associated infections between 2006 and 2007. They found *Enterobacter* species to be the eighth most common cause of healthcare-associated infections (5% of all infections) and the fourth most common gram-negative cause of these infections [82].

Enterobacter cloacae subsp. *cloacae* (*E. cloacae*) occurs in the intestinal tracts of humans and animals, in hospital environments, the skin, in water, sewage, soil, meat. Nitrogen-fixing strains have been isolated from the roots of rice plants. *E. amnigenus* has been mostly isolated from water, but some strains were isolated from clinical specimens from the respiratory tract, wounds and feces. *E. asburiae* strains were isolated from clinical specimens, mostly urine, respiratory tract, feces, wounds, and blood [79,80].

7.4. Origin of the Use of Fecal Indicator Bacteria

Historically, the design and use of indicators of fecal pollution comes from the end of the 19th to beginning of the 20th century. In 1880, von Fritsch described *Klebsiella pneumoniae* and *K. pneumoniae* subsp. *rhinoscleromatis* (*Klebsiella rhinoscleromatis*) as micro-organisms characteristically found in human feces [83]. In 1885, Escherich described several microorganisms in the feces of newborn and suckling babies. This included a motile, rod-shaped microorganism that caused milk to clot, which was named "*Bacterium coli commune*". He observed that within a few weeks after birth, this bacterium became the dominant organism in the infant colon [6]. Also in 1885, Percy and Grace Frankland started the first routine bacteriological examination of water in London, using Robert Koch's solid gelatin media to count bacteria [83]. In 1891, Percy and Grace Frankland came up with the concept that organisms characteristic of sewage must be identified to provide evidence of potentially dangerous pollution [83]. In 1892, Schardinger proposed that since "*Bacterium coli*" was a characteristic component of the fecal flora, its presence in water could be taken as an indication of the presence of fecal pollution and therefore of the potential presence of enteric pathogens [6]. Soon after the description of "*Bacterium coli*", other bacteria were isolated from stools and water—*Klebsiella* in 1882 and *Enterobacter* in 1890 [6]. By 1893, the "Wurtz method" of enumerating "*Bacterium coli*", by direct plating water samples on litmus lactose agar, was being used by sanitary bacteriologists. This was based on the concept of acid and gas production (detected by the Durham tube) from lactose as a diagnostic feature [6]. In 1905, MacConkey described his now famous MacConkey's broth, which was diagnostic for lactose-fermenting bacteria tolerant of bile salts. Coliforms were already considered to be a heterogeneous group of organisms, many of which were not of fecal origin. The origins of the critical observation that "*Bacterium coli*" was largely fecal in origin while other coliforms were not, could be claimed by Winslow and Walker in 1907 [83].

Various classification schemes for coliforms have emerged. The earliest were those of MacConkey in 1909, which recognized 128 different coliform types, while Bergey and Deehan in 1908, identified 256. By the early 1920s, differentiation of coliforms had come to a series of correlations that suggested that indole production, gelatin liquefaction, sucrose fermentation and the Voges–Proskauer reaction were among the more important tests for determining fecal contamination. These developments culminated in the IMViC (Indole, Methyl Red, Voges–Proskauer and Citrate) tests for the differentiation of so-called fecal coliforms, soil coliforms and intermediates [83].

7.5. Fecal Indicator Bacteria

7.5.1. Coliforms

Total coliforms are Gram-negative, oxidase-negative, non-sporeforming rods, that ferment lactose with gas production at 35–37 °C, after 48h, in a medium with bile salts and detergents [1,4,6,57,84]. When the test of coliforms is carried out with environmental waters, several species of the four *Enterobacteriaceae* genera *Escherichia*, *Klebsiella*, *Enterobacter* and *Citrobacter* give positive results and therefore are coliforms according to this definition. However, the environmental significance of these four genera is very disparate as discussed in the present text. Therefore, total coliform counts are not necessarily a measure of fecal pollution and indeed can have no relation with this cause [1,4,6,84].

Fecal coliforms (or thermotolerant coliforms) are traditionally defined as coliforms that ferment lactose at 44.5 °C in a medium with bile salts [1,4,57,84]. The range of species detected by the experimental procedure is much lower than that of total coliforms. With environmental polluted waters, only *E. coli*, and *K. oxytoca* and *K. pneumoniae* gave positive results in the test [85].

Traditional tests for total and fecal coliforms are carried out either by the multiple-tube fermentation technique or by filtration through membrane. The multiple-tube fermentation technique is used for medium or highly contaminated waters, and the filtration through membrane for low or very low contaminated waters. Filtration through membrane is a very sensitive technique since can detect one (culturable) cell in 500 or even 1,000 mL of water. However, both methods take several days to complete and do not detect viable but non-culturable bacteria [3,57,86]. These limitations stimulate the discovery of alternative methods, faster and, if possible, less prone to false negative results such as those caused by the viable but non-culturable bacteria.

The detection of β-D-galactosidase activity (at 37 °C) is usually a good marker for total coliforms in environmental waters, since most of these bacteria display this enzymatic activity [1,3,57,87–91]. Most *Escherichia*, *Citrobacter*, *Enterobacter*, *Klebsiella* and *Raoultella* strains have galactosidase. *Hafnia*, *Serratia* and *Yersinia* also possess this enzymatic activity. Most *Proteus*, *Salmonella* and *Edwardsiella* strains do not display β-galactosidase [92–95]. Ca. 10% of the coliform strains isolated from the environment do not have an active formic hydrogenolyase (cleaves formate with the formation of CO_2) and therefore do not produce gas being undetected by the traditional techniques but are detected by the assay of β-galactosidase activity [57,96,97].

β-galactosidase cleaves lactose in glucose and galactose, and can be detected by using colored or fluorescent markers that change color after enzyme action, such as XGAL (5-bromo-4-chloro-3-indol-β-galactopyranoside) and ONPG (O-nitrophenyl-β-D-galactopyranoside) or MUGAL (4-methylumbelliferyl-β-D-galactopyranoside), respectively [57, 96,97].

In environmental waters, the presence of *Aeromonas* or *Vibrio cholerae* can be a source of false positives in the β-D-galactosidase assay, since these bacteria have galactosidase, but are not coliforms [93,95,96,98]. Additionally, in particular environments, such as estuaries, β-galactosidase activity can overestimate total coliform count due to UV-stimulated enzymatic activity in certain bacteria such as *E. coli* [86].

The detection of β-D-glucuronidase activity (at 44.5 °C) is, generally, a good marker for fecal coliforms in environmental polluted waters and very specific for *E. coli* [13,85,87–91,97,99–101]. In Gram-negative bacteria, this enzymatic activity if found in most *E. coli* strains and in some *Salmonella* and *Shigella* strains [92–95,97,102]. *Aeromonas, Citrobacter, Enterobacter*, non-*coli Escherichia*, *Hafnia*, *Klebsiella*, *Proteus*, *Serratia*, *Vibrio*, *Yersinia*, and most *Salmonella* strains do not display β-glucuronidase activity [93–95,101,102].

β-D-glucuronidase activity can be detected by using colored or fluorescent markers that change color after enzyme action, such as XGLUC (5-bromo-4-chloro-3-indoxyl-β-D-glucuronide), IBDG (indoxil-β-glucuronide), and MUGLU (4-methylumbelliferyl-β-D-glucuronide), respectively [57,97,100].

The presence of this enzyme in some strains of *Bacteroides*, *Flavobacterium*, *Staphylococcus*, *Streptococcus*, in anaerobic corynebacteria and *Clostridium*, has also been reported [93,95–97,102]. β-D-glucuronidase activity in fecal bacteria other then *E. coli* (*Bacteroides*, bifidobacteria, clostridia, enterococci and *Lactobacillus*) is very limited [61]. Although all these glucuronidase positive bacteria could lead to false positive detections in the fecal coliform test, experimental results for environmental polluted waters indicate a significant correlation between fecal coliform detection using conventional techniques and the glucuronidase assay, suggesting that false positives are not significant [85,96].

The detection of total coliforms and fecal coliforms by enzymatic methods are much less time consuming than traditional techniques. With fluorescent markers and the use of a spectrofluorimeter the detection of coliforms can be performed in minutes [57,101]. However, in very low contaminated waters, enzymatic methods might not be able to detect coliform cells. Moreover, on-line monitoring of glucuronidase activity is currently too insensitive to replace culture based detection of *E. coli*. Nevertheless, on-line enzymatic methods can be a valuable complementary tool for high temporal resolution monitoring. More research is needed in order to enhance sensitive and lower detection limits of available on-line glucuronidase techniques.

The seminal work of Leclerc *et al.* [103] clarified the diversified roles that coliforms have in the environment and the real meanings of the tests on total coliforms and fecal coliforms. It was shown that *Enterobacteriaceae* encompass three groups of bacteria with very different roles in the environment. Group I harbored only *E. coli*. Since this species usually do not survive for long periods outside this environment (but see topic 10), it was considered a good and reliable indicator of fecal pollution (both animal and human). Group II, the "ubiquitary" group, encompassed several species of *Klebsiella* (*K. pneumoniae* and *K. oxytoca*), *Enterobacter* (*Enterobacter cloacae* subsp. *cloacae*, *E. aerogenes*) and *Citrobacter* (*C. amalonaticus*, *C. koseri* and *C. freundii*). These bacteria live in the animal and human gut, but also in the environment, and are easily isolated from the soil, polluted water and plants. Their presence in polluted waters does not necessarily indicate fecal pollution. Finally Group III was composed of *Raoultella planticola*, *R. terrigena*, *Enterobacter amnigenus* and *Kluyvera intermedia* (*Enterobacter intermedius*), *Serratia fonticola*, and the genera *Budvicia*, *Buttiauxella*, *Leclercia*, *Rahnella*, *Yersinia*, and most species of *Erwinia* and *Pantoea*. These bacteria live in fresh waters, plants and small animals. They grow at 4 °C, but not at 41 °C. They are not indicators of fecal pollution, although can be detected in the total coliform test. Leclerc *et al.* concluded that: (1) in the enterobacteria, *E. coli* is the only true and reliable indicator of fecal pollution in environmental waters; (2) the traditional total coliform test should be abandoned, since it can detect bacteria that have no connection with fecal pollution; (3) the detection of fecal coliforms must be carried out at 44.5 °C, and positive results confirmed by identification to species levels in order to exclude false positives such as *K. pneumoniae*.

7.5.2. Streptococci and Enterococci

Fecal streptococci also belong to the traditional indicators of fecal pollution. Fecal streptococci are Gram-positive, catalase-negative, non-sporeforming cocci that grow at 35 °C in a medium containing bile salts and sodium azide. Cells hydrolyze esculin [1,4,57]. Azide is a strong inhibitor of the respiratory chain. Since streptococci are one of the very few bacteria that have no respiratory chain, the test is very specific for this group, and false positives are rarely found [104,105].

Fecal enterococci (*E. faecalis*, *E. faecium*, *E. avium* and *E. gallinarum*) are fecal streptococci that grow in the presence of 6.5% NaCl at 45 °C.

Selective media use these particular characteristics in order to separate enterococci from the other streptococci [104,105].

Several studies [104,106] have reported on the microbiological composition of human and animal (cattle, chicken, deer, dog, fowl, goose, and swine) feces. *E. faecalis* and *E. faecium* were present in human and animal feces. However, whereas human feces almost have only these two enterococci, in the animals others species co-occur, like *E. avium*, *E. cecorum*, *E. durans*, *E. gallinarum* and *E. hirae*. It was concluded that in urban areas where contamination with dog and chicken feces is not likely, the best marker for human fecal pollution was *E. faecalis*.

The intestinal enterococci group has been used as an index of fecal pollution. In human feces, the numbers of intestinal enterococci are generally about an order of magnitude lower than those of *E. coli* (Table 6). Most species do not grow in environmental waters. In this milieu, fecal enterococci are able to survive longer, are more resistant to drying and chlorination, than *E. coli* [1,84].

However, caution should be taken with interpreting the results obtained by the enterococci procedure in water analysis. Enterococci and other group D-streptococci are present in many foods, especially those of animal origin. The isolation of *E. faecalis* and *E. faecium* was used to indicate fecal contamination of food. However, enterococci are now also considered as normal parts of the food microflora and not only as indicators for poor hygiene [104]. In addition, agricultural soils and crops with added with manure also harbor enterococci [105].

7.5.3. The use of ratios between indicator counts

The ratio of counts of fecal coliforms to fecal streptococci has been proposed as a means to differentiating between contamination from human and animal sources. Ratios greater than 4 have been suggested to indicate a human source whereas ratios less than 0.7 suggest an animal source. This results from the fact that streptococcal concentrations in human feces are generally less than coliforms (Table 6). In contrast, in animal feces fecal streptococci generally outnumber fecal coliforms (Table 7). In urban sewage, fecal streptococci tend to be present in concentrations 10–100 times less than fecal coliforms [65].

Geldreich [107] summarized the information available on the fecal coliforms to fecal streptococci ratios in the feces of warmblooded animals, and reported the following values: human feces, 4.3; cattle, sheep, and poultry, from 0.104 to 0.421; and wild animals (including rabbits, field mice, chipmunks, and birds), 0.0008 to 0.043. Fecal coliforms to fecal streptococci ratios for the feces of wild animals appear to be at least 10-fold lower than those of domestic livestock.

Table 7. Bacteria in the feces of farm and domestic warm-blooded animalsa.

Animal	Log$_{10}$ cells/g wet weight feces		
	Fecal coliforms	**Fecal streptococci**	***Clostridium perfringens***
Chicken	5.4	6.1	2.3
Duck	7.5	7.7	-
Horse	4.1	6.8	< 0
Pig	6.5	7.9	3.6
Sheep	7.2	7.6	5.3
Turkey	5.5	6.4	-
Cat	6.9	7.4	7.4
Dog	7.1	9.0	8.4

[a] Adapted from [83].

Doran and Linn [108] reported a study of the runoff from a cow-calf pasture in eastern Nebraska (USA), monitored during a three-year period. It was concluded that the fecal coliforms to fecal streptococci ratio in pasture runoff was useful in identifying the relative contributions of cattle and wildlife and in evaluating the effects of cattle management and distribution on runoff water quality. Ratios below 0.05 were indicative of wildlife sources and ratios above 0.1 were characteristic of grazing cattle. Fecal coliforms to fecal streptococci ratios of diluted cattle waste in excess of 1 were interpreted as the result of differential aftergrowth and die-off between fecal coliforms and fecal streptococci. Ratios between 0.7 and 4.0 may indicate situations where cattle are localized close to sampling or outflow points.

However, the interpretation of this ratio should be cautious. It has been observed a shift in the ratio with time and distance from the fecal pollution source. This resulted from the fact that both in surface and groundwaters, fecal streptococci are more persistent than fecal coliforms. Therefore increasing the distance from the pollution point and with passing time, the ratio tends to decrease without a change in the nature of the pollution source [65]. For these reasons, this ratio has been considered by some authors as too unreliable to be useful in characterizing pollution sources [65,84].

The ratio of fecal enterococci to fecal streptococci differs among vertebrate species. Humans have a predominance of enterococci, whereas animals contain appreciable amounts of streptococci. However, since enterococci are also present in animals and are more persistent in the environment than other fecal streptococci, the identification of the enterococci and streptococci species present in polluted waters, and the concomitant calculation of this ratio is generally considered unreliable as an indicator of the source of fecal pollution [65].

7.5.4. Limitations of coliform and enterococcus counts as indicator of fecal pollution

An extreme case of uselessness of the determination of total and fecal coliforms and enterococci in the assessment of fecal pollution has been demonstrated by several authors studying the microbiology of pulp and paper mill effluents.

Caplenas and Kanarek [109] reported a study of pulp and paper mills located in Wisconsin (USA). Fresh water supplies, re-cycled water within mills, treated effluent wastewater and waters receiving effluent wastes downstream, were assessed for the presence of fecal coliforms and *Klebsiella*. Wastewaters prior to treatment contained fecal coliforms and *Klebsiella*. Up to 84% of the fecal coliforms (detected by the standard test procedure) were indeed *Klebsiella*. In treated effluent wastewaters this value reached 90%. Treatment of the wastewater lowered the concentration of "true" fecal bacterial contamination, but since *Klebsiella* grew rapidly in the wastewaters, fecal coliform counts were high, although no true fecal contamination was involved. The source of *Klebsiella* was traced to the early pulping stages in the mills. *Klebsiella* maintains a wood, bark or soil reservoir. It was concluded that: (1) *Klebsiella* are ubiquitous in the pulp and paper mill industry processing stages; (2) the standard procedure for fecal coliform estimation is useless to assess the microbiological quality of the effluents of these industries; (3) The assay of *E. coli* should replace the fecal coliform detection procedure.

Gauthier *et al.* [110] and Gauthier and Archibald [58] reported two studies of seven pulp and paper mills in Ontario and Quebec, Canada. Total and fecal coliforms and enterococci were detected in nearly all the biotreatment, biosolids (sludges), and in-mill water system samples. In the mill samples, the majority of the fecal coliforms (detected by the standard test procedure) were *K. pneumoniae*, *Raoultella terrigena* and *Raoultella planticola*, with *E. coli* in minority. *E. faecalis* and *E. faecium* were detected in relatively large numbers in most samples from all of the seven mills examined. Other coliforms such as *Enterobacter* spp. and *Citrobacter freundii* were occasionally recovered from total and fecal coliform tubes. Biofilms established in the piping, tanks, and machinery where thermal and pH conditions permit were the most likely source of these bacteria. Analyses using two independent *Salmonella* detection/enumeration methods showed no detectable *Salmonella* cells in the sludges and final effluents of the five mills tested. It was concluded that for these particular systems, the determination of total and fecal coliforms and enterococci is useless and have no relationship with real fecal pollution. These studies also demonstrated the importance of checking the identities of bacteria causing the positive results in the tests. Both *Escherichia* and *Klebsiella* can give positive results in the fecal coliform test, but their ecological meaning is opposite.

Another important case of failure of the use of coliforms to detect fecal pollution was the 1993 *Cryptosporidium* outbreak in Milwaukee (USA).

Cryptosporidium parvum, a protozoan parasite that causes gastrointestinal illness, is transmitted by ingestion of oocysts excreted in human or animal feces. Typical modes of transmission include person to person, animal to person, by exposure to contaminated food or water [111,112].

From 1990 to 2000, at least 10 cryptosporidiosis outbreaks associated with contaminated drinking water were reported in the USA. In 1993, an estimated 403,000 residents of the greater Milwaukee area (Wisconsin, population, ca. 1.61 million) became ill when an ineffective filtration process led to the inadequate removal of *Cryptosporidium* oocysts in one of two municipal water-treatment plants [111,112].

It was the largest waterborne disease outbreak in documented USA history. Over the span of approximately two weeks, people became ill with stomach cramps, fever, diarrhea and dehydration caused by the pathogen. More than half the people who received residential drinking water from the southern water-treatment plant became ill, which was twice the rate of illness among people whose residential drinking water came mainly from the northern water-treatment plant. Over 54 deaths were attributed to this outbreak, mostly among the elderly and immunocompromised people, such as AIDS patients [111,112].

Standard microbiological water analysis was ineffective in detecting this parasite. Indeed, throughout the period from February to April, samples of treated water from both plants were negative for coliforms. The origin of the contamination was determined as water from Lake Michigan. No specific source of the *Cryptosporidium* was ever identified but runoff from abnormally heavy spring rains most likely carried the parasite to the lake [111,112].

7.5.5. Clostridium perfringens

Sulphite-reducing clostridia, namely *Clostridium perfringens*, are spore-forming Gram-positive, non-motile, anaerobic, sulfite-reducing rods. *C. perfringens* is present in higher numbers in the feces of some animals, such as dogs, than in the feces of humans and less often in the feces of many other warm-blooded animals. The numbers excreted in feces are normally substantially lower than those of *E. coli.*

Clostridium spores are exceptionally resistant to unfavorable conditions in water environments, including UV irradiation, temperature and pH extremes, and disinfection processes, such as chlorination. Although clostridia probably do not growth in surface waters, the high resistance of their spores makes their presence ubiquitous in environmental waters [1,4,84,113].

The presence of chlorine in water rapidly inactivates indicator bacteria such as *E. coli* and coliforms, but it leaves the most resistant pathogens almost unaffected for several hours. This creates a false sense of security by providing negative coliform and negative *E. coli* results to authorities responsible for water testing. *Giardia* cysts, *Crystosporidium* oocysts, and human enteric viruses all have higher resistance to disinfectants and constitute a major public health risk if distribution system integrity is breached. *C. perfringens* spores are less affected by the residual concentrations of chlorine. Testing for the spores of this bacterium can probably provide an added margin of safety in the evaluation of treatment [114].

7.5.6. Correlations between parameters used to assess fecal pollution

In environmental waters, several studies have reported significant correlations between indicators of fecal pollution and between indicators and pathogenic gastrointestinal bacteria.

Charriere *et al.* [115] reported a study of deep aquifer waters (raw waters and piped chlorinated waters) in Normandy, France. In heavily contaminated raw waters and in slightly contaminated treated waters, fecal coliforms and enterococci were correlated.

Martins *et al.* [116] reported a study of 60 public outdoor swimming pools in Sao Paulo city, Brazil. Total coliforms, fecal coliforms and fecal streptococci levels increased with number of bathers and water temperature, and decreased with chlorine levels. All these indicators were significantly correlated with each other.

Ferguson *et al.* [117] reported a study carried out in Georges River, in the Sydney region, Australia. In the water column, concentrations of fecal coliforms, fecal streptococci, *C. perfringens* spores were all positively correlated with each other. Isolation of *Salmonella* spp. were most frequent during rainfall and sewage overflow events. In the water column, 55% of samples contained *Salmonella* when fecal coliform densities exceeded 2,000 CFU/100 mL.

Medema *et al.* [118] reported a study of seven different fresh water sites normally used for triathlon competitions. Sites were small rivers, channels, lakes and harbors, and were influenced by sewage effluents and agricultural run-off. When data from all triathlons were pooled, geometric mean densities of fecal coliforms and *E. coli*, of *E. coli* and fecal enterococci and of fecal coliforms and fecal enterococci, were significantly correlated.

Polo *et al.* [119] reported a study of water samples obtained from 213 beaches, eight rivers and 14 freshwaters in north-eastern Spain. In freshwaters and heavily contaminated seawaters, *Salmonella* and fecal coliforms were correlated, while in less contaminated seawaters, the highest correlation was with between *Salmonella* and fecal streptococci.

Byamukama *et al.* [99] reported a study of the microbiology of Nakivubo channel, Uganda. This channel receives raw sewage from slums, industrial effluents, and discharges from a sewage treatment plant and from a complex of slaughterhouses. Water from eight sampling sites was assessed for the presence of total and fecal coliforms, *E. coli* and sulphite-reducing clostridia. All microbiological parameters were significantly correlated.

Noble *et al.* [120] reported a comparative determination of total coliforms, fecal coliforms and enterococci in 108 sites along the southern California coastline, USA. Results by traditional and enzymatic methods and from all three parameters were correlated.

Harwood *et al.* [121] reported a study on indicator and pathogenic microorganisms carried out in six wastewater reclamation facilities in the USA. Data from disinfected effluent (reclaimed water) samples were analyzed separately (by facility) and as a pooled data set (all facilities). Significant correlations between indicator organism concentrations were observed in the pooled data sets, namely for total and fecal coliforms.

Cabral and Marques [85] reported a microbiological study of a polluted river (Febros) in the Great Oporto area, northwest Portugal. Total and fecal coliforms, fecal streptococci and enterococci were all significantly correlated with each other.

Touron *et al.* [122] reported a study carried out in the Seine estuary, France. Water was sampled at nine stations (along an upstream/downstream transect of 156 km), during nine years, for fecal coliforms, *E. coli*, enterococci and *Clostridium perfringens* spores. At the upstream part of the estuary (at Poses), *Salmonella* and fecal coliforms, and *E. coli* and enterococci counts were correlated. At the mouth of the estuary (at Honfleur), significant correlation was found for *Salmonella* and enterococci counts. No significant correlation between concentrations of any combination of indicator organism and pathogen was observed.

Wilkes *et al.* [59] reported a comparative study on the presence and concentration of several pathogenic and indicator bacteria in the surface water of a Canadian river. Surface water was collected within the South Nation River basin in eastern Ontario, from the river proper, and from several lower stream order tributaries. Using data aggregated during the entire multi-year study, significant correlations were found among all indicator bacteria - total and fecal coliforms, *E. coli*, *Enterococcus*, and *C. perfringens*.

However, in others studies no correlation was found between the different fecal indicator bacteria. Garrido-Pérez *et al.* [123] reported a study of the bathing seawater quality in 18 Spanish beaches near the Strait of Gibraltar. Sample locations were selected as a single point located in the area of highest bather density of each beach. No significant correlation was found between fecal coliforms and *Clostridium perfringens* counts in the bathing seawater.

8. FECAL INDICATOR CHEMICAL COMPOUNDS

Several chemical substances have been used as markers of fecal pollution in environmental waters. Caffeine is present in several beverages and in many pharmaceutical products. It is excreted in the urine of individuals who have ingested the substance. The main source of caffeine in domestic wastewaters is excretion following consumption of coffee, tea, soft drinks, or medication. Levels of caffeine in domestic wastewater have been measured to be between 20 and 300 μg/L. Levels in receiving waters are much lower due to significant dilution. Due to its high solubility, low octanol-water partition coefficient, insignificant volatility and clear anthropogenic origin, the presence of caffeine in environmental waters can be a good marker for human fecal pollution [124–129].

However, relationships between fecal indicator bacteria and caffeine are variable. Wu *et al.* [129] reported a study carried out in the Rochor Canal and Marina Bay, Singapore. In Rochor Canal, the highest concentration of caffeine (1.35 ng/mL) was found at downstream, and the lowest (0.68 and 0.37 ng/mL) were determined at middle and upstream points. At Marina Bay, the concentration of caffeine was in the range of 0.41–0.96 ng/mL. Fecal coliform concentrations were very high, exceeding 5,000 CFU/100 mL. Caffeine and fecal coliform concentrations were significantly correlated in Rochor Canal samples, but no significant correlation was observed in Martina Bay water samples.

The use of caffeine as marker of fecal pollution has additional important limitations. Caffeine is often present in the urban environment from numerous plant species debris as well as from human "dumping" of coffee wastes. In addition, the current analytical methods used are relatively complex and expensive [124].

Coprostanol (5β-cholestan-3β-ol) is a fecal stanol that is formed by indigenous bacteria present in the gut of humans and higher animals, during catabolism of cholesterol. It is the main stanol present in human feces (24 to 89% of total steroids) and in domestic wastewater. Based on these facts, it has been proposed as a chemical indicator of human fecal pollution. Feces from pigs and cats also contain coprostanol, but at much lower levels. Additional fecal stanols, such as 24-ethylcoprostanol, were found to be predominant in herbivores, such as cows, horses, and sheep, suggesting potential use of this chemical as an indicator of fecal pollution from these sources. Reported half-lives of coprostanol in aerobic conditions are generally lower than 10 days at 20 °C. Thus, the presence of coprostanol in an aerobic environment can be considered an indication of recent fecal input to the waters. [124,128,130].

Isobe *et al.* [130] reported a study carried out in the Mekong Delta (Vietnam) and in the Tokyo metropolitan area. During the wet season in the Mekong Delta, higher bacterial densities were observed in rivers, probably due to the higher bacterial inputs from soil particles with runoff. In Tokyo, higher bacterial densities were usually observed during summer, followed by those in the typhoon aftermath and winter. Significant correlations between the concentrations of *E. coli* and coprostanol (log scale) were found in all surveys. It was concluded that the determination of coprostanol can improve standard microbiological assays of fecal pollution.

Reports from several regions throughout the world indicate variable quantitative relationships between fecal coliform densities and coprostanol concentrations. In the Derwent Estuary and Sydney region (Australia), coprostanol concentration of 400 ng/L corresponded to 1,000 CFU of fecal coliforms/100 mL. In the Mekong Delta, during the wet season, this coliform density corresponded to 30 ng coprostanol/L, and in the dry seasons, 100 ng/L. In Tokyo metropolitan area, these values were 30 ng coprostanol/L, in summer, and 100 ng coprostanol/L, in a typhoon aftermath. These differences were interpreted as a result of differences in water temperature and soil particle concentration [130].

Fecal sterol analysis, although expensive and complex, has resolved problems of source attribution in urban and rural environments not possible with use of traditional fecal indicator bacteria [124]. These chemical indicators are especially useful in environments which allow survival and growth of fecal bacteria. For instance, in tropical regions, characterized by high temperatures and frequent rainstorms that facilitate erosion of soils, fecal bacteria can proliferate in environmental waters reaching densities that are not representative of real sewage inputs in the environment [130].

9. SOURCES OF FECAL BACTERIAL POLLUTION OF ENVIRONMENTAL WATERS

9.1. Sources of Surface and Groundwater Contamination

Determinations carried out in the sewage systems of urbanized areas have confirmed the presence of high numbers of intestinal bacteria. Treatment of sewage reduces the concentration of these bacteria by 1–2 logs, but effluent still contains high levels of intestinal bacteria (Table 8). Effluents from sewage treatment plants can be a source of contamination of surface waters with fecal bacteria.

Table 8. Typical concentrations of selected bacteria in raw and treated domestic wastewatera.

Bacterial group	Raw sewage (cells/ml)	Treated effluent (cells/ml)
Salmonella	$10^{-1} - 10^{1}$	$10^{-1} - 10^{1}$
Total coliforms	$10^{4} - 10^{6}$	$10^{3} - 10^{5}$
Fecal coliforms	$10^{3} - 10^{5}$	$10^{2} - 10^{4}$
Enterococci	$10^{3} - 10^{4}$	$10^{1} - 10^{3}$
Clostridium perfringens	$10^{2} - 10^{3}$	$10^{1} - 10^{2}$

[a] Adapted from Medema *et al.* [131].

Septic tanks, cesspools, latrines and other on-site systems are widely used for wastewater storage and treatment. The water percolating from these facilities contains bacteria that may contaminate groundwater supplies.

Many farmers use cellars, tanks or landfills to store manure. Water leaching from these storage sites may also contaminate groundwater, especially during periods of rainfall. The application of animal manure to agricultural lands as fertilizer is common practice throughout the world. Bacteria present in the manure may leach into the groundwater.

The potential for bacteria present in human and animal wastes to contaminate water in nearby wells needs special attention [131]. An important source of contamination of surface and ground waters is runoff water from agricultural and pasture lands, and urban areas.

Fecal bacteria enter surface water by direct deposit of feces and by overland runoff. The movement of animal wastes into surface waters can be a major factor contributing to the pollution of available water in many regions. Over one-third of the land area of USA is used for grazing livestock and receives 50% of all livestock wastes.

In a study reported by Doran and Linn [108], runoff from a cow-calf pasture in eastern Nebraska was monitored during a three-year period. Rainfall runoff from the grazed area contained 5 to 10 times more fecal coliforms than runoff from the fenced, ungrazed area. However, fecal streptococci counts were higher in runoff from the ungrazed area and reflected the contributions from wildlife.

Urban and suburban areas are dominated by impervious cover. During storms, rainwater flows across these impervious surfaces, mobilizing contaminants. The pollutants carried in runoff originate from a variety of urban and suburban nonpoint sources. Contaminants commonly found in stormwater runoff include fecal and pathogenic bacteria. Stormwater transports pollutants to water bodies such as lakes and streams [132].

Enterococci and *E. coli* can be found in high numbers in most storm drains and creeks. In Southern California (USA), Ferguson *et al.* [133] found high levels of enterococci (*Enterococcus faecalis, Enterococcus faecium, Enterococcus hirae, Enterococcus casseliflavus* and *Enterococcus mundtii*) in intertidal sediments in a seasonal river, and near a storm drain outlet.

9.2. Survival in Surface Water

Most intestinal bacteria that contaminate environmental waters are not able to survive and multiply in this environment. Survival rates vary widely among fecal bacteria introduced in environmental waters. Pathogenic enteric bacteria and *E. coli* display low survival rates (Table 9).

Table 9. Reduction times for fecal bacteria in surface watersa.

Bacterial group	Time for 50% reduction in concentration (days)
Total coliforms	0.9
E. coli	1.5–3
Enterococci	0.9–4
Clostridium perfringens	60 – > 300
Salmonella	0.1–0.67
Shigella	1

[a] Adapted from Medema *et al.* [131].

Table 10. Disappearance rates of fecal bacteria in groundwatersa.

Bacterial group	Disappearance rate (per day)
E. coli	0.063–0.36
Fecal streptococci	0.03–0.24
Clostridium bifermentans spores	0.00
Salmonella enterica subsp. *enterica* serovar Typhimurium	0.13–0.22

[a] Adapted from Medema *et al.* [131].

The ability of fecal bacteria to survive in environmental waters generally increases as the temperature decreases. Others factors that influence survival include dissolved organic carbon concentration, sunlight intensity and the ability to enter the viable but non-culturable state [131].

In a comparative study on the survival of 10 different coliform species (*E. coli, Citrobacter freundii, Citrobacter youngae, Klebsiella pneumoniae, K. oxytoca, Enterobacter amnigenus, Enterobacter cloacae* subsp. *cloacae*, and *Pantoea agglomerans* (*Enterobacter agglomerans*)) inoculated in sterilized river water with different concentrations of dissolved organic carbon, Boualam *et al.* [134] found that only *C. freundii, K. pneumoniae* and *E. cloacae* subsp. *cloacae* remained cultivable after 96 hours of incubation. In a posterior study, using the same bacteria and medium, Boualam *et al.* [135] found that after 28 days, only *C. freundii* and *E. cloacae* subsp. *cloacae* survived.

Baudišová [136] reported a comparative study on the survival of total coliforms, fecal coliforms and *E. coli*, in sterile and non-sterile river water. In sterile water, all bacteria survived for many months. However, in non-sterile conditions (closer to true environmental conditions), the elimination rate of all bacteria was considerably faster. Total coliforms survived the longest and *E. coli* the shortest.

9.3. Survival in Groundwater

Survival of bacteria in groundwater is influenced by several factors, namely the survival in soil, since in order to reach the groundwater bacteria have to percolate through the soil. Generally, survival in soil (and concomitantly in groundwater) is enhanced by low temperatures, high soil humidity, neutral or alkaline soil pH and the presence of organic carbon [131].

10. WHICH INDICATORS OF FECAL POLLUTION SHOULD BE USED?

Several fecal indicator bacteria in environmental waters are in current use. From these stand out fecal coliforms, *E. coli* and enterococci [1,6,137]. In environmental waters, most fecal coliform strains are *E. coli.*

In particular situations, the presence of *E. coli* is definitively not associated with fecal pollution. These situations were firstly detected in some African countries, namely Nigeria, Ivory Coast and New Guinea (although not in others, such as Uganda) [42,99].

Recent studies carried out in temperate zones indicated that *E. coli* can persist in secondary, nonhost habitats, outside the hot tropical areas, and become naturalized in these habitats. Byappanahalli *et al.* [138] reported that *E. coli* could be isolated from coastal temperate forest soils in Indiana (USA). The aquatic alga *Cladophora glomerata* (L.) from several Lake Michigan beaches was shown to harbor high densities of *E. coli* [139].

Ishii *et al.* [140] reported a study on the survival of *E. coli* in temperate riverine soils of northern Minnesota (USA). Viable *E. coli* populations were repeatedly isolated from northern temperate soils in three Lake Superior watersheds. Seasonal variation in the population density of soilborne *E. coli* was observed; the greatest cell densities were found in the summer to fall, and the lowest numbers, occurred during the winter to spring months. Horizontal, fluorophore-enhanced repetitive extragenic palindromic PCR (HFERP) DNA fingerprint analyses indicated that identical soilborne *E. coli* genotypes, overwintered in frozen soil and were present over time, and that these strains were different from *E. coli* strains obtained from wildlife commonly found in the studied habitats or river water. Soilborne *E. coli* strains had HFERP DNA fingerprints that were unique to specific soils and locations. In laboratory studies, naturalized *E. coli* strains had the ability to grow and replicate to high cell densities, in nonsterile soils when incubated at 30 or 37 °C and survived longer than 1 month when soil temperatures were lower than 25 °C. It was concluded that these *E. coli* strains became naturalized, autochthonous members of the soil microbial community.

In a latter paper, Ksoll *et al.* [141] studied epilithic periphyton communities at three sites on the Minnesota shoreline of Lake Superior (USA). Fecal coliform densities increased up to 4 orders of magnitude in early summer, and decreased during autumn. HFERP DNA fingerprint analyses indicated that waterfowl (geese, terns, and gulls) were the major primary source of periphyton *E. coli* strains that could be identified. Periphyton and sewage effluent were also major potential sources. Several periphyton *E. coli* isolates were genotypically identical, repeatedly isolated over time. Inoculated *E. coli* rapidly colonized natural periphyton in laboratory microcosms and persisted for several weeks, and some cells were released to the overlying water. It was concluded that *E. coli* had became a naturalized member of the bacterial periphyton communities.

The presence, persistence, and possible naturalization of *E. coli* in these habitats can confound the use of fecal coliforms as a reliable indicator of recent fecal contamination of environmental waters. Future studies should consider other nonhost habitats as potential sources of fecal coliform bacteria in aquatic environments.

Considering these limitations, it appears to be advisable, in order to check the microbiological quality of drinking water, to complement the determination of *Escherichia coli* with the assay of enterococci. This rationale that has been followed, for many years, in the making of the drinking-water legislation in the European Union.

However, for many developing countries, where limited financial resources are the norm and reality, the routine determination of these two parameters can be difficult to implement. In these circumstances, it appears common sense that is better to determine a (good) parameter, such as *Escherichia coli*, than have no analysis done.

In this context, USA legislation emerges as a pragmatic approach to the problem. According to the American legislation, total coliforms are the routine parameter to be determined. Only when these determinations are repeatedly positive, it is mandatory to assess fecal coliforms [142,143]. Although total coliforms are not necessarily fecal bacteria, the rationale behind this system is correct, since: (1) a positive test in fecal coliforms (which is our target) is necessarily positive in the total coliform procedure; (2) the inverse is not necessarily true; (3) total coliforms are easily and cheaply assayed in waters.

As an alternative to the determination of both *E. coli* and enterococci, the assay of ammonia in environmental waters can be useful and complement the determination of fecal coliforms.

Ammonia is one of the key molecules in the nitrogen cycle. The presence of ammonia in surface waters can be due to direct contamination by agricultural fertilizers, and/or to microbial degradation of proteins, nucleic acids and urea, implying therefore the presence of a considerable concentration of organic matter in the water. Ammonia is rapidly oxidized in the environment and is typically found in natural waters at concentrations less than 0.1 mg/L. Concentrations significantly above this indicate gross contamination by fresh sanitary waste, where ammonia levels are typically very high (tens or hundreds of mg/L) [84].

Espigares *et al.* [144] reported a comparative study of chemical and microbiological indicators (total and fecal coliforms, fecal streptococci and sulphite-reducing clostridia) in a stretch of the Guadalquivir River (Spain) and its affluents. Total coliforms were correlated with fecal coliforms, but were not correlated with fecal streptococci and clostridia. Fecal coliforms were correlated with the other indicators. Fecal streptococci and sulphite-reducing clostridia were correlated with the other indicators except for total coliforms. All these microbiological indicators were correlated with dissolved oxygen (negatively), dissolved organic carbon and ammonia (positively). Cabral and Marques [85] in a study of a polluted river (Febros) in the Great Oporto area (northwest Portugal) found that ammonia was significantly correlated with all the microbiological assayed parameters—total and fecal coliforms, fecal streptococci and enterococci. These correlations are most probably due to the carry-over of organic matter in wastewaters, and to a high microbial ammonification activity [85].

Simple and rapid in-field tests and automated and continuous systems are available for the assay of ammonia in environmental waters. More studies are needed in order to confirm the use of ammonia as a reliable parameter in a preliminary screening for emergency fecal pollution outbreaks.

11. CONCLUSIONS

1) Safe drinking water for all is one of the major challenges of the 21st century.
2) Microbiological control of drinking water should be the norm everywhere.
3) Routine basic microbiological analysis of drinking water should be carried out by assaying the presence of *Escherichia coli* by the culture methods. On-line monitoring of glucuronidase activity is currently too insensitive to replace culture based detection of *E. coli* but is a valuable complementary tool for high temporal resolution monitoring. Whenever financial resources are available, coliform determinations should be complemented with the quantification of enterococci.
4) More studies are needed in order to check if ammonia is reliable for a preliminary screening for emergency fecal pollution outbreaks.

5) Financial resources should be devoted to a better understanding of the ecology and behavior of human and animal fecal bacteria in environmental waters.

REFERENCES

1. WHO (World Health Organization). Guidelines for Drinking-water Quality, Incorporating 1st and 2nd Addenda, Volume 1, Recommendations, 3rd ed; WHO: Geneva, Switzerland, 2008.
2. Fenwick, A. Waterborne Diseases—Could they be Consigned to History? Science **2006**, 313, 1077–1081.
3. George, I; Crop, P; Servais, P. Use of β-D-Galactosidase and β-D-Glucuronidase Activities for Quantitative Detection of Total and Faecal Coliforms in Wastewater. Can. J. Microbiol **2001**, 47, 670–675.
4. Grabow, WOK. Waterborne Diseases: Update on Water Quality Assessment and Control. Water SA **1996**, 22, 193–202.
5. Seas, C; Alarcon, M; Aragon, JC; Beneit, S; Quiñonez, M; Guerra, H; Gotuzzo, E. Surveillance of Bacterial Pathogens Associated with Acute Diarrhea in Lima, Peru. Int. J. Infect. Dis **2000**, 4, 96–99.
6. Medema, GJ; Payment, P; Dufour, A; Robertson, W; Waite, M; Hunter, P; Kirby, R; Anderson, Y. Safe drinking water: an ongoing challenge. In Assessing Microbial Safety of Drinking Water Improving Approaches and Method; WHO & OECD, IWA Publishing: London, UK, 2003; pp. 11–45.
7. Farmer, JJ; Hickam-Brenner, FW. The Genus Vibrio and Phtotobacterium. In The Prokaryotes: An Evolving Electronic Resource for the Microbiological Community-electronic release 314,, 3th ed; Dworkin, M, Falkow, S, Rosenberg, E, Eds.; Springer-Verlag: New-York, NY, USA, 2003.
8. Farmer, JJ; Janda, JM; Brenner, FW; Cameron, DN; Birkhead, KM. Genus Vibrio. In Bergey's Manual of Systematic Bacteriology, 2nd ed; Brenner, DJ, Krieg, NR, Staley, JT, Eds.; Springer: New York, NY, USA, 2005; Volume 2, Part B, pp. 494–546.
9. DSMZ (Deutsche Sammlung von Mikroorganismen und Zellkulturen GmbH). Bacterial Nomenclature Up-to-Date (Approved Lists; Validation Lists); DSMZ: Braunschweig, Germany, 2010; List 05/2010.
10. Sack, DA; Sack, RB; Nair, GB; Siddique, AK. Cholera. Lancet **2004**, 363, 223–233.
11. Todar, K. Vibrio cholerae and asiatic cholera. Todar's Online Textbook of Bacteriology. 2009. Available online: http://www.textbookofbacteriology.net/cholera.html (assessed on 4 May 2010).
12. Ali, M; Emch, M; Yunus, M; Sack, RB. Are the Environmental Niches of Vibrio cholerae 0139 Different from those of Vibrio cholerae 01 El Tor? Int. J. Infect. Dis **2001**, 5, 214–219.
13. Ramamurthy, T; Yamasaki, S; Takeda, Y; Nair, GB. Vibrio cholerae 0139 Bengal: Odyssey of a Fortuitous Variant. Microbes Infect **2003**, 5, 329–344.
14. Arbora, S; Crum, S. Medical Geography and Cholera in Perú; The Geographer's Craft Project, Department of Geography, The University of Colorado at Boulder: Boulder, CO, USA, 1996. Available online: http://www.colorado.edu/geography/ gcraft/warmup/cholera/cholera_f.html (assessed on 4 May 2010).
15. Glass, RI; Claeson, M; Blake, PA; Waldmann, RJ; Pierce, NF. Cholera in Africa: Lessons on Transmission and Control for Latin America. Lancet **1991**, 338, 791–795.
16. Glass, RI; Libel, M; Brandling-Bennet, AD. Epidemic Cholera in the Americas. Science **1992**, 256, 1524–1525.
17. Tamplin, ML; Parodi, CC. Environmental Spread of Vibrio cholerae in Peru. Lancet **1991**, 338, 1216–1217.

18. Steinberg, EB; Greene, KD; Bopp, CA; Cameron, DN; Wells, JG; Mintz, ED. Cholera in the United States, 1995–2000: Trends at the End of the Twentieth Century. J. Infect. Dis **2001**, 184, 799–802.
19. Meibom, KL; Blokesch, M; Dolganov, NA; Wu, C-Y; Schoolnik, GK. Chitin Induces Natural Competence in Vibrio cholerae. Science **2005**, 310, 1824–1827.
20. Waldor, MK; Mekalanos, JJ. Lysogenic Conversion by a Filamentous Phage Encoding Cholera Toxin. Science **1996**, 272, 1910–1914.
21. Hervio-Heath, D; Colwell, RR; Derrien, A; Robert-Pillot, A; Fournier, JM; Pommepuy, M. Occurrence of Pathogenic Vibrios in Coastal Areas of France. J. Appl. Microbiol **2002**, 92, 1123–1135.
22. Alam, M; Hasan, NA; Sadique, A; Bhuiyan, NA; Ahmed, KU; Nusrin, S; Nair, GB; Siddique, AK; Sack, RB; Sack, DA; Huq, A; Colwell, RR. Seasonal Cholera Caused by Vibrio cholerae Serogroups O1 and O139 in the Coastal Aquatic Environment of Bangladesh. Appl. Environ. Microbiol **2006**, 72, 4096–4104.
23. Bhanumathi, R; Sabeena, F; Isac, SR; Shukla, BN; Singh, DV. Molecular Characterization of Vibrio cholerae O139 Bengal Isolated from Water and the Aquatic Plant Eichhornia crassipes in the River Ganga, Varanasi, India. Appl. Environ. Microbiol **2003**, 69, 2389–2394.
24. Munro, PM; Colwell, RR. Fate of Vibrio cholerae O1 in Seawater Microcosms. Water Res **1996**, 30, 47–50.
25. Alam, M; Sultana, M; Nair, GB; Sack, RB; Sack, DA; Siddique, AK; Ali, A; Huq, A; Colwell, RR. Toxigenic Vibrio cholerae in the Aquatic Environment of Mathbaria, Bangladesh. Appl. Environ. Microbiol **2006**, 72, 2849–2855.
26. Chaiyanan, S; Chaiyanan, S; Huq, A; Maugel, T; Colwell, RR. Viability of the Nonculturable Vibrio cholerae O1 and O139 System. Appl. Microbiol **2001**, 24, 331–341.
27. Molloy, S. Biofilms in the Time of Cholera. Nature Rev. Microbiol **2007**, 5, 906–907.
28. Le Minor. The genus Salmonella. In The Prokaryotes: An Evolving Electronic Resource for the Microbiological Communityelectronic release 314,, 3th ed; Dworkin, M, Falkow, S, Rosenberg, E, Eds.; Springer-Verlag: New York, NY, USA, 2003.
29. Popoff, MY; Le Minor, LE. Genus Salmonella. In Bergey's Manual of Systematic Bacteriology, 2nd ed; Brenner, DJ, Krieg, NR, Staley, JT, Eds.; Springer: New York, NY, USA, 2005; Volume 2, Part B, pp. 764–799.
30. Tindall, BJ; Grimont, PAD; Garrity, GM; Euzéby, JP. Nomenclature and Taxonomy of the Genus Salmonella. Int. J. Syst. Evolut. Microbiol **2005**, 55, 521–524.
31. Ben Aissa, R; Al-Gallas, N; Troudi, H; Belhadj, N; Belhadj, A. Trends in Salmonella enterica Serotypes Isolated from Human, Food, Animal, and Environment in Tunisia, 1994–2004. J. Infect **2007**, 55, 324–339.
32. Arvanitidou, M; Kanellou, K; Vagiona, DG. Diversity of Salmonella spp. and Fungi in Northern Greek Rivers and their Correlation to Faecal Pollution Indicators. Environ. Res **2005**, 99, 278–284.
33. Kapperud, G; Rørvik, LM; Hasseltvedt, V; Høiby, EA; Iversen, BG; Staveland, K; Johnsen, G; Leitão, J; Herikstad, H; Andersson, Y; Langeland, G; Gondrosen, B; Lassen, J. Outbreak of Shigella sonnei Infection Traced to Imported Iceberg Lettuce. J. Clin. Microbiol **1995**, 33, 609–614.
34. Strockbine, NA, Maurelli, AT. Genus Shigella. In Bergey's Manual of Systematic Bacteriology, 2nd ed; Brenner, DJ, Krieg, NR, Staley, JT, Eds.; Springer: New York, NY, USA, 2005; Volume 2, Part B, pp. 811–823.
35. Emch, M; Ali, M; Yunus, M. Risk Areas and Neighborhood-Level Risk Factors for Shigella dysenteriae 1 and Shigella flexneri. Health Place **2008**, 14, 96–105.
36. Germani, Y; Sansonetti, PJ. The Genus Shigella. In The Prokaryotes: An Evolving Electronic Resource for the Microbiological Communityelectronic release 314,, 3th ed; Dworkin, M, Falkow, S, Rosenberg, E, Eds.; Springer-Verlag: New York, NY, US, 2003.
37. Hale, TL. Genetic Basis of Virulence in Shigella Species. Microbiol. Rev **1991**, 55, 206–224.
38. Todar, K. Shigella and shigellosis. Todar's Online Textbook of Bacteriology. 2009. Available online: http://www.textbookofbacteriology.net/Shigella.html (assessed on 1 September 2010).

39. Tetteh, GL; Beuchat, LR. Survival, Growth, and Inactivation of Acid-Stressed Shigella flexneri as Affected by pH and Temperature. Int. J. Food Microbiol **2003**, 87, 131–138.
40. Chompook, P; Todd, J; Wheeler, JG; von Seidlein, L; Clemens, J; Chaicumpa, W. Risk Factors for Shigellosis in Thailand. Int. J. Infect. Dis **2006**, 10, 425–433.
41. Faruque, SM; Khan, R; Kamruzzman, M; Yamasaki, S; Ahmad, QS; Azim, T; Nair, GB; Takeda, Y; Sack, DA. Isolation of Shigella dysenteriae type 1 and S. flexneri Strains from Surface Waters in Bangladesh: Comparative Molecular Analysis of Environmental Shigella Isolates versus Clinical Strains. Appl. Environ. Microbiol **2002**, 68, 3908–3913.
42. Bettelheim, KA. The genus Escherichia. In The Prokaryotes: An Evolving Electronic Resource for the Microbiological Communityelectronic release 314,, 3th ed; Dworkin, M, Falkow, S, Rosenberg, E, Eds.; Springer-Verlag: New York, NY, USA, 2003.
43. Scheutz, F; Strockbine, NA. Genus Escherichia. In Bergey's Manual of Systematic Bacteriology, 2nd ed; Brenner, DJ, Krieg, NR, Staley, JT, Eds.; Springer: New York, NY, USA, 2005; Volume 2, Part B, pp. 607–623.
44. World Health Organization. Enterotoxigenic Escherichia coli (ETEC). Diarrhoeal Diseases. Available online: http://www.who.int/vaccine_research/diseases/diarrhoeal/ en/index4 .html (assessed on 4 September 2010).
45. Health Canada. Guidelines for Canadian Drinking Water Quality: Guideline Technical Document. Bacterial Waterborne Pathogens. Current and Emerging Organisms of Concern; Health Canada: Ottawa, ON, Canada, 2006.
46. World Health Organization. Enterohaemorrhagic Escherichia coli (EHEC). Fact sheet N°125; Available online: http://www.who.int/mediacentre/factsheets/fs125/en/ (assessed on 4 September 2010).
47. Enteroinvasive Escherichia coli (EIEC); US Department of Health & Human Services, U.S. Food and Drug Administration. Available online: http://www.fda.gov/Food/ FoodSafety/ FoodborneIllness/FoodborneIllnessFoodbornePathogensNaturalToxins/BadBugBook/ucm071298.htm (assessed on 4 September 2010).
48. WHO (World Health Organization). Pathogenic Mycobacteria in Water: A Guide to Public Health Consequences, Monitoring and Management; Pedley, S, Bartram, J, Rees, G, Dufour, A, Cotruvo, J, Eds.; IWA Publishing: London, UK, 2004.
49. Holtzman, A; Aronson, T; Froman, S; Berlin, OGW; Dominguez, P; Kunkel, KA; Overturf, G; Stelma, G; Smith, C; Yakrus, M. The Isolation and Identification of Mycobacterium avium Complex (MAC) Recovered from Los Angeles Potable Water, a Possible Source of Infection in AIDS Patients. Int. J. Environ. Health Res **1994**, 4, 63–72.
50. Von Reyn, CF; Maslow, JN; Barber, TW; Falkinham, JO; Arbeit, RD. Persistent Colonisation of Potable Water, a Source of Mycobacterium avium Infection in AIDS. Lancet **1994**, 343, 1137–1141.
51. Gião, MS; Azevedo, NF; Wilks, SA; Vieira, MJ; Keevil, CW. Persistence of Helicobacter pylori in Heterotrophic Drinking Water Biofilms. Appl. Environ. Microbiol **2008**, 74, 5898–5904.
52. Hulten, K; Han, SW; Enroth, H; Klein, PD; Opekun, AR; Gilman, RH; Evans, DG; Engstrand, L; Graham, DY; El-Zaatari, FA. Helicobacter pylori in the Drinking Water in Peru. Gastroenterology **1996**, 110, 1031–1035.
53. Handfield, M; Simard, P; Couillard, M; Letarte, R. Aeromonas hydrophila Isolated from Food and Drinking Water: Hemagglutination, Hemolysis, and Cytotoxicity for a Human Intestinal Cell Line (HT-29). Appl. Environ. Microbiol**1996**, 62, 3459–3461.
54. Daskalov, H. The Importance of Aeromonas hydrophila in Food Safety. Food Control **2006**, 17, 474–483.
55. Chauret, C; Volk, C; Creason, R; Jarosh, J; Robinson, J; Warnes, C. Detection of Aeromonas hydrophila in a Drinking-Water Distribution System: A Field and Pilot Study. Can. J. Microbiol **2001**, 47, 782–786.
56. Fernández, MC; Beatriz, N; Giampaolo, SB; Ibañez, M; Guagliardo, V; Esnaola, MM; Conca, L; Valdivia, P; Stagnaro, SM; Chiale, C; Frade, H. Aeromonas Hydrophila and its Relation with Drinking Water Indicators of Microbiological Quality in Argentine 2000. Genetica **2000**, 108, 35–40.

57. George, I; Servais, P. Sources et Dynamique des Coliformes dans le Basin de la Sein; Rapport de Synthèse; Programme PIREN-Seine 1998–2001, Sources et dynamique des coliformes dand le bassin de la Seine; C. N. R. S.: Paris, France, 2002.
58. Gauthier, F; Archibald, F. The Ecology of "Faecal Indicator" Bacteria Commonly Found in Pulp and Paper Mill Water Systems. Water Res **2001**, 35, 2207–2218.
59. Wilkes, G; Edge, T; Gannon, V; Jokinen, C; Lyautey, E; Medeiros, D; Neumann, N; Ruecker, N; Topp, E; Lapena, DR. Seasonal Relationships Among Indicator Bacteria, Pathogenic Bacteria, Cryptosporidium Oocysts, Giardia Cysts, and Hydrological Indices for Surface Waters Within an Agricultural Landscape. Water Res **2009**, 43, 2209–2223.
60. Saarela, M; Lähteenmäki, L; Crittenden, R; Salminen, S; Mattila-Sandholm, T. Gut Bacteria and Health Foods—The European Perspective. Int. J. Food Microbiol **2002**, 78, 99–117.
61. Hawksworth, G; Drasar, BS; Hill, MJ. Intestinal Bacteria and the Hydrolysis of Glycoside Bonds. J. Med. Microbiol**1971**, 4, 451–459.
62. Hopkins, MJ; Sharp, R; Macfarlane, GT. Variation in Human Intestinal Microbiota with Age. Digest Liver Dis **2002**, 34(Suppl.2), S12–S18.
63. Leclerc, H; Moriamez, JC. Étude Quantitative de la Flore Fécale de l'Adulte et du Nourrison Alimenté Artificiellement. Path. Bio **1980**, 28, 217–226.
64. Wilson, M. Microbial Inhabitants of Humans Their Ecology and Role in Health and Disease; Cambridge University Press: Cambridge, UK, 2005.
65. Sinton, LW; Finlay, RK; Hannah, DJ. Distinguishing Human from Faecal Contamination in Water: A Review. New Zealand J. Marine Freshwater Res **1998**, 32, 323–348.
66. Biavati, B; Mattarelli, P. The family Bifidobacteriaceae. In The Prokaryotes: An Evolving Electronic Resource for the Microbiological Communityelectronic release 314,, 3th ed; Dworkin, M, Falkow, S, Rosenberg, E, Eds.; Springer-Verlag: New York, NY, USA, 2003.
67. Hippe, H; Andreesen, JR; Gottschalk, G. The genus Clostridium—Nonmedical. In The Prokaryotes: An Evolving Electronic Resource for the Microbiological Communityelectronic release 314,, 3th ed; Dworkin, M, Falkow, S, Rosenberg, E, Eds.; Springer-Verlag: New York, NY, USA, 2003.
68. Rainey, FA; Hollen, BJ; Small, A. Genus Clostridium. In Bergey's Manual of Systematic Bacteriology, 2nd ed; De Vos, P, Garrity, GM, Jones, D, Krieg, NR, Ludwig, W, Rainey, FA, Schleifer, K-H, Whitman, WB, Eds.; Springer: New York, NY, USA, 2009; Volume 3, pp. 738–828.
69. Smith, LDS. The genus Clostridium—Medical. In The Prokaryotes: An Evolving Electronic Resource for the Microbiological Communityelectronic release 314,, 3th ed; Dworkin, M, Falkow, S, Rosenberg, E, Eds.; Springer-Verlag: New York, NY, USA, 2003.
70. Bezirtzoglou, E; Panagiou, A; Savvaidis, I; Maipa, V. Distribution of Clostridium perfringens in Polluted Lake Environments. Anaerobe **1997**, 3, 169–172.
71. Švec, P; Devriese, LA. Genus Enterococcus. In Bergey's Manual of Systematic Bacteriology, 2nd ed; DE Vos, P, Garrity, GM, Jones, D, Krieg, NR, Ludwig, W, Rainey, FA, Schleifer, K-H, Whitman, WB, Eds.; Springer: New York, NY, USA, 2009; Volume 3, pp. 594–607.
72. Pinto, B; Pierotti, R; Canale, G; Reali, D. Characterization of 'Faecal Streptococci' as Indicators of Faecal Pollution and Distribution in the Environment. Lett. Appl. Microbiol **1999**, 29, 258–263.
73. Bergey's Manual of Determinative Bacteriology, 9th ed; Holt, JG; et al. (Eds.) Williams & Wilkins. Baltimore, MD, USA, 1994, pp. 175–190.
74. Gordon, DM; FitzGibbon, F. The Distribution of Enteric Bacteria from Australian Mammals: Host and Geographical Effects. Microbiology **1999**, 145, 2663–2671.
75. Frederiksen, W; Søgaard, P. The genus Citrobacter. In The Prokaryotes: An Evolving Electronic Resource for the Microbiological Communityelectronic release 314,, 3th ed; Dworkin, M, Falkow, S, Rosenberg, E, Eds.; Springer-Verlag: New York, NY, USA, 2003.
76. Frederiksen, W. Genus Citrobacter. In Bergey's Manual of Systematic Bacteriology, 2nd ed; Brenner, DJ, Krieg, NR, Staley, JT, Eds.; Springer: New York, NY, USA, 2005; Volume 2, Part B, pp. 651–656.
77. Grimont, F; Grimont, PAD; Richard, C. The Genus Klebsiella. In The Prokaryotes: An Evolving Electronic Resource for the Microbiological Communityelectronic release 314,,

3th ed; Dworkin, M, Falkow, S, Rosenberg, E, Eds.; Springer-Verlag: New York, NY, USA, 2003.

78. Grimont, PAD; Grimont, F. Genus Klebsiella. In Bergey's Manual of Systematic Bacteriology, 2nd ed; Brenner, DJ, Krieg, NR, Staley, JT, Eds.; Springer: New York, NY, USA, 2005; Volume 2, Part B, pp. 685–693.
79. Grimont, F; Grimont, PAD. The Genus Enterobacter. In The Prokaryotes: An Evolving Electronic Resource for the Microbiological Communityelectronic release 314,, 3th ed; Dworkin, M, Falkow, S, Rosenberg, E, Eds.; Springer-Verlag: New York, NY, USA, 2003.
80. Grimont, PAD; Grimont, F. Genus Enterobacter. In Bergey's Manual of Systematic Bacteriology, 2nd ed; Brenner, DJ, Krieg, NR, Staley, JT, Eds.; Springer: New York, NY, USA, 2005; Volume 2, Part B, pp. 661–669.
81. Lockhart, SR; Abramson, MA; Beekmann, SE; Gallagher, G; Riedel, S; Diekema, DJ; Quinn, JP; Doern, GV. Antimicrobial Resistance Among Gram-negative Bacilli Causing Infections in Intensive Care Unit Patients in the United States between 1993 and 2004. J. Clin. Microbiol **2007**, 45, 3352–3359.
82. Hidron, AI; Edwards, JR; Patel, J; Horan, TC; Sievert, DM; Pollock, DA. NHSN Annual Update: Antimicrobial Resistant Pathogens Associated with Healthcare-Associated Infections: Annual Summary of Data Reported to the National Healthcare Safety Network at the Centers for Disease Control and Prevention, 2006–2007. Infect. Control Hosp. Epidemiol **2008**, 29, 996–1011.
83. Ashbolt, NJ; Grabow, OK; Snozzi, M. Indicators of microbial water quality. In Water Quality: Guidelines, Standards and Health; Fewtrell, L, Bartram, J, Eds.; World Health Organization (WHO), IWA Publishing: London, UK, 2001; pp. 289–316.
84. Payment, P; Waite, M; Dufour, A. Introducing parameters for the assessment of drinking water quality. In Assessing Microbial Safety of Drinking Water Improving Approaches and Method; WHO & OECD, IWA Publishing: London, UK, 2003; pp. 47–77.
85. Cabral, JP; Marques, C. Faecal Coliform Bacteria in Febros river (Northwest Portugal): Temporal Variation, Correlation with Water Parameters, and Species Identification. Environ. Monit. Assess **2006**, 118, 21–36.
86. Villarino, A; Toribio, AL; Brena, BM; Grimont, PAD; Bouvet, OMM. On the Relationship Between the Physiological State of Bacteria and Rapid Enzymatic Assays of Faecal Coliforms in the Environment. Biotechnol. Lett **2003**, 25, 1329–1334.
87. George, I; Petit, M; Servais, P. Use of Enzymatic Methods for Rapid Enumeration of Coliforms in Freshwaters. Lett. Appl. Microbiol **2000**, 88, 404–413.
88. George, I; Petit, M; Theate, C; Servais, P. Distribution of Coliforms in the Seine River and Estuary (France) Studied by Rapid Enzymatic Methods and Plate Counts. Estuaries **2001**, 24, 994–1002.
89. Nelis, H; van Poucke, S. Enzymatic Detection of Coliforms and Escherichia coli Within 4 Hours. Water, Air, and Soil Pollution **2000**, 123, 43–52.
90. Rompré, A; Servais, P; Baudart, J; de-Roubin, M-R; Laurent, P. Detection and Enumeration of Coliforms in Drinking Water: Current Methods and Emerging Approaches. J. Microbiol. Method **2002**, 49, 31–54.
91. Van Poucke, SO; Nelis, HJ. Rapid Detection of Fluorescent and Chemiluminescent Total Coliforms and Escherichia coli on Membrane Filters. J. Microbiol. Method **2000**, 42, 233–244.
92. Kämpfer, P; Rauhoff, O; Dott, W. Glycosidase Profiles of Members of the Family Enterobacteriaceae. J. Clin. Microbiol**1991**, 29, 2877–2879.
93. Kilian, M; Bülow, P. Rapid Diagnoses of Enterobacteriaceae. I. Detection of Bacterial Glycosidases. Acta Pathol. Microbiol. Scandinavica **1976**, 84, 245–251.
94. Muytjens, HL; van der Ros-van de Repe, J; nan Druten, HAM. Enzymatic Profiles of Enterobacter sakazakii and Related Species with Special Reference to the α-Glucosidase Reaction and Reproducibility of the Test System. J. Clin. Microbiol **1984**, 20, 684–686.
95. Tryland, I; Fiksdal, L. Enzyme Characteristics of β-D-Galactosidase- and β-D-Glucuronidase-Positive Bacteria and their Interference in Rapid Methods for Detection of Waterborne Coliforms and Escherichia coli. Appl. Environ. Microbiol **1998**, 64, 1018–1023.

96. Köster, W; Egli, T; Ashbolt, N; Botzenhart, K; Burlion, N; Endo, T; Grimont, P; Guillot, E; Mabilat, C; Newport, L; Niemi, M; Payment, P; Prescott, A; Renaud, P; Rust, A. Analytical methods for microbiological water quality testing. In Assessing Microbial Safety of Drinking Water Improving Approaches and Method; WHO & OECD, IWA Publishing: London, UK, 2003; pp. 237–292.
97. Manafi, M; Kneifel, W; Bascomb, S. Fluorogenic and Chromogenic Substrates Used in Bacterial Diagnostics. Microbiol. Rev **1991**, 55, 335–348.
98. Geissler, K; Manafi, M; Amorós, I; Alonso, JL. Quantitative Determination of Total Coliforms and Escherichia coli in Marine Waters with Chromogenic and Fluorogenic Media. J. Appl. Microbiol **2000**, 88, 280–285.
99. Byamukama, D; Kansiime, F; Mach, RL; Farnleitner, AH. Determination of Escherichia coli Contamination with Chromocult Coliform Agar Showed a High Level of Discrimination Efficiency for Differing Faecal Pollution Levels in Tropical Waters of Kampala, Uganda. Appl. Environ. Microbiol **2000**, 66, 864–868.
100. Eccles, JP; Searle, R; Holt, D; Dennis, PJ. A Comparison of Methods Used to Enumerate Escherichia coli in Conventionally Treated Sewage Sludge. J. App. Microbiol **2004**, 96, 375–383.
101. Farnleitner, AH; Hocke, L; Beiwl, C; Kavka, GG; Zechmeister, T; Kirschner, AKT; Mach, RL. Rapid Enzymatic Detection of Escherichia coli Contamination in Polluted River Water. Lett. Appl. Microbiol **2001**, 33, 246–250.
102. Edberg, SC; Kontnick, CM. Comparison of β-Glucuronidase-Based Substrate Systems for Identification of Escherichia coli. J. Clin. Microbiol **1986**, 24, 368–371.
103. Leclerc, H; Mossel, DAA; Edberg, SC; Struijk, CB. Advances in the Bacteriology of the Coliform Group: their Suitability as Markers of Microbial Water Safety. Ann. Rev. Microbiol **2001**, 55, 201–234.
104. Klein, G. Taxonomy, Ecology and Antibiotic Resistance of Enterococci from Food and the Gastro-Intestinal Tract. Int. J. Food Microbiol **2003**, 88, 123–131.
105. Kühn, I; Iversen, A; Burman, LG; Olsson-Liljequist, B; Franklin, A; Finn, M; Aarestrup, F; Seyfarth, AM; Blanch, AR; Vilanova, X; Taylor, H; Caplin, J; Moreno, MA; Dominguez, L; Herrero, IA; Möllby, R. Comparison of Enterococcal Populations in Animals, Humans, and the Environment—An European Study. Int. J. Food Microbiol **2003**, 88, 133–145.
106. Wheeler, AL; Hartel, PG; Godfrey, DG; Hill, JL; Segars, WI. Potentital of Enterococcus faecalis as a Human Fecal Indicator for Microbial Source Tracking. J. Environ. Qual **2002**, 31, 1286–1293.
107. Geldreich, EE. Fecal Coliform and Fecal Streptococcus Density Relationships in Waste Discharges and Receiving Waters. Crit. Rev. Environ. Control **1976**, 6, 349–369.
108. Doran, JW; Linn, DM. Bacteriological Quality of Runoff Water from Pasteureland. Appl. Environ. Microbiol **1979**, 37, 985–991.
109. Caplenas, NJ; Kanarek, MS. Thermotolerant Non-Fecal Source of Klebsiella pneumoniae: Validity of the Fecal Coliform Test in Recreational Waters. Am. J. Public Health **1984**, 74, 1273–1275.
110. Gauthier, F; Neufeld, JD; Driscoll, BT; Archibald, FS. Coliform Bacteria and Nitrogen Fixation in Pulp and Paper Mill Effluent Treatment Systems. Appl. Environ. Microbiol **2000**, 66, 5155–5160.
111. Corso, PS; Kramer, MH; Blair, KA; Addiss, DG, Davis, JP, Haddix, AC. Cost of Illness in the 1993 Waterborne Cryptosporidium Outbreak, Milwaukee, Wisconsin. Emerg. Inf. Dis **2003**, 9, 426–431.
112. MacKenzie, WR; Hoxie, NJ; Proctor, ME; Gradus, MS; Blair, KA; Peterson, DE; Kazmierczak, JJ; Addiss, DG; Fox, KR; Rose, JB; Davis, JP. A Massive Outbreak in Milwaukee of Cryptosporidium Infection Transmitted Through the Public Water Supply. N. Engl. J. Med **1994**, 331, 161–167.
113. Vilanova, X; Manero, A; Cerdà-Cuéllar, M; Blanch, AR. The Composition and Persistence of Faecal Coliforms and Enterococcal Populations in Sewage Treatment Plants. J. Appl. Microbiol **2004**, 96, 279–288.

114. Araujo, M; Sueiro, RA; Gómez, MJ; Garrido, MJ. Enumeration of Clostridium perfringens Spores in Groundwater Samples: Comparison of Six Culture Media. J. Microbiol. Methods **2004**, 57, 175–180.
115. Charriere, G; Mossel, DAA; Beaudeau, P; Leclerc, H. Assessment of the Marker Value of Various Components of the Coli-Aerogenes Group of Enterobacteriaceae and of a Selection of Enterococcus spp. for the Official Monitoring of Drinking Water Supplies. J. Appl. Bacteriol **1994**, 76, 336–344.
116. Martins, MT; Sato, MIZ; Alves, MN; Stoppe, NC; Prado, VM; Sanchez, PS. Assessment of Microbiological Quality for Swimming Pools in South America. Water Res **1995**, 29, 2417–2420.
117. Ferguson, CM; Coote, BG; Ashbolt, NJ; Stevenson, IM. Relationships Between Indicators, Pathogens and Water Quality in an Estuary System. Water Res **1996**, 30, 2045–2054.
118. Medema, GJ; van Asperen, IA; Havelaar, AH. Assessment of the Exposure of Swimmers to Microbiological Contaminants in Fresh Waters. Water Sci. Technol **1997**, 35, 157–163.
119. Polo, F; Figueras, MJ; Inza, I; Sala, J; Fleisher, JM; Guarro, J. Relationship Between Presence of Salmonella and Indicators of Faecal Pollution in Aquatic Habitats. FEMS Microbiol. Lett **1998**, 160, 253–256.
120. Noble, RT; Leecaster, MK; McGee, CD; Weisberg, SB; Ritter, K. Comparison of Bacterial Indicator Analysis Methods in Stormwater-Affected Coastal Waters. Water Res **2004**, 38, 1183–1188.
121. Harwood, VJ; Levine, AD; Scott, TM; Chivukula, V; Lukasik, J; Farrah, SR; Rose, JB. Validity of the Indicator Organism Paradigm for Pathogen Reduction in Reclaimed Water and Public Health Protection. Appl. Environ. Microbiol **2005**, 71, 3163–3170.
122. Touron, A; Berthe, T; Gargala, G; Fournier, M; Ratajczak, M; Servais, P; Petit, F. Assessment of Faecal Contamination and the Relationship Between Pathogens and Faecal Bacterial Indicators in an Estuarine Environment (Seine, France). Mar. Pollut. Bull **2007**, 54, 1441–1450.
123. Garrido-Pérez, MC; Anfuso, E; Acevedo, A; Perales-Vargas-Machuca, JA. Microbial Indicators of Faecal Contamination in Waters and Sediments of Beach Bathing Zones. Int. J. Hyg. Environ. Health **2008**, 211, 510–517.
124. Ashbolt, N; Fujioka, R; Glymph, T; McGee, C; Schaub, S; Sobsey, M; Toranzos, G. Pathogen indicators, and indicators of fecal contamination. In Report of the Experts Scientific Workshop on Critical Research Needs for the Development of New or Revised Recreational Water Quality; EPA 823-R-07–006; U.S. Environmental Protection Agency, Office of Water, Office of Research and Development: Warrenton, VA, 2007; Chapter 2,; pp. 35–56.
125. Ferreira, AP. Caffeine as an Environmental Indicator for Assessing Urban Aquatic Ecosystems. Cad. Saúde Pública**2005**, 21, 1884–1892.
126. Gardinali, PR; Zhao, X. Trace Determination of Caffeine in Surface Water Samples by Liquid Chromatography-Atmospheric Pressure Chemical Ionization–Sass spectrometry (LC-APCI-MS). Environ. Int **2002**, 28, 521–528.
127. Peeler, KA; Opsahl, SP; Chanton, JP. Tracking Anthropogenic Inputs Using Caffeine, Indicator Bacteria, and Nutrients in Rural Freshwater and Urban Marine Systems. Environ. Sci. Technol **2006**, 40, 7616–7622.
128. Scott, TM; JB; Rose, TM; Jenkins, S; Farrah, R; Lukasik, J. Microbial Source Tracking: Current Methodology and Future Directions. Appl. Environ. Microbiol **2002**, 68, 5796–5803.
129. Wu, J; Yue, J; Hu, R; Yang, Z; Zhang, L. Use of Caffeine and Human Pharmaceutical Compounds to Identify Sewage Contamination. World Acad. Sci. Eng. Technol **2008**, 44, 438–442.
130. Isobe, KO; Tarao, M; Chiem, NH; Minh, LY; Takada, H. Effect of Environmental Factors on the Relationship between Concentrations of Coprostanol and Fecal Indicator Bacteria in Tropical (Mekong Delta) and Temperate (Tokyo) Freshwaters. Appl. Environ. Microbiol **2004**, 70, 814–821.
131. Medema, GJ; Shaw, S; Waite, M; Snozzi, M; Morreau, A; Grabow, W. Catchment characteristics and source water quality. In Assessing Microbial Safety of Drinking Water

Improving Approaches and Method; WHO & OECD, IWA Publishing: London, UK, 2003; pp. 111–158.
132. EPA. Source Water Protection Practices Bulletin. Managing Stormwater Runoff to Prevent Contamination of Drinking Water; Office of Water (4606); United States Environmental Protection Agency (EPA): Washington, DC, USA, 2009.
133. Ferguson, DM; Moore, DF; Getrich, MA; Zhowandai, MH. Enumeration and Speciation of Enterococci Found in Marine and Intertidal Sediments and Coastal Water in Southern California. J. Appl. Microbiol **2005**, 99, 598–608.
134. Boualam, M; Mathieu, L; Fass, S; Cavard, J; Gatel, D. Relationship between Coliform Culturability and Organic Matter in Low Nutritive Waters. Water Res **2002**, 36, 2618–2626.
135. Boualam, M; Fass, S; Saby, S; Lahoussine, V; Cavard, J; Gatel, D; Mathieu, L. Organic Matter Quality and Survival of Coliforms in Low-Nutrient Waters. J. AWWA **2003**, 95, 119–126.
136. Baudišová, D. Evaluation of Escherichia coli as the Main Indicator of Faecal Pollution. Water Sci. Technol **1997**, 35, 333–336.
137. Edberg, SC; Rice, EW; Karlin, RJ; Allen, MJ. Escherichia coli: The Best Biological Drinking Water Indicator for Public Health Protection. J. Appl. Microbiol **2000**, 88, 106S–116S.
138. Byappanahalli, MN; Whitman, RL; Shively, DA; Sadowsky, MJ; Ishii, S. Population Structure, Persistence, and Seasonality of Autochthonous Escherichia coli in Temperate, Coastal Forest Soil from a Great Lakes Watershed. Environ. Microbiol **2006**, 8, 504–513.
139. Ishii, S; Yan, T; Shively, DA; Byappanahalli, MN; Whitman, RL; Sadowsky, MJ. Cladophora (Chlorophyta) spp. Harbor Human Bacterial Pathogens in Nearshore Water of Lake Michigan. Appl. Environ. Microbiol **2006**, 72, 4545–4553.
140. Ishii, S; Ksoll, WB; Hicks, RE; Sadowsky, MJ. Presence and Growth of Naturalized Escherichia coli in Temperate Soils from Lake Superior Watersheds. Appl. Environ. Microbiol **2006**, 72, 612–621.
141. Ksoll, WB; Ishii, S; Sadowsky, MJ; Hicks, RE. Presence and Sources of Fecal Coliform Bacteria in Epilithic Periphyton Communities of Lake Superior. Appl. Environ. Microbiol **2007**, 73, 3771–3778.
142. Hecq, P; Hulsmann, A; Hauchman, FS; McLain, JL; Schmitz, F. Drinking water regulations. In Analytical Methods for Drinking Water; Quevauviller, P, Thompson, KC, Eds.; John Wiley: London, UK, 2006; Chapter 1.
143. Total Coliform Rule: A Handbook for Small Noncommunity Water Systems Serving less than 3,300 Persons; EPA 816-B-06–001; Office of Water (4606M), EPA: Washington, DC, USA, 2006.
144. Espigares, M; Coca, C; Fernández-Crehuet, M; Moreno, O; Gálvez, R. Chemical and Microbiologic Indicators of Faecal Contamination in the Guadalquivir (Spain). Eur. Water Pollut. Control **1996**, 6, 7–13.

CHAPTER 6

Drinking water contamination and treatment techniques

S. Sharma[1], A. Bhattacharya[1]

[1] Reverse Osmosis Division, Central Salt and Marine Chemicals Research Institute, G. B. Marg, Bhavnagar 364002, Gujarat, India

ABSTRACT

Water is of fundamental importance for life on earth. The synthesis and structure of cell constituents and transport of nutrients into the cells as well as body metabolism depend on water. The contaminations present in water disturb the spontaneity of the mechanism and result in long/short-term diseases. The probable contaminations and their possible routes are discussed in the present review. Continued research efforts result in some processes/technologies to remove the contaminations from water. The review includes concepts and potentialities of the technologies in a comprehensible form. It also includes some meaningful hybrid technologies and promising awaited technologies in coming years.

Keywords: Water Contaminants, Purification, Hybrid technology

1. INTRODUCTION

Availability of fresh water, the nature's gift controls the major part of the world economy. The adequate supplies of water are necessary for agriculture, human consumption, industry as well as recreation. Ironically, sometimes, natural or added contaminations rob us of the gift and making us confront a lot more challenging world. It is a well-known fact that fresh water is an important necessity for our health. With the advancement of technology and industrial growth, fresh water resources all over the world are threatened. One-sixth of the world population suffers from the freshwater unavailability situation (Elimelech 2006). It is seen that developed countries suffer most from chemical discharge problems, whereas developing countries from agricultural sources. Contaminated water causes problems to health and leads to waterborne diseases which can be prevented by taking measures even at the household level. Providing safe water for all is a challenging task. Continued research efforts in this area for more than few decades result in many processes/technologies (Shannon et al. 2008).

Water contamination is a common problem to all over the world. These may be geological or anthropogenic (man-made) (Fawell and Nieuwenhuijsen 2003). Higher levels of contaminants in drinking water are seldom to cause acute health effects. Of course it depends on individual susceptibility and mode of contact with the body.

The types and concentrations of natural contaminates depend on the nature of the geological materials through which the groundwater flows and quality of the recharge water. Groundwater moving through sedimentary rocks and soils may pick up a wide range of compounds, such as magnesium, calcium, and chloride, arsenate, fluoride, nitrate, and iron; thus, the effect of these natural contaminations depends on their types and concentrations. The natural occurring elements present at unacceptable levels can contaminate water as well (Liu et al.2005; Charles et al. 2005; Rukah and Alsokhny 2004; Mulligan et al. 2001; Ghrefat et al. 2014; Meenakshi and Maheshwari 2006).

Other contaminants are man-made by-products of industry, and agriculture, including heavy metals like mercury, copper, chromium, lead, and hazardous chemicals, dyes and compounds like insecticides and fertilizers. Improper storing or disposing of household chemicals such as paints, synthetic detergents, solvents, oils, medicines, disinfectants, pool chemicals, pesticides, batteries, gasoline and diesel fuel can lead to ground water contamination (Kass et al. 2005; Anwar 2003) According to UN report 2003 (UN WWAP 2003) every day 2 million tons of sewage, industrial and agricultural waste are discharged into the world's water.

The microbial contaminants include pathogens like bacteria, viruses, and parasites such as microscopic protozoa and worms. These living organisms can be spread by human and animal wastes knowing or unknowingly.

Some contaminants can be easily identified by assessing color, odor, turbidity and the taste of the water. However, most cannot be easily detected and require testing to reveal whether water is contaminated or not. Thus, the contaminants may result in unappealing taste or odor and staining as well as health effects.

Color of the drinking water is a physical characteristic that cannot be noticed unless it is one of high concentration. For example, if ground water containing a high iron concentration, it gives a reddish appearance; similarly, high tannin concentration makes the water look brown. Generally, it is measured by comparing a water sample to a color standard. One color unit has no effect on the water and usually not detectable while 100 color units could be compared to the color of light tea (Ligor and Buszewski 2006). Odor is also an indicator for the presence of some contamination though odor-free water is not necessarily safe for drinking purpose. Also, some contaminant odors are noticeable even when present in extremely small amounts.

On the other hand, the presence of clays, silts or sand, or organic, algae, and leaf particles results in turbidity. The turbidity may shield bacteria, preventing disinfection chemicals from attacking and destroying the cells. The presence of organic materials in conjunction with chlorine can form trihalomethanes and other potentially harmful chemicals. Generally, surface water sources have higher turbidity compared to groundwater sources. The turbidity of a surface water source can vary greatly from 1 to 200 NTU (NTU: nephelometric turbidity unit). The immunity in turbidity level is different from children to adult people.

Types of contaminants

Basically, the contaminants are four types associated with water pollution

- Inorganic contaminants
- Organic contaminants
- Biological contaminants
- Radiological contaminants.

2. INORGANIC CONTAMINANTS

The presence of contaminants can also be measured by its chemical parameters. Hardness of the drinking water is a naturally occurring contaminate, which basically depends on the geographical status. It is caused by significant amounts of calcium or magnesium components; the hardness is classified into carbonate or non-carbonate hardness depending on what molecules are combined with calcium or magnesium. If they are combined with carbonate ions (CO_3^{-2}), the hardness is termed as 'carbonate hardness'; if combined with other ions, it is non-carbonate hardness. Generally, 300–400 mg/L hardness is suitable for drinking purpose. Prolonged exposure to water containing salts (TDS >500 mg/L) can cause kidney stone, etc.

Apart from carbonate/noncarbonated hardness, there are several inorganic substances (viz. fluoride, arsenic, lead, copper, chromium, mercury, antimony, cyanide) that contaminate water resource. They can get into drinking water from natural sources, industrial processes, as well as from plumbing systems (EPA US 2006; Nriagu 1988).

Sources of fluoride may be geological or anthropogenic. Weathering of fluoride-bearing minerals (fluorite, fluorspar, cryolite, fluorapatite, ralstonite and others) on the earth's crust can lead to higher fluoride levels in groundwater. The over exploitation of ground water also aggravates the problem of fluoride concentration in the water even more. Further, the anthropogenic sources of fluorides are certain pharmaceutical products (for treating hyperthyroidism), medicines, tooth pastes, insecticides, disinfectants, preservatives, super phosphate fertilizer, vitamin supplements and others. Their effects are especially harmful to develop children and the elderly people. Fluoride is known to cause dental and skeletal fluorosis. It is also associated with Alzheimer's disease and other forms of dementia (Susheela 1999; Fawell et al. 2006; WHO 2008). Fluoride enters the brain and enables aluminum to cross the blood–brain barrier, resulting in increased risk for these diseases (Ram Gopal and Ghosh 1985). Excessive fluoride (4.0 mg/L by EPA) concentrations have been reported in ground waters of more than 20 developed and developing countries including India where 19 states are facing acute fluorosis problems (Eswar and Devaraj 2011). *Arsenic* (MCL 0.01 mg/L) (EPA US 2006) enters in drinking water supplies from natural deposits in the earth or from agricultural and industrial practices (Smith et al. 2000). Arsenic contamination is by far the biggest mass poisoning case in the world, especially in India and Bangladesh (Chatterjee et al. 1995; Khan et al. 2003). Arsenic contamination of drinking water causes a disease called arsenicosis (Chen et al. 1988). Non-cancer effects can include thickening and discoloration of the skin, stomach pain, nausea, vomiting, diarrhea, numbness in hands and feet, partial paralysis, and blindness. Arsenic has been linked to cancer of the skin, bladder, lungs, kidney, nasal passages, liver, and prostate (Yoshida et al. 2004). The toxicity and excretion of arsenic compounds and their metabolites highly depend on the oxidation states [viz. Arsenite (As III) and Arsenate (As V)] and degree of methylation of arsenicals. It is seen that As(III) is ten times more toxic than As(V) (Pontius et al. 1994).

Mercury (MCL 0.002 mg/L) (EPA US 2006) gets into drinking water from agricultural runoff as well as seepage from landfills and some factories. The presence of mercury in water causes impairment of brain functions, neurological disorders, and retardation of growth in children, abortion and disruption of the endocrine system (Clarkson 1992; Counter and Buchanan 2004). *Copper* (MCL 1.3 mg/L) (EPA US 2006) can enter into the water through natural deposits in rock and soil, but more often as a result of corrosion in household plumbing. In short term, exposure leads to mild gastrointestinal distress, but long-term exposure can lead to permanent liver or kidney damage (Semple et al. 1960; Manuel et al. 1998). *Chromium* (MCL 0.1 mg/L) (EPA US 2006) occurs naturally in the ground and is often used in electroplating of metals and leather industries. Generally, it gets into water from runoff from old mining operations and improper waste disposal from these industries. A high level of exposure of chromium causes liver and kidney damage, dermatitis and respiratory problems (Zhang et al. 1997; Ray (Arora) and Ray 2009). *Lead* (MCL 0.015 mg/L) (EPA US 2006) is an increasing problem in cities with older water systems. Water slowly corrodes the lead in municipal water systems which can cause a wide range of developmental difficulties for children and high blood pressure and kidney ailments in older and adults (Needleman et al. 1990). *Antimony* (MCL 0.006 mg/L) (EPA US 2006) occurs naturally in the ground and originates from flame retardant industry. It is also used in ceramics, glass, batteries, fireworks and explosives. It may get into drinking water through natural weathering of rock as well as through industrial and municipal waste disposal or from manufacturing processes. It affects cholesterol, glucose in blood levels (Cooper and Harrison 2009a, b; Public Health Service, US 1992). *Nitrate* (MCL 10 mg/L as Nitrogen) (EPA US 2006) contamination comes through fertilizers. It is found in sewage and wastes from human and/or farm animals and generally gets into drinking water from these activities. Excessive levels of nitrate in drinking water have caused serious illness because of nitrate conversion to nitrite in the body and interferes oxygen transport in the blood. The symptoms include shortness of breath and blueness of the skin (Gupta et al. 2000). *Asbestos* (MCL 7 million fibers/lit) (EPA US 2006) is a mineral that forms minute fibers in the environment. Asbestos fibers in water have been linked to an increase in the risk of certain cancers and regulated by EPA because of asbestos exposure from water (EPA US 2009a, b; Bull

2007). *Selenium* (MCL 0.05 mg/L) (EPA US 2006) contamination comes through mainly food and soils. It is used in electronics, photocopy operations, manufacture of glass, chemicals, drugs, and as a fungicide and feed additive. Exposure to high levels of selenium over a long period of time has resulted in a number of adverse health effects, including a loss of feeling and control in the arms and legs (Olson 1986; Fan and Kizer 1990). *Barium* (MCL 2 mg/L) (EPA US 2006) occurs naturally in some aquifers that serve as sources of ground water. It generally gets into drinking water after dissolving from naturally occurring minerals in the ground. It may damage heart and cardiovascular system, and is associated with high blood pressure in laboratory animals such as rats exposed to high levels during their lifetimes (Brenniman et al. 1979; Wones et al. 1990). Drinking water that meets the EPA standard is associated with little to none of this risk and is considered safe with respect to Barium. *Beryllium* (MCL 0.004 mg/L) (EPA US 2006) generally gets into water from runoff from mining operations, discharge from processing plants and improper waste disposal. Beryllium compounds have been associated with damage to the bones and lungs and also may increase the risk of cancer in humans who are exposed over long periods of time (Cooper and Harrison 2009a, b). *Cyanide* (MCL 0.2 mg/L) (EPA US 2006) usually gets into water as a result of improper waste disposal. It has been shown to damage the spleen, brain and liver of humans fatally poisoned with cyanide (Ronald 1991).

3. ORGANIC CONTAMINANTS

The major anthropogenic sources of organic contamination are pesticides, domestic waste, and industrial wastes, etc. Contamination through organic materials can cause serious health problems like cancers, hormonal disruptions, and nervous system disorder (Ram et al. 1990; Harvey et al. 1984). Trihalomethanes (THMs) are formed when chlorine in the treated drinking water combines with naturally occurring organic matter.

- Pesticides contaminate through agricultural as well as public hygienic sources (Damalas and Eleftherohorinos 2011; Younes and Galal-Gorchev 2000). The adverse environmental effects of pesticides used in agriculture and public health are due to an improper handling and application procedure (WHO 2010). Pesticides are designed to interact with various chemical processes in the pest's living body chemistry. Unfortunately, doing this, all pesticides may interact with the metabolism of non-targeted living organism. Mostly, pesticides damage the liver and nervous system. Tumor formation in the liver has also been reported (Bolognesi 2003). Environmental agencies have fixed their MCL's (EPA US 2009a, b). Some of the pesticides with their MCLs are in the ensemble (Table 1).

- Volatile organic chemicals (VOCs) include solvents and organic chemicals like toluene benzene, styrene, trichloroethylene (TCE) and vinyl chloride, etc., degreasers, adhesives, gasoline additives, and fuel additives (Wehrmann et al. 1996). These VOCs cause chronic health effects like cancer, central nervous system disorders, liver and kidney damage, reproductive disorders, and birth defects (Brown et al. 1984).
- Dyes constitute one of the largest groups of organic compounds that represent an increasing environmental concern. The release of this contaminated water into the environment is a considerable source of non-esthetic pollution and eutrophication, which can originate dangerous byproducts through oxidation, hydrolysis, or other chemical reactions taking place in the wastewater phase (Pagga and Bruan 1986; Prevot et al. 2001).
- Apart from the above, compounds present in the water have the potential to cause known or suspected adverse ecological or human health effects. These compounds are termed 'Emerging Organic Contaminants' (Pal et al. 2010, 2014; Stuart et al. 2012; Lapworth et al. 2012). It includes pharmaceuticals (viz. ciprofloxacin, erythromycin, tetracycline, codeine, salbutamol, carbamazepine, paracetamol, ibuprofen, salicylic acid, Tamiflu, chemotherapy drugs such as 5-flurourcil, ifosfamide) industrial compounds (viz. chlorinated solvents, petroleum hydrocarbons, including the polyaromatic hydrocarbons, the fuel oxygenate methyl tertiary butyl ether, plasticizers/resins bisphenols, adipates and phthalates), personal care products (viz. *N,N* diethyl meta toluamide, alkyl esters of *p*-hydroxy benzoic acid, triclosan), fragrances (viz. tonalide, galaxolide), water treatment by products (viz. trihalomethanes, haloacetic acids, *N*-nitroso dimethyl amine), plasticizers, flame retardants as well as surfactants. Mostly, they are endocrine disruptors, carcinogenic.
- These are commonly derived from a variety of municipal, agriculture and industrial sources and pathways. The pharmaceuticals (viz. antibiotics, analgesic and anti-inflammatory) come from hospital effluents and/or chemical manufactures. There are reports of lowest predicted no-effect concentration (PNEC) values of emerging organic contaminants. Some of them are listed in Table 2 (Pal et al. 2010, 2014).

Table 1 Some of the pesticides with their maximum contamination level (MCL) (Adapted from EPA, US Protection agency)

Pesticides	**Nature**	**Maximum contamination level (MCL), μg/L**
Carbofuran	Nematicide	40
Dalapon	Herbicide	200
Dibromochloropropane	Nematocide	0.2
Dinoseb	Insecticide/miticide	7
Dioxin	Herbicide	0.0003
Diquat	Herbicide	20
Endothall	Algicide	100
Ethylene dibromide	Insecticide	0.05
Glyphosate	herbicide	700
Methoxychlor	Insecticide	40
Oxamyl	Insecticide	200
Pentachlorophenol	Fungicide	1
Picloram	Herbicide	500
Simazine	Herbicide	4
Toxaphene	Insecticide	3

Table 2 Lowest predicted no effect concentration (PNEC) values for some of the emerging organic contaminants (Pal et al. 2010, 2014)

Compounds	**Lowest PNEC (ng/l)**	**Compounds**	**Lowest PNEC (ng/l)**
Antibiotics			
Trimethoprim	1000	Bisphenol A (making plastics)	60–150
Ciprofloxacin	20		
Sulfamethaoxazole	20,000		

Analgesic and anti-inflammatory			
Naproxen	37,000	PPOS (protective coatings, surfactants)	1100
Ibuprofen	5000		
Ketoprofen	15.6×10^6		
Diclonofenac	10,000		
Beta blockers			
Propranolol	500	Fipronil (termiticide)	250
Atenolol	10×10^6		
Blood lipid regulators			
Clofibric acid	12000	NP1EO (surfactant)	330
Gemfibrozil	100,000		
Benzafibrate	100,000		
Hormones			
Estriol	0.8	4MBC (sun screen)	560
Estrone	18		
Sucralose (sugar substitute)	93×10^4	DEET (mosquito repellent)	$5–24 \times 10^6$

4. BIOLOGICAL CONTAMINANTS

Biological contamination of water is caused by the presence of living organisms, such as algae, bacteria, protozoan or viruses. Each of these can cause distinctive problems in water (Daschner et al. 1996; Ashbolt 2004). Algae are in general single celled and microscopic. These are quite abundant and depend on nutrients (viz. Phosphorus) in water. The nutrients are generally from domestic run-off or industrial pollution. The excess algae growth is not only imparted taste and odor problems in water; it clogs filters, and produces unwanted slime growths on the carriers. Sometimes, they [viz. blue-green algae (Anabaena, Aphanizomenon and Microcystis)] are capable of liberating toxins and they damage the liver (hepatotoxins), nervous system (neurotoxins) and skin (Hitzfeld et al. 2000; Rao et al. 2002).

Bacteria are also microscopic single celled. There are numerous pathogenic bacteria and can be contaminated with water (Inamori and Fujimoto 2009). They can result in typhoid, dysentery, cholera and gastroenteritis. Some non-pathogenic bacteria (viz. sulfur, crenothrix iron bacteria), although not harmful, may cause taste and odor problems (Nwachcuku and Gerba 2004; Rusin et al. 1997). Similarly, Protozoans are also single-celled and microscopic organisms. Some protozoans (like Giardia and Cryptosporidium) are commonly found in rivers, lakes, and streams contaminated with animal feces or which receive wastewater from sewage treatment plants. These may cause diarrhea, stomach cramps, nausea, fatigue, dehydration and headaches. Viruses are the smallest living organisms capable of producing infection and causing diseases. Hepatitis and polio viruses are commonly reported in the contaminated water.

5. RADIOLOGICAL CONTAMINANTS

Radiological contaminants are caused by radioactive elements. Sources of radioactive material could be soils or rocks the water moves through or some industrial waste. Erosion of natural deposits of certain minerals (radioactive) may emit radiations (like α, β). Radiological elements (viz. U^{226}, Ra^{226}, Ra^{228} and Rn^{228}) tend to be a greater problem in groundwater than in surface water. All types of radiological contamination increase the risk of cancer (Alireza et al. 2010; Haki et al. 1995). Some of the radioactive contaminants with their MCLs are listed in Table 3.

Table 3 Radioactive contaminants and their MCLs (adapted by EPA, US)

Contaminants	**MCL**
Alpha particles	15 (pCi/L)
Beta particles and photon emitters	4 mrems/year
Radium 226 and Radium 228 (combined)	5 pCi/L
Uranium	30 ug/L

6. SOLVING APPROACHES

The famous saying of Minora Shirota's statement is "Prevent disease rather than treat disease: a healthy intestine leads to a long life, and deliver health benefits to as many people as possible at an affordable price" (Heasman and Mellentin 2001). This philosophy, elaborated almost half a century ago, is becoming more valid now than ever before. The need of science-based solutions for uncontaminated water provisioning results in several water treatment methods to counter the problem. Of course, the suitable technology is based on raw water characteristics (i.e., the nature and extent of contamination), infrastructure (i.e., power, manpower, availability of chemicals), affordability/cost as well as acceptability. Some of the common water purification methods are sedimentation or settling, boiling/distillation, chemical treatment (precipitation/ coagulation/ adsorbents), disinfection and filtration. The processes and techniques in mitigating the contaminations are as follows.

7. PRECIPITATION AND COAGULATION

Precipitation is a technique of removing one or more substances from a solution by adding reagents so that insoluble solids appear. The 'solubility' rules the technique, i.e., when the product of ion concentrations (in simple) in the solution over the solubility product of the respective solid, the precipitation occurs. It is one of the simple methods to purify water. The chemicals are added to form particles which settle and remove contaminants from water. The treated water is reused whereas the settled portion is dewatered and disposed of. The technique is used in softening of water as well as to remove impurities like phosphorus, fluoride, arsenic, ferrocyanide and heavy metals, etc. (EPA US 2000; Matlock et al. 2002; Eikebrokk et al. 2006).

8. SOFTENING OF WATER

The presence of Ca/Mg in terms of carbonate, bicarbonate, chloride and sulfate results in hardness of water. Addition of proper chemical forms precipitation and makes it soft.

Addition of $Ca(OH)_2$ forms precipitation with bicarbonate and sulfate in water.

$$Ca(HCO_3)_2 + Ca(OH)_2 \rightarrow 2CaCO_3\downarrow + 2H_2O,$$

$$MgSO_4 + Ca(OH)_2 \rightarrow Mg(OH)_2\downarrow + CaSO_4$$

Addition of Na-aluminate forms precipitation of hydroxide with sulfate and chloride in water. Actually, Na-aluminate forms sodium hydroxide with water, and with sulfate/chloride it forms hydroxide.

$$MgSO_4/Cl_2 + Na_2Al_2O_4 + 4H_2O \rightarrow Mg(OH)_2\downarrow + Na_2SO_4/NaCl + 2Al(OH)_3$$

Formation of aluminum hydroxide aids in floc formation, sludge blanket conditioning and silica reduction.

Softening of water is also feasible by simple boiling

$$Ca(HCO_3)_2 + heat \rightarrow CaCO_3\downarrow + H_2O + CO_2$$

9. REMOVAL OF HEAVY METALS

Heavy metals (e.g., Ba, Cd, Pb, Hg, Ni, Cu) typically precipitated from waste water as sulfates, sulfides, hydroxides, and carbonates (Matlock et al. 2002). Metal co-precipitation during flocculation with iron and aluminum salts is also possible for some metals (e.g., As, Cd, Hg, Cr). The following reaction represents as chromium co-precipitation in terms of hydroxides or sulfates

$$H_2Cr_2O_7 + 6FeSO_4 + 6H_2SO_4 \rightarrow Cr_2(SO_4)_3\downarrow + 3\ Fe_2(SO_4)_3 + 7H_2O,$$
$$Cr_2(SO_4)_3 + 3Ca(OH)_2 \rightarrow 2Cr(OH)_3\downarrow + 3CaSO_4$$

10. REMOVAL OF ARSENIC

Arsenic removal with coagulants, viz. Alum [$Al_2(SO_4)_3 \cdot 18H_2O$] ferric chloride ($FeCl_3$) and ferric sulfate [$Fe_2(SO_4)_3 \cdot 7H_2O$] is effective (Harper and Kingham 1992; Fields et al. 2000). In these cases, arsenic (V) can be more effectively removed than arsenic (III). The microparticles and negatively charged arsenic ions are attached to the flocs by electrostatic attachment during the process. The possible steps of coagulation and co-precipitation are as follows:

Alum dissolution:

$$Al_2(SO_4)_3 \cdot 18H_2O \rightarrow 2Al^{+3} + 3SO^{-2}_4 + 18H_2O$$

Aluminum precipitation (acidic):

$$2Al^{+3} + 6H_2O \rightarrow 2Al(OH)_3\downarrow + 6H^+$$

Co-precipitation (non-stoichiometric, non-defined product):

$$H_2AsO_4^- + Al(OH)_3 \rightarrow Al{-}As\ (complex) + Other\ Products.$$

Similar reactions take place in case of ferric chloride and sulfate with the formation of Fe–As complex as an end product which is removed by the process of sedimentation and filtration (Mok and Wai 1994; Hering et al.1997). The efficient removal depends on pH range.

11. REMOVAL OF PHOSPHORUS

The removal of phosphates is generally done by coagulant, i.e., by mixing coagulant into waste water (Xie et al. 2005). The most commonly used multivalent metal ions are Ca, Al, and Fe.

$$10\ Ca^{2+} + 6\ PO_4^{3-} + 2OH^- \rightarrow Ca_{10}(PO_4) \times 6(OH)_2\downarrow$$

$$Al^{3+} + H_nPO_4^{3-n} \rightarrow AlPO_4 + n\ H^+,$$

$$Fe^{3+} + H_nPO_4^{3-n} \rightarrow FePO_4 + n\ H^+.$$

12. REMOVAL OF FLUORIDE

Precipitation of fluoride species into chemically stable form is the most effective option for the removal of fluoride (in terms of Ca, Mg, Al) from effluent streams (Dahi 1997; Lawrence et al. 2005). Among all metal fluorides, CaF_2 is less soluble in water. Consequently, removal of fluoride from the effluents by converting it into CaF_2 has become the most widely used method of treatment. $CaCl_2$, limes, may be used for this purpose, but $CaCl_2$ is preferred with respect to lime due to its higher solubility and the lower ratio of additive to effluent.

$$2\ HF + Ca(OH)_2 \rightarrow CaF_2 + 2H_2O$$

$$CaCl_2 + 2HF \rightarrow CaF_2\downarrow + 2HCl$$

The reaction of hydrofluoric acid and ammonium fluoride with the aluminum treatment agent is as follows:

$$3\ HF + AlO_2 \rightarrow AlF_3\downarrow + H_2O + OH^-,$$

$$3NH_4F + AlO_2 \rightarrow AlF_3\downarrow + 3NH_4^+ + O_2,$$

$$6NH_4HF_2 + 4\ AlO_2 \rightarrow 4\ AlF_3\downarrow + 6NH_4^+ + 2O_2 + 2H_2O + 2OH^-.$$

Inorganic flocculants have the potential in different separations (Gray et al. 1995; Jiang and Graham 1998), but they are used in very large quantities. These leaves large amounts of sludge and strongly affected by pH changes, whereas polymeric flocculants cause the formation of large cohesive aggregates (flocs) and inert to pH changes. Both natural and synthetic polymers are useful for this purpose. Generally, synthetic polymers (viz. polyacrylamide, polyethylene oxide, poly (diallyl dimethyl ammonium chloride), poly (styrene sulfonic acid) are highly effective flocculants at small dosages and have high tailor ability but poor shear stability, whereas though natural polymers (viz. starch, guar gum, alginate, glycogen, dextran) are biodegradable and effectively shear stable (Brostow et al. 2009).

13. REMOVAL OF DYES

Dyes are non-biodegradable, and precipitation with $CaCO_3$ can be one of the approaches to remove them from the water (Hoffmann et al. 1995; Reife and Freeman 1996).

As a whole, the precipitation technique has the features

Benefits:

- Simple process,
- Effective for the removal of As, Cd, Ba, Cd, Cr, Pb, Hg, Se, Ag, etc.,
- It is also applicable to remove natural organic matter (NOM) or dissolved organic carbon (DOC).

Limitations:

- Requires continuous supply of huge chemicals,
- Handling of by-products,
- Disposal of coagulation/precipitation sludge is a concern.

14. DISTILLATION

It is the most common separation technique (http://www.msue.msu.edu; Veil. 2008). In this separation technique, the mixed components in water are separated by the application of heat. It is based on the differences in boiling points of the individual components. The boiling point characteristics depend on the concentrations of the components present.

Thus, the distillation process depends on the vapor pressure characteristics of liquid mixtures. The basic principle described as the input of heat energy raises vapor pressure. When the vapor pressure reaches its surrounding pressure, the liquid mixture boils and distillation occurs because of the differences of volatility in the mixture.

This process results in a separation between water and inorganic substances, such as lead, calcium, magnesium, etc. are also destroying bacteria. However, organics with boiling points lower than 100 °C cannot be removed efficiently and can actually become concentrated in the product water. Distilled water purification technology was originally developed for industrial purpose. However, it came eventually for home use. Since, this process is not very effective in removing organic chemicals so the carbon filter system must be added to make the water really safe to drink. The carbon filters require regular changing because they can quickly become breeding grounds for bacterial growth.

Although distilled water is safe, it is not healthy as this contains no nutrient minerals, which are essential for the drinking purpose. This type of water purification technology is also very slow. Adding to that, the cost of a carbon filter and the result is an unwieldy system of water purification.

Benefits:

- Removes a broad range of contaminants (toxic chemicals, heavy metals, bacteria, viruses, parasites),
- Continuous,
- Does not rely on physical barriers (filters),
- Does not require additional disinfecting process.

Limitations:

- It consumes an enormous amount of energy both in terms of cooling and heating requirements,
- Some contaminants can be carried into the condensate,
- Requires careful maintenance to ensure purity,
- The process is not very effective which are of lower volatility (viz. organics) compared to water.

15. ADSORPTION

In this physical process, dissolved contaminants adhere to the porous surface of the solid particles (Jiuhui 2008). It is the surface phenomena and the outcome of surface energy. With the material, all the bonding requirements of the constituent atoms of the material are filled with other atoms. However, atoms on the surface of the adsorbent are not wholly surrounded by other adsorbent atoms and physical attractive force results. It can be physisorption (originates from vanderwaals forces) and chemisorption (originates from co-valent forces).

The adsorbent systems are added directly to the water supply or via mixing basin. Adsorbents combine chemical and physical processes to remove the compounds that impart color, taste, and odor to water. In principle, all microporous materials can be used as adsorbents. However, those with well controlled and highly microporous are the most preferred (Yang 1997). The porous solids, e.g., activated carbon, silica gels, aluminas, zeolites, etc. contain many cavities or pores with diameters as small as a fraction of a nanometer is useful (Ali and Gupta 2007; Qu 2008).

The isotherms are the quantitative interrelation between the adsorbate and adsorbent. The three most well-known isotherms are Freundlich, Langmuir and Linear. The most commonly used for the water contaminants is Freundlich and it is expressed as:

$$\frac{x}{m} = KC_e^{1/n}$$

where x is the mass of solute adsorbed, M is the mass of adsorbent, C_e is the equilibrium concentration of solute, and K and n are experimental constants.

16. ACTIVATED CARBON

The most commonly used adsorbent is activated carbon—a substance which is quite similar to common charcoal. Actually, the active carbon is much more efficient because of its high porous character. The high porous character is generated by treating carbon to steam and high temperature (1300 °C) with or without oxygen in the presence of inorganic salts (physical method). The carbon may be of petroleum coke, bituminous coal, lignite, wood products, and coconut/peanut shells. At high temperature, parts of carbon are oxidized in CO_2 and steam. The gases are evacuated and micro fractures and pores are generated in the carbon structure. It dramatically increases the carbon surface area, making a useful material for the removal of contaminants (Baudu et al. 1991; Yang and Benton 2003). In some cases, the carbonaceous matter may be treated with a chemical activating agent such as phosphoric acid, zinc chloride and the mixture carbonized at an elevated temperature, followed by the removal of activating agent by water washing (chemical method).

Active carbon uses the physical adsorption process, whereby Vanderwaals attractive forces pull the solute contamination out of the solution and onto its surface. The efficiency of the adsorption depends on the nature of the carbon particle and pore size, surface area, density and hardness as well as the nature of the contaminants (concentration, hydrophobicity, polarity and solubility of the contaminant and contaminant attraction to the carbon surface).

There are two different forms of activated carbon in common use, granular activated carbon (GAC) and powdered activated carbon (PAC). Physically, the two differ as their names suggested by particle size and diameter. The reusability of the carbon is done primarily with the GAC as PAC particles are too small to be reactivated.

Benefits:

- Activated charcoal is effective for trapping carbon-based impurities (volatile organic chemicals), chlorine (including cancer-causing by-product trihalomethanes) as well as colors and odors,
- Very cost effective,
- Long life (high capacity).

Limitations:

- In GAC scheduled filter replacements, it is important to eliminate the possibility of 'channeling' which reduces the contact between the contaminant and the carbon. Therefore, it reduces efficiency, and the accumulation of bacteria in the filter,
- Frequent filter changes often required,
- Can generate carbon fines

17. ACTIVATED ALUMINA

Activated alumina consists mainly of aluminum oxide (Al_2O_3) spherical beads, highly porous and exhibits tremendous surface area. The surface area of activated alumina is in the range 345–415 m^2/g. It does not shrink, swell, soften or disintegrate when immersed in water. It can exist in three forms, viz. activated alumina sorbent, activated alumina desiccant and activated alumina catalyst carrier. The granulated alumina has the internal active surface of the alumina.

In this process, contaminated water is passed through a cartridge or canister of activated alumina. The contaminant adsorbs on the alumina (Chen et al. 1987). As the physical adsorption has a particular limit, the cartridge of activated alumina must be replaced periodically.

Benefits:

- Tailoring of activated alumina is possible by varying the activation process and dopant variation,
- Effective in removing As^{5+}, PO_4^{3-}, Cl^-, and F^- from water,
- Removal of Se, Sb, Pb and Bi from the water is also possible.

Limitations:

- The method is not very much capable of reducing levels of other contaminants of health concern. It needs another support.

18. ZEOLITE

Zeolites are aluminosilicates with an Si/Al ratio between 1 and infinity. It has a tetrahedral network of silicon and oxygen atoms, and some of the silicon atoms are replaced by aluminum to form alumino-silicates. The adsorptive property of zeolite is considered due to the crystalline nature of the materials. The channels in it are of extended honeycomb and cavities. Zeolites have the surface area 1–20 m^2/g. Synthetic Zeolites are manufactured by hydrothermal processes in a temperature range of 90–100 °C, an autoclave followed by ion exchange with certain cations (Na^+, Li^+, Ca^{2+}, K^+, NH_4^+) (Rahman et al. 2012). The high cation exchange and molecular sieve properties, such as zeolites, have been widely used as adsorbents. The water softening process is by exchanging Na^+ with the Ca^{2+}/Mg^{2+} in water, as follows:

$$\text{Na–Zeolite} + Ca^{2}+/Mg^{2+} \rightarrow \text{Ca/Mg–Zeolite} + Na^{+}$$

Natural zeolites in the waste-water treatment are very useful (Margeta et al.2013; Kalló 2001). Many natural zeolites (e.g., Clinoptilolite, mordenite, phillipsite, chabazite) show selective separation towards NH_4^+ and also for transition metals (e.g., Cu^{2+}, Ag^+, Zn^{2+}, Cd^{2+}, and Hg^{2+}) (Jafarpour et al. 2010; Karapınar 2009).

Benefits:

- Recharging of zeolite is feasible by exchanging the cations with the initial one and thus reuse is also feasible,
- Removes NH_4^+ and heavy metal removal of inorganic anions (nitrates, phosphates, arsenates, chromates and fluorides) as well as radionuclides (e.g., 137Cs, 90Sr, 60Co, 45Ca, 51Cr, 111mCd, 110mAg) is also possible,
- Removal of organics and other humic substances (including humic, fulvic acid, and humin) and odor is also possible,
- Microorganisms capturing (the large surface area of the zeolites is accessible for adhering microorganisms. This makes selecting a suitable material for biofilter for removal of pathogenic microorganisms),
- By the zeolites, the permeable reactive barriers (PRB) can be prepared in the waste disposal site, so that contaminations could not spread in the ground water.

Limitations:

- As zeolites are used as softener in detergent formulations and insoluble, they lead to increase in sewage sludge mass.

19. SILICA GEL

Silica gel is an amorphous hard glass-like granules or beaded material made of silicon dioxide (SiO_2). Basically, it is a naturally occurring mineral which is purified and processed. Silica gel is a high capacity adsorbent with fine pores on the surface and can be used especially as desiccant, moisture-proof, rust inhibitor as well as catalysis (Heckel and Seebach 2000).

Generally, it is formed by two routes: (1) polymerizing silicic acid, and (2) aggregation of particles of colloidal silica. Silicic acid, $Si(OH)_4$, has a strong tendency to polymerize and form a network of siloxane (Si–O–Si), leaving a minimum number of uncondensed Si–O–H groups. The aggregation is by Van der Waals forces or by cations bridging as coagulants. Commercial silica is prepared through the first route by mixing a sodium silicate solution with a mineral acid, such as sulfuric or hydrochloric acid. The reaction produces a concentrated dispersion of finely divided particles of hydrated SiO_2, known as silica hydrosol or silicic acid.

$$Na_2SiO_3 + 2HCl + nH_2O \rightarrow 2NaCl + SiO_2 \cdot nH_2O + H_2O$$

The hydrosol, on standing, polymerizes into a white jelly-like precipitate, which is silica gel. The resulting gel is washed, dried, and activated. Various silica gels with a wide range of properties, such as surface area, pore volume, and strength, can be made by varying the silica concentration, temperature, pH, and activation temperature (Iler 1979). Two common types of silica gel are known as regular-density and low-density silica gels, although they have the same densities (true and bulk). The regular-density gel has a surface area of 750–850 m^2/g and an average pore diameter of 22–26 A°, whereas the respective values for the low density gel are 300–350 m^2/g and 100–150 A°).

Because of its large pore volume and mesoporosity, silica gel is used as desiccant. The modified silica gel (modified by the impregnation) with a high-molecular weight quaternary amine (triethyl octadecyl ammonium iodide) has been used for the concentration of heavy metals (Cs, Ag, Hg, Cu, Cd, etc.) for water purification (Tzvetkova and Nickolov 2012; Bowe et al. 2003; Bowe and Martin. 2004).

Benefits:

- Silica gel is non-toxic, non-corrosive material,
- It has high adsorption capacity because of very high surface area and porosity.

Limitations:

- Preparative aspects needed very precise control,
- Modification is needed to remove the contaminants.

20. ION EXCHANGE

The coulombic attractive force between ions and charged functional groups is more commonly classified as ion exchange. It is a typical reversible chemical reaction where an ion from a solution is exchanged for a similarly charged ion attached to an immobile solid particle.

The selectivity coefficient controls the preference for ions of particular resins and is expressed as follows:

$K_{B^+}^{A^+} = \frac{\{\bar{A}\}\{B^+\}}{\{A^+\}\{\bar{B}\}}$ for the exchange of Ā in solution for B^+ on the resin:

$$A^+ + \bar{B} \leftrightarrow B^+ + \bar{A}$$

The barred terms indicate location on the resin (solid phase) as opposed to solution phase. The superscript and subscript on the selectivity coefficient indicate the direction of the reaction.

Ion exchange materials are insoluble substances containing loosely held ions, capable of exchanging particular ions within them with ions in a solution that is passed through them. Many natural substances like proteins, cellulose, living cells and soil particles exhibit ion exchange properties, which play an important role in the way the function in nature. Synthetic ion-exchange polymers can be made in two forms, viz. beaded polymer matrix (resins) and membranes.

21. ION EXCHANGE RESINS

Ion exchange resins are very small polymer matrix (beads), with a diameter of about 0.6–1.0 mm. The ion exchange resins can be manufactured in one of the two physical structures, gel and porous. The gel resins are crosslinked polymers having no porous structure, while porous resins have considerable external and pore surfaces (microporous, mesoporous and macroporous) where ions can attach. The porous polymer matrices contain invisible water inside the pores of the beads, measured as "humidity" or "moisture content". The functional groups (ions) can be attached on the polymer matrix which cannot be removed or displaced.

Based on their functional groups attached on polymer matrix, the ion exchange resins are two types: cation and anion exchange resins, which further subdivided into four categories-

- Strongly acidic (typically, sulfonic acid groups, e.g., sodium polystyrene, sulfonate, etc.,
- Strongly basic (quaternary amino groups, for example, trimethylamonium group,
- Weakly acidic (mostly, carboxylic acid groups),
- Weakly basic (primary, secondary, and/or ternary amino groups, e.g., polyethylene amine).

Cation exchange resins (Fig. 1) exchange cations like calcium, magnesium, radium, and anion resins, used to remove anions like nitrate, arsenate, arsenite, or chromate from waste solution/water (Alexandratos 2009; Calmon 1986). Regeneration can be possible using sodium chloride. In case of cation resins, sodium ion displaces the cation from the exchange site; whereas in case of anion resins, the chloride ion displaces the anion from the exchange site. Resins can be designed to show a preference for specific ions, so that the process can be easily adapted to a wide range of different contaminants.

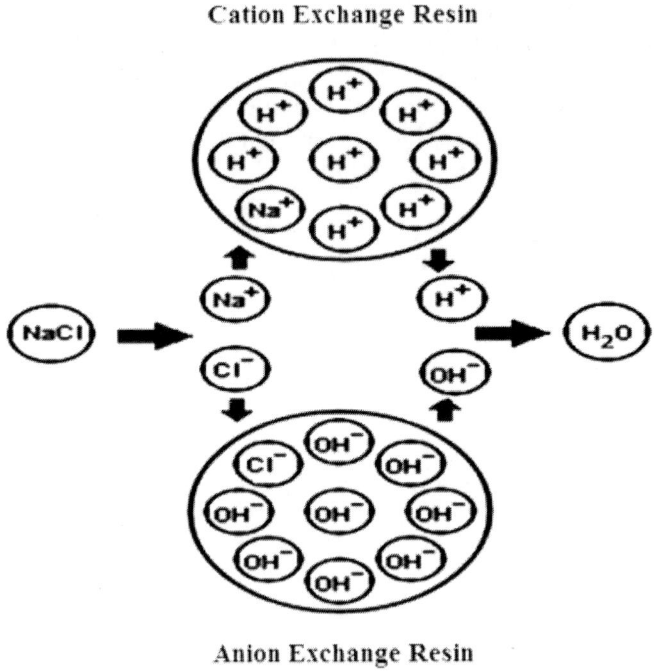

Figure. 1 Schematic diagram regarding the behavior cation exchange and anion exchange resin

The mode of preparation of ion exchange resins is through suspension polymerization technique containing the monomers, cross-linkers and initiators. Various types of polymeric beads like styrene, MMA, MAA, DVB, etc. can be prepared by this technique, varying the ratio of monomers, diluents, the stabilizer, concentration and the agitation rate is dispersed by agitation in a liquid phase, usually water, in which the monomer droplets are polymerized while they are dispersed by continuous agitation also known as pearl polymerization technique (Penlidis et al. 1997).

The most important issue in the practical operation of suspension polymerization is the control of the final particle size distribution. The size of the particles depends on monomer type, monomer purity, interfacial tension, stabilizer concentration, agitation condition in the reactor (degree of agitation) design of reactor/stirrer.

Benefits:

- Simple and low running cost technique,
- Technique is very useful in separating components/contaminations (cations and anions) from dilute solutions/wastes and in water purification, etc.,
- Useful for the recovery of expensive materials from industrial waste (e.g., precious metals),
- Recycling components present in the solutions and/or regenerating chemicals,
- Capability to handle hazardous wastes,
- Simple regeneration process and well-maintained resins last for many years.

Limitations:

- Limitation on the concentration in the effluent to be treated,
- Ion exchange resin-treated water contains sodium, which cannot be recommended for the diet requiring low sodium intake,
- Generation of waste (sodium wastewater) as a result of ion exchange regeneration,
- Ion exchange resins do not remove organic compounds or biological contaminants,
- If resin is not sanitized or regenerated regularly, bacterial colonies proliferate on resin surfaces and can contaminate drinking water.

The ion exchange membranes are discussed in the following part.

Apart from the above, interests are growing to develop different low cost adsorbents. For this purpose, numerous agro-waste biomaterials are found suitable, viz. rice-husk, soyabean hulls, coconut shells, rice straw, sugarcane bagasse, tea leaves, petiolar felt-sheath of palm trees, etc. (Ahluwalia and Goyal 2007; Tee and Khan 1988; Low et al. 1993; Mustafiz et al. 2002). These are useful for removal of heavy metal ions (Pb^{2+}, Ni^{2+}, Cd^{2+}, Zn^{2+}, etc.) in low concentrations. Biosorption is a rapid phenomenon of passive metal uptake sequestration of non-growing biomass (Beveridge and Doyle 1989). Biomass of *Aspergillusniger, Penicillum Chrysogenoum, Rhizopusnigricans, Ascophyllumnodosum, Sargassumnatans, Chlorella fusca, Oscillatoriaanguistissima, Bacillus firmus and Streptomyces sp.* has also the potential to sequester metal ions by forming metal complexes from solution and obviates the necessity to maintain special growth-supporting conditions.

22. MEMBRANE WATER TREATMENT

Membrane technology is one of the innovative ideas of water treatment. Over here, a semipermeable membrane is used for the removal of water impurities. There are two types of membrane water treatment technologies, namely pressure-driven (e.g., reverse osmosis) and electrically driven (electro-membrane) (Charcosset 2009).

23. REVERSE OSMOSIS

The two processes (viz. osmosis and reverse osmosis) are the regulator of life. Though they are termed as concentration and pressure driven simultaneously, both are controlled by thermodynamic function, i.e., 'chemical potential' of the systems. It is essentially a driving force expressed as a change in the free energy of the system as a result of the change in the composition of the system. Though literally the two signify just the opposite process, thermodynamically they are similar. Under isothermal operating condition, the tendency for material transport is always in the direction of lower chemical potential for both the processes. In osmosis, the flow is occurring solvent to solution side through a semipermeable membrane, whereas in reverse osmosis the flow is a solution for solvent. In both cases, only solvent molecules migrate from one side to another. The schematic diagram of osmosis and reverse osmosis is presented in Fig. 2.

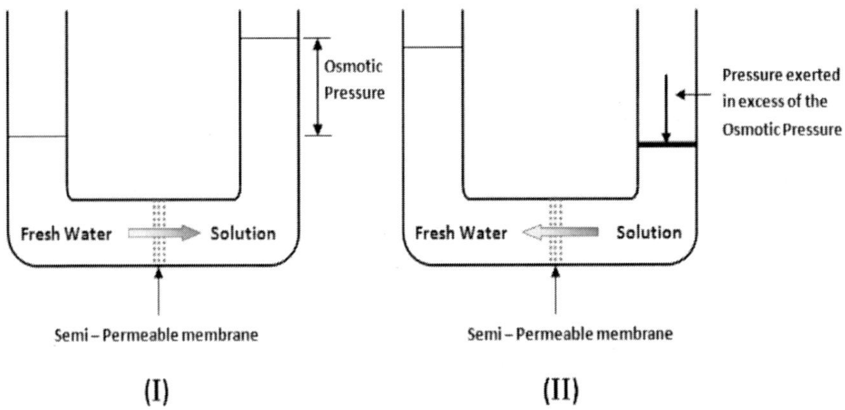

Figure. 2 Schematic diagram of osmosis and reverse osmosis

The main two characteristics of a membrane process are flux and rejection. If an RO membrane is considered as permeating water only, the water and solute flux can be written as:

$$Jw = A(\Delta P - \Delta \Pi)$$

$$Js = B\Delta cs$$

where *A* is the permeability coefficient, and ΔPΔP and ΔΠΔΠ are the hydraulic pressure and osmotic pressure difference across the membrane and *B* is the solute permeability coefficient and ΔC_S is the solute concentration difference across the membrane.

The microfiltration and ultrafiltration membranes have a pore size in the range of >10 and 1–100 nm, respectively, whereas in the case of nanofiltration and reverse osmosis membranes are in the range of ~1 and <1 nm. Size selective separation operates in case of micro and ultrafiltration, whereas the size and charge selective separation operate in the latter two.

The membranes are generally based on natural and synthetic polymers (cellulose acetate, cellulose triacetate, polysulfone, polyamide, etc.). The most popular RO membrane is thin film composite membranes (i.e., polyamide layer on asymmetric polysulfone) (Cadotte and Peterson 1981). The polyamide layer is formed by interfacial polymerization of diamine and acyl halide and shows the charge holding capacity in it (Fig. 3).

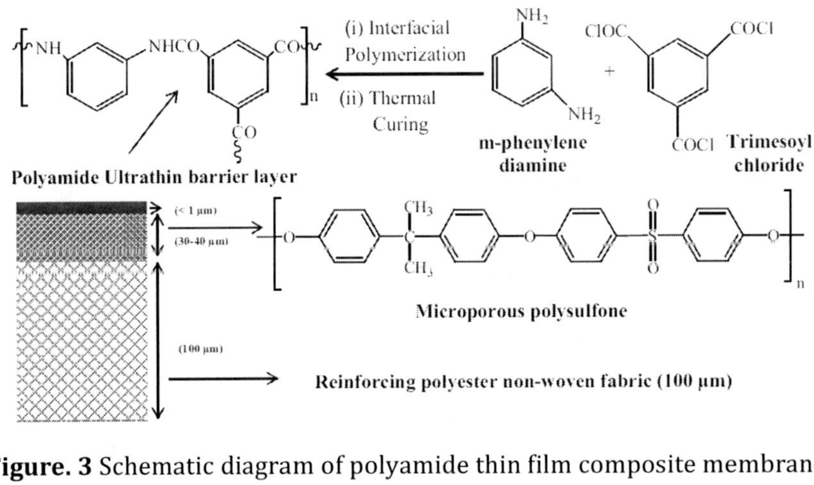

Figure. 3 Schematic diagram of polyamide thin film composite membrane

Reverse osmosis (RO) is one of the most effective types of water treatment and widely used water purification processes in the world. It is usually used for home water treatment to remove salts (Bhattacharya and Ghosh 2004), chemical toxins (Pawlak et al. 2005), organic contaminants (Bhattacharya et al. 2008), dyes (Nataraj et al. 2009), pesticides (Bhattacharya 2006) and microbes (Park and Hu 2010). In reverse osmosis, the raw water is forced (with pressure) through a dense membrane filter that prevents passing of impurities.

Benefits:

- No phase changes and thus requirement of low energy,
- Eco-friendly as they do not produce or use any harmful chemicals; compactness and space requirements are less compared to distillation, and can be designed according to the requirement,
- Ability to remove almost all kinds of contaminates like Cl^-, NO_3^-, F^-, $SO_4^=$, Pb^{2+}, Na^+, K^+, Mg^{2+}, organics as well as microorganisms,
- No alteration in the taste and smell of water and effective removal of microbes and toxins.

Limitations:

- The purified water obtained after reverse osmosis treatment is devoid of useful minerals,
- Membrane may become clogged after prolonged use and, hence, requires periodical replacement of the membrane.

24. ELECTRODIALYSIS MEMBRANE TREATMENT

Electrodialysis (ED) is electric potential-driven membrane-based separation process. The basic principle of the membrane separation is similar to ion exchange reactions (Xu 2005; Strathmann 2010a). The charged groups are attached to the polymer backbone of the membrane material and it is obvious that the fixed charge groups partially or completely exclude ions of the same charge from the membrane, i.e., an anionic membrane with fixed positive groups excludes positive ions, but is freely permeable to negatively charged ions whereas cationic membrane with fixed negative groups excludes negative ions but is freely permeable to positively charged ions.

Since the membrane is of ion selective, it separates or rejects opposite charge ions, useful in removal, or separation of electrolytes (Koter and

Warszawski 2000; Strathmann 2010b). The schematic diagram is presented in Fig. 4.

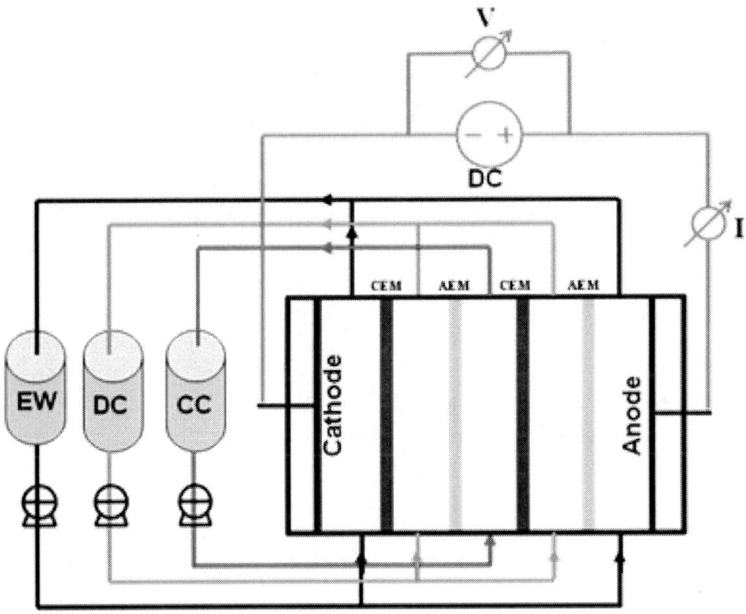

Figure. 4 Schematic diagram of electrodialysis. *DC* diluted chamber, *EW* electrode wash, *CC* concentrated chamber, *CEM* cation exchange membrane, *AEM* anion exchange membrane

The ion transportation depends on the current efficiency in the particular system. Generally, the current efficiencies >80 % are required in commercial stacks to minimize energy operating costs. The low current efficiencies result in water splitting in the dilute or concentrate streams, shunt currents between the electrodes, or back-diffusion of ions from the concentrate to the dilute. The current efficiency is calculated according to the following equation (Shaffer and Mintz 1980).

$$\xi = zFQ_f(C_{inlet} - C_{outlet})/NI$$

where ξ is the current utilization efficiency, z is the charge of the ion, F is the Faraday constant 96485 Amp-s/mol, Q_f is the dilute flow, L/s C_{inlet} is the dilute ED cell inlet concentration, mol/L, C_{outlet} is the ED cell outlet concentration, mol/L, N is the number of cell pairs, and I is the current, Amps.

Apart from their chemical structure (cation and anion), the commercial ion-exchange membranes can be divided, according to their structure and preparation procedure, into two major categories, homogeneous and heterogeneous and depending on the degree of heterogeneity of the ion-exchange membranes, these can be further classified into different types: mono polar (cation/anion) ion-exchange, amphoteric ion-exchange, bipolar ion-exchange, inter-polymer membranes.

The ion-exchange membranes are very similar to normal ion-exchange resins in terms of chemical structure as well as of high selectivity and low resistivity. The difference between membranes and resins arises largely from the mechanical requirement of the membrane process. Thus, it is generally not possible to use sheets of material that have been prepared in the same way as a bead resin. However, the most common solution to this problem is the preparation of membrane with a backing of a stable reinforcing material that gives the necessary strength and dimensional stability. The preparation method of ion-exchange membranes can be summarized in three different steps, viz. Polymerization or polycondensation of monomers; at least one of them must contain a moiety that either is or can be made anionic or cationic groups, respectively, introduction of anionic or cationic moieties into a preformed solid film such as styrene-DVB-based membrane, and introduction of anionic or cationic moieties into a polymer, such as Polysulfone, followed by dissolving the polymer and casting it onto a film.

Benefits:

- Non-pollution, safety and reliability,
- Completely eliminated the chemicals for regeneration,
- Effective for complete removal of dissolved ionic particles (cation and anions), heavy metals, etc.,
- Ability to treat feed water with higher SDI, TOC and silica concentrations.

Limitations:

- Removal of low-molecular weight ionic contaminations,
- Non-charged, higher molecular weight, and less mobile ionic species cannot be significantly removed by the process,
- Large membrane areas are required to satisfy capacity requirements for low concentration (and sparingly conductive) feed solutions.

25. CATALYTIC PROCESSES

Catalytic processes are typically achieved by the following three methods: hydrogenation of nitrate, photocatalytic and electrocatalytic.

25.1 Hydrogenation of nitrate

The hydrogenation via catalytic method is one of the promising techniques for removal of nitrate from water. It needs very active catalysts because the reaction is performed preferably at an ambient/low temperature. The reaction scheme shows that nitrate is reduced to the desired products involving NO_2^-, NO, N_2O and N_2. The undesired byproduct NH_4^+ is also formed by a side reaction due to over hydrogenation (Soares et al. 2010; Mikami et al. 2006; Berndt et al. 2001). Supported bimetallic catalyst (viz. Pd/Cu, Pd/In, and Pd/Sn) has emerged as efficient catalysts for nitrate hydrogenation (Gao et al. 2003; Mikami et al. 2003; Deganello et al. 2000). Apart from Pd, the other metals (e.g., Cu, In, Sn, Co) serve as the role of promoter for the first reduction step to convert NO_3^- into NO_2^- (Soares et al. 2008; Pintar et al. 2004; Arino et al. 2004; Qi et al. 2006). It is seen in the schematic reaction, below that N_2 and ammonium (NH_4^+) are the stable end products of the catalytic reduction process. N_2 is not harmful, but the second one is considered a hazardous aquatic pollutant. That is why target is to convert NO_3^- into N_2as an end product.

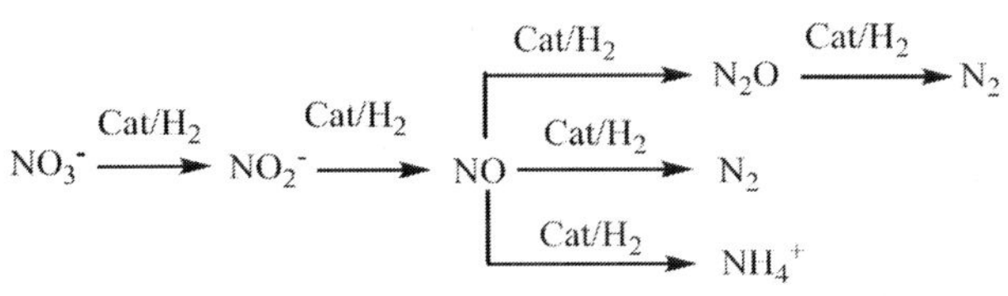

Benefits:

- The method can be of single operation mode,
- Selectivity of catalyst can counter the formation of ammonia ions,
- Addition of other chemicals can be avoided.

Limitations:

- Increase in pH in the reaction medium forms ammonia in dissolved condition, which is more harmful than nitrate.

25.2 Photocatalytic method

The method is based on the acceleration of photodegradation of organic pollutants, pathogens, green algae, and substances in the presence of catalyst (Esplugas et al. 2002; Pera-Titus et al. 2004; Akira et al. 2000; Eggins et al. 1997; Bekbolet et al. 1998; Ibhadon and Fitzpatrick 2013; Gaya and Abdullah 2008; Chong et al. 2010). In response to UV light, when they excited charge separation followed by scavenging e–s and holes by surface adsorbed species. The heterogeneous photocatalysts employing semiconductor catalysts (TiO_2, ZnO, Fe_2O_3,) have shown their efficiency in degrading a wide range of pollutants in water. Metal oxides are more suitable, since they are more resistant to poisoning and deactivation.

Upon UV-irradiation, photocatalytic reactions are initiated by the absorption of illumination with photo-energy equal to or greater than the band gap of the semiconductor. It results in electron–hole (e^-/h^+) pairs as shown in Fig. 5. Thus, it participates in the redox reaction with the adsorbed pollutant species in water. Apart from the reaction, the semiconductor also oxidizes water to produce OH•, a powerful oxidant, which rapidly reacts with the pollutants in the water (Teoh et al. 2012).

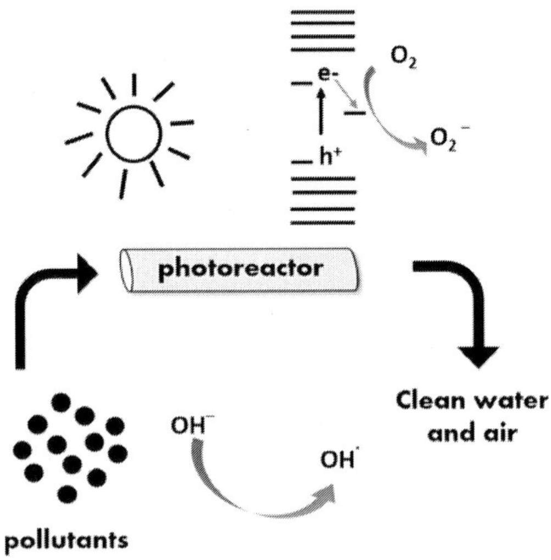

Figure. 5 Schematic diagram of the photocatalytic arrangement

To improve the catalytic activity using visible light, various approaches are also developed, viz. addition of dopants, stoichiometry of catalytic metal oxides and mixed metal oxides, particle size and shape. TiO_2 doped with nitrogen showed excellent photo catalytic activities compared to unmodified TiO_2 nanoparticles in both degradation of chemicals and bactericidal reaction (Daneshvar et al. 2007).

Benefits:

- Reusability of the catalyst as it is unchanged during the process,
- Reactions can occur in ambient condition as well as no consumable chemicals are required,
- Operational process is simple
- It is good enough to treat low concentration of pollutants.

Limitations:

- Post-separation of the semiconductor catalysts after water treatment is important and failing results in catalyst poisoning.
- The catalysts with their fine particle size and large surface area to volume ratio create a strong catalyst agglomeration tendency during the operation.

25.3 Electrocatalytic oxidation

In the electrocatalysis, the oxidation occurs through surface mediator on the anodic surface. The rate of oxidation depends on temperature, pH and diffusion rate of generating oxidants in indirect electrolysis (Mohana and Balasubramanian 2006). This is somewhat different from electrolysis where direct oxidation of pollutants takes place and rate of oxidation depends on electrode activity, pollutants diffusion rate and current density.

The electrocatalysis through metal oxide (MO_x) electrode can be shown (Comninellis 1994) as follows:

$$MO_x + H_2O \rightarrow MO_x(\cdot OH)$$

In the presence of organics (*R*) present in waste water, the physiosorbed active oxygen ($^{\cdot}OH$) involves in complete combustion of organics (1) and chemisorbed active oxygen in the form of MO_{x+1} (2) does the selective oxidation

$$R + MO_x(\cdot OH) \rightarrow CO_2 + H^+ + e + MO_x + 1 \qquad (1)$$

$$R + MO_x + 1 \rightarrow RO + MOx \qquad (2)$$

The key role in the electrocatalytic process is electrocatalytic material. Ru/Pb/Sn oxide and Pb/PbO_2 coated with Ti is used in the dye oxidation (Mohana and Balasubramanian 2006; Morsia et al. 2011). Pt, TiO_2, IrO_2, PbO_2, several Ti-based alloys and boron-doped diamond (BDD) electrodes are employed for the removal of effluents containing various organics, viz. phenols, pharmaceuticals, alcohols, carboxylic acids, anionic surfactants and pesticides (Comninellis and Nerini 1995; Radovici et al. 2009; Klavarioti et al. 2009; Canizares et al. 2008; Louhichi et al. 2008; Ventura et al. 2002). $Pt(acac)_2$ onto ruthenium nanoparticles is used for the removal of formic acid (Chen et al. 2009).

Electrocatalytic reduction is largely used for NO_3^- removal. In this regard, the development of electrodes (viz. Ti–Rh, Ti/IrO_2–Pt, PPy–Graphite, Carbon cloth–Rh, Pd–Sn activated carbon fiber (Tucker et al. 2004; Li et al. 2010; Zhang et al. 2005; Peel et al. 2003; Wang et al. 2006) is interesting direction.

Benefits:

- High pollutant degradation, easy control and low cost,
- It can be easily controlled by putting on/off the power,
- Environmentally compatible since there is little or no need for additional chemicals,
- It has the potential to eliminate different types of pollutants as well as bulk volume,
- It operates at low temperature and pressure compared to nonelectrochemical methods; thus, the volatilization and discharge of un-reacted wastes can also be avoided.

Limitations:

- High operating cost due to the high energy consumption during operation,
- Electrode fouling may also occur on the surface of the electrodes,
- It needs, conducting nature of the effluent. Sometimes the addition of an electrolyte may be necessary,
- The use of metal ions resulted in an effluent with a higher toxicity than that of the initial effluent. Thus, this approach requires a separation step to recover the metallic species (Martinez-Huitle and Ferro 2006).

Phytoremediation

It signifies the removal of pollutants from the environment by the use of plants. The technology involves different mechanisms, viz. phytoextraction, rhizofiltration, phytostabilization, phytotransformation/ phytodegradation (Rai 2009). Phytoextraction involves metal accumulation into the harvestable parts of the roots and the above ground shoot. Rhizofiltration indicates the absorption, precipitation and concentration of toxic metals from polluted effluents. Phytostabilization is a process in which mobility of heavy metals is reduced through the use of tolerant plants, whereas phytotransformation/phytodegradation is the process in which contaminants can be eliminated via phytodegradation or phytotransformation by plant enzymes or enzyme co-factors.

The history of the particular study, including the uptake of toxic metals Hg, As, and other metals, begins in the 70's (Dolar et al. 1971) and other metals (Stanley 1974). In this regard, macrophytes water hyacinth (*Eichhornia-crassipes*) (Zhu et al. 1999); pennywort (*Hydrocotyle umbellate* L) (Dierberg et al. 1987); duckweeds (*Lemna minor* L.) (Rai 2007a) and water velvet (*Azollapinnata*) (Rai 2007a, b) are considered the biological filters and play the important role in the maintenance of the aquatic ecosystem. The floating plants *Lemna minor* (Zayed et al. 1998), *Eichhorniacrassipes* (Zhu et al. 1999) and *Pistiastratiotes*and *Salviniaherzogii* (Maine et al. 2001, 2004) show good potential in accumulating the metals directly from industrial effluents.

Benefits:

- Cost effective,
- Eco-friendly.

Limitations:

- Seasonal growth of the plants,
- Biomass disposal.

Vegetated filter strips

The filter strips are meant as land areas of either planted or indigenous vegetation, situated between a potential, pollutant-source area and a surface-water body that receives runoff. Vegetated filter strips (viz. grassed filter strips, filter strips, and grassed filters) are vegetated surfaces that are designed to treat sheet flow from adjacent surfaces (Dillaha et al.1989; Delgado et al. 1995). The run-off usually carries sediment, organics, plant nutrients and pesticides.

The trapped plant nutrients and pesticides may be easily degraded or transformed by biological and chemical processes. Cole et al. (1997) report the removal of chloropyrifos (62–99 %), dicamba (90–100 %), 2,4D (89–98 %), and mecoprop (89–95 %) using Bermuda grass buffer. On the other hand, atrazine (98 %) and pyrethroid (100 %) removal is possible using vegetated drainage ditch (Moore et al. 2001)

Benefits:

- Trap sediments,
- Capture nutrients both through plant uptake and adsorption of soil particles,
- Promote transformation and degradation of pollutants into less toxic forms,
- Removal of pathogens is possible.

Limitations:

- The design is important,
- Proper vegetation is necessary,

Biologically active carbon filtration

Biologically active carbon is another prospective process with this bioremediation technique. The process utilizes granulated activate carbon (GAC) as its water filtration. The microbial (bacterial) colonization is possible over the GAC media particles form 'biofilm' (Scholz and Martin 1997). Actually, it is described as a 'porous tangled mass of slime matrix (Weber et al. 1978). It consists of microbial cells, either immobilized on the surface of the GAC (substratum) or embedded in an extracellular microbial organic polymer matrix (Ghosh et al. 1999; Lawrence and Tong 2005). Bacteria and fungi cells in the biofilms secrete extracellular polymeric substances to form a cohesive, stable matrix in which cells are held in dense agglomeration (Branda et al. 2005; Lazarova and Manem 1995). The extracellular matrix is composed of polysaccharides, proteins, nucleic acids and lipids (Goodwin and Forster 1985). The activity of the biofilm relates to the physiological modifications associated with the promotion of certain genes (Dagostino et al. 1991), or changes the bacteria cell surrounding to increase the local concentration of nutrients, oxygen and enzymes (Ghosh et al. 1999) or limit the invasion of toxic or inhibiting substances (Blenkinsopp and Costerton 1991).

Most of the dissolved organic chemical removal occurs through physical adsorption in the GAC media. Apart from the adsorption, biodegradation can also operate.

Benefits:

- It can avoid chemical disinfection water treatment processes,
- Because of the microbial biodegradation of organic substrates on the GAC media, the service life can be extended,
- Bacterial regrowth is less possible,
- Eliminates the need for coagulant in source filtration processes (Hillis 2000).

Limitations:

- The control of the growth of the process is necessary.

Magnetic separation

In the magnetic separation process, the high-gradient magnetic separation (HGMS) is a commonly used process (Hoffmann and Franzreb 2004a, b; Ditsch et al. 2005; Okada et al. 2005). In this case, device comprising bed of magnetically susceptible wires placed inside an electromagnet is used. There are various influencing factors, viz. nature of impurities, concentration, size, magnetic susceptibility, spacing design, and intensity of magnetic field and its orientation, magnetic field strength.

Generally, there are three categories of separators based on magnet type, viz. permanent magnet, electromagnet and superconducting magnet. The permanent magnet (ferro-magnets of iron-based, nickel, cobalt or rare earth element) is having magnetic fields of less than 1 T, though trend is to improve the magnetic field strength by the development of materials and shape design parameters (Ormerod and Constantinides 1997; Zhu and Halbach 2001; Iwashita et al. 2008). The electromagnetic-based device consists of a solenoid of electrically conducting wires which can generate a magnetic field of 2–4 T within their cavity on passage of electric current (Li et al. 2007) (Timoshenko and Ugarov 1994). The third category of magnetic separators generates the highest intensity magnetic field from 2 to 10 T (Selvaggi et al. 1998; Yan et al. 1996).

The magnetically assisted water purification can be primarily classified into the following type depending on the difference in adoption of physical processes, viz. direct purification, seeding and separation by magnetic flocculant, and magnetic sorbent application in organic and inorganic contaminants including radionuclides. In the direct purification technique, there is no carrier magnetic component utilized for the separation. The basic properties of ions or solid response to the magnetic field are utilized for purification. In this method, the anti-scaling technique is most commonly practiced. In the area of anti-scale magnetic treatment, the most common constituents of scale are $CaCO_3$, $CaSO_4{\cdot}2H_2O$ and silica, $BaSO_4$, $SrSO_4$, $Ca_3(PO_4)_2$ and ferric and aluminum hydroxides (Busch and Busch 1997; Gabrielli et al. 2001; Fathia et al. 2006; Jianxin et al. 2007). In the magnetic flocculant separation, coagulant cation [viz. Fe(III)] forms an insoluble precipitate under applied magnetic field. It is an effective means of lowering significantly both the oil and suspended solids of water effluent streams (Kakihara et al. 2004; Nishijima and Takeda 2007). Ions (polymerise as polyhydroxycomplexes, or nitroso-hydroxy, or hydroxy-carbonato or halogenohydroxo-carbonato complexes), which are difficult to coagulate magnetic sorbents, are utilized for waste water purification.

Benefits:

- Useful for the separation of pollutants,
- Magnetic pre-treatment improvises purification RO membrane filters,
- Calcium carbonate scale formation in heat exchanger can be reduced,
- Promotes the homogeneous precipitation of calcium carbonate scales.

Limitations:

- Not fully sufficient.

Disinfection

The disinfection methods are classified as physical and chemical methods. In physical treatment UV, solar radiation, and ultrasound are included, whereas chlorine, iodine, ozone are included in chemical treatment (Kerwick et al. 2005).

The features of the treatments are described in the following:

By ultraviolet radiation:

In the ultraviolet treatment, the water to be treated passes through germicidal ultraviolet (UV) light configured inside a low-pressure lamp. As the water passes the ultraviolet purifier, the biological contaminants are exposed to UV light, which damages the genetic components of the microbes. The microbes are killed this way using ultraviolet water treatment (Hijnen et al. 2006; Bergmann et al. 2002). They are the pioneers of using UV in water purifiers. A major drawback of this water treatment type is that it is ineffective in removal of dissolved chemicals and other particulate matter.

Benefits:

- Ability to destroy or make inactive many pathogenic microorganisms,
- It has no effect on minerals in water,
- Ability to degrade some organic contaminants,
- No additional toxic and nontoxic chemicals are introduced.

Limitations:

- Not suitable for water with high levels of suspended solids, turbidity, color or soluble organic matter,
- Without electricity, it could not operate.

It is employed by solar radiation also. It is very useful to inactivate pathogens, especially diarrhea. The contaminated water is to fill into transparent plastic bottles and expose to the full sunlight for 6 h. The UV-A radiation (wavelength 320–400 nm) of the sunlight destroys the pathogen.

Benefits:

- Easy to use as well as inexpensive,
- Good bacterial and viral disinfection,
- No toxic chemicals except plastic bottles,
- Does not require constant attention to use,
- No effect for minerals in water.

Limitations:

- Dependent on climatic condition,
- Toxicity can come from poor quality of plastic bottles,
- Need turbidity of 30NTU or less,
- Less effective against bacterial spores and cysts stage of some parasites.

By ultrasound

Ultrasound is the cyclic sound pressure with a frequency greater than the upper limit of human hearing. The ultrasound is used in many different fields by penetrating the medium, measuring the reflection signature or supplying focussed energy. The mechanical vibration of the waves can be caused to damage cellular structures of bacteria. Thus, it can be useful to disinfect water. However, the regrowth of the microorganisms is also possible. Thus, combination of this and chemical disinfectant is the best technique.

Benefits:

- Easy to use,
- Does not require constant attention.

Limitations:

- Regrowth of microorganism is also feasible,
- Not fully self-sufficient.

By ozone

Ozone, O_3 is an unstable form of oxygen and protective layer of UV-radiation. But in drinking water, it makes an effective disinfectant (VonGunten 2003a, b). It readily gives up oxygen and thus a powerful oxidizing agent. Ozone is made by passing oxygen through UV-light or a 'cold' electrical discharge. The very high oxidation potential of ozone is easy enough to insert oxygen into the bonds of organic compounds to form aldehydes and ketones. It is effective for killing the biological contaminants (viz. pathogens) than that of chemical disinfection method like chlorination. Actually, the ozone oxidizes the organics in bacterial membrane, which weakens the cell wall and leads to cellular rupture. This exposes the organism to the external environment, which causes almost immediate death of the cell. Ozone also improves the clarity (clarifying iron, sulfur and manganese). The soluble Fe(II) and Mn(II) which are not filtered in the normal condition transformed to insoluble Fe(III) and Mn(VII) with ozone treatment and, thus, filtration is possible. It also reduces odor problems and concentrations of sulfur and other dissolved chemicals. The main advantage of ozone is that it leaves no disinfectant residual in the water. To use ozone as disinfectant, it is generated and immediately applied on site. The limitations of using ozone as disinfectant are a significant air pollutant, explosive, and an irritant to skin, eyes, respiratory tract and mucous membrane. It can produce carcinogens if little bromine is there in the water.

By chlorine

The most common strong oxidant in the form of chlorine and its compounds, viz. chloramine or chlorine oxide are used in disinfection technique. Chlorine is well to do against bacteria and protozoa that form cysts (viz. Giardia lamblia and Cryptospordium) (Gala-Gorchev 1996); Melvin et al. 1967). Handling of chlorine gas is dangerous, thus the use of sodium and calcium hypochlorite is the trend. It releases free chlorine in water. Electrolytic method is another mode to get chlorine solution. The free chlorine is released when dissolved in water. The limitation of using chlorine that reacts with natural organic compounds in the water forms potentially harmful chemical by-products, such as trihalomethanes and haloacetic acids. They are shown to cause cancer (Univ. Florida Report 1998). The maximum allowable annual average level of trihalomethane and haloacetic acids is 80 and 60μg/L, respectively. They tend to increase with pH, temperature, time and the level of organics in water. One way to decrease the level of trihalomethane and haloacetic acids is to reduce the organics (EPA, US 2012). Thus, it is preferable to use after the removal of organic compounds from water. EPA has suggested 'enhanced coagulation' (i.e., the process by increasing the feed rate of coagulants, adding better coagulants, possibly ferric coagulants) process to remove the organics for controlling trihalomethane and trihaloacetic acids. The advantage of using chloramine is that it will not form THMs or haloacetic acids, but it results in nitrification, as ammonia is a nutrient for bacterial growth, with nitrates being generated as by-products

$$Cl_2+H_2O \rightarrow HOCl+HCl$$

$$HOCl \rightarrow HCl+[O]$$

Benefits:

- Simple method,
- Availability of inexpensive chemicals/cheaply available,

Limitations:

- Excess of chlorine produces characteristic unpleasant taste and odor and irritating effect on mucous membrane.

However, the usual practice is bleaching powder in place of chlorine. The mechanistic process is as follows:

$$Ca(OCl)Cl+H_2O \rightarrow Ca(OH)_2+Cl_2$$

$$Cl_2+H_2O \rightarrow HOCl+HCl$$

$$HOCl \rightarrow HCl+[O]$$

Bleaching powder should be used only in calculating amount because excess of it will give a bad taste and disagreeable odor, while the lesser amount of it will not sterilize the water completely (Snoeyink and Jenkins 1980).

Benefits:

- Simple method,
- Availability of inexpensive chemicals.

Limitations:

- It requires continuous supply of the chemicals and trained personal so that the chlorine is at effective levels.

By iodine

Similar to chlorine, it is also a good oxidizing agent. It is effective against many varieties of pathogenic organisms including spores, cysts, viruses, etc. in a short time (Punayani et al. 2006). Compounds-based formulations (organic iodide compounds: bisglycine hydroiodide, potassium tetraglycine triiodide, etc., iodophores: combination of iodine and solubilizing compounds, i.e., non-ionic surfactants, iodine incorporated resins, cross-linked copolymer of styrene and divinylbenzene with I_3^-, iodine with polyvinyl pyrrolidone) are there to regulate the release of iodine.

The mechanistic way of inhibition of protein function is forming *N*-iodo compounds, i.e., reacting with basic –NH functions of amino acids and nucleotides. Thus, important positions for H-bonds are blocked, resulting in a lethal change in protein structure. The –SH groups in the cytoplasm are oxidised. Thus, the ability to make disulphide bonds in protein formation is lost. The addition of olefinic double bonds of unsaturated fatty acids may also be the reason to decrease the fluidity of cell membranes.

Benefits:

- Effective against many varieties of pathogenic organisms including spores, cysts, viruses, etc. in a short time,
- Eliminates the chances of disease caused due to deficiency of iodine,

- Ammonia and other nitrogeneous substance have no pronounced effect on the efficiency.

Limitations:

- Higher concentrations are required compared to chlorine,
- Costly than chlorine per unit of germicidal effectiveness,
- Taste and slight color produced can affect palatability and esthetic quality.

By hydrogen peroxide

Though it is known for high oxidative and biocidal efficiency, the use in drinking water disinfection is not available, but coupled with ozone, UV-radiation it can be used (Andreozzi et al. 1999). The disinfection mechanism is based on the decomposition of peroxide, i.e., the release of free oxygen radicals. The free radicals have both oxidizing and disinfecting abilities.

Contrary to other chemical substances, it does not produce residues or gases. The limitations of peroxide are: it can irritate the eyes, skin, and lung. Skin exposure causes painful blisters, burns and skin whitening.

Apart from the above, recent trend is there to employ Ag, Au, Cu, Zn, titanium nanoparticles supported in solid matrix. Due to the bactericidal effect, the water passed in the matrix will be free from bacterial contamination (Li et al. 2008; Savage and Diallo 2005).

26. HYBRID TECHNOLOGIES

In true sense, no technologies independently counter all the problems. The development of technology is a dynamic process that moves forward slowly and recommendations are made based on the best science available at that time. However, with new research and new results, the flaws of existing technologies may be removed. That is why the concept of the combinations of technologies or in other sense hybrid technologies has come. Scientists and technologists have orchestrated according to the requirement. Let us discuss with the synergistic RO technology first. In the RO technology, feed pretreatment is vital for RO to avoid problems, i.e., fouling, damaging the membranes. Conventional pretreatment steps include chemicals addition, i.e., acid, coagulant/flocculant, disinfection. Coagulation and flocculation (coagulants–flocculants) are dealt in water treatment process. Chlorine treatment is treated as disinfection process and commonly employed. But chlorination shortens the stability of the membrane and, thus, dechlorination treatment (viz. sodium bi sulfite) is required. In media filtration, water is treated by passing through granular media like pumice, anthracite, gravels, etc. that can be used in combination. Cartridge filter (made up of papers, woven wire, cloth) is used as the last pretreatment step to retain particles in the size range 1–10μm. To check the quality, 'Silt Intensity Index' or SDI parameter is important. Actually, SDI considers the ratio of two flow measurements, one at the beginning, and the other at the end by passing feed water through a 0.45 μm filter paper in dead end mode at constant pressure (Saha and Bhattacharya 2010).

Similarly, the pretreatment step coagulation is coupled with ion exchange treatment of water. The coupled electrodeionization technology based on electrodialysis and ion exchange results in a process which effectively deionizes water, while the ion exchange membranes are continuously regenerated by the electric current in the unit. This electrochemical regeneration replaces the chemical regeneration of conventional ion exchange systems. Recently, hybrid technologies like ED-RO or ED-RO and distillation have been developed for the water purification and these processes offer many advantages over the traditional technologies (Saha and Bhattacharya 2010; Makwana et al. 2010). The ED-RO technologies desalinate the brackish water with high recoveries along with zero discharge and reduced energy consumption. ED-RO is a high recovery system since RO concentrates can be recycled through the ED system to reduce the feed flow rate, pre-treatment cost and the reduced amount of effluent. Thus, coupling of the technologies/processes offers a solution to an increasing important issue in water treatment as well as for water conservation. To get better results, UV is typically used as a final purification stage in terms of removing contaminants bacteria and viruses.

27. AWAITED/COMING TECHNOLOGIES

The torch of the scientific quest along with the traditional technologies has now been handed over to the nanotechnologists of the twenty-first century, to whom a major challenge is to transform this into the field. Nanotechnology refers to technologies involving particles on the approximate size scale of a few to hundreds of nanometers in diameter. The elevated surface area to mass ratio, a common characteristic of nanoparticles, makes it promising. In terms of applicability, three approaches are there, viz. individual nanoparticles (Watlungton 2005), binding the nanoparticles to a powder/granule form and nanoparticles onto membranes/polymers. As individual zero valent iron particles (Can-

Bao and Zhang 1997), palladium-coated iron particles and palladium-coated gold nanoparticles (Michael et al. 2005) are very promising in terms of permeable reactive barriers and photocatalysts. The nanoscopic materials such as carbon nanotubes and alumina fibers embedded in zeolite filtration membranes (Valli et al. 2010), TiO_2/Al_2O_3 membranes (Zhang et al. 2006), carbon nanotubes, wrapped around a carbon block filter structure (Cooper et al. 2007) have the capacity to remove the impurities from water. Nanoreactive membranes are able to decompose pollutants such as 4-nitrophenol (Dotzauer et al. 2006) and bind metal ions (Hollman and Bhattacharya 2004) in water solution. Polysulfone membranes impregnated with silver nanoparticles are found to be effective in bacteria and virus removal (Zodrow et al. 2009). Superchlorination is another advanced technique to get clean and disinfected water. It signifies that extra dose of chlorine oxidize organics kill and remove algae and pathogens from the water within the short-contact time. HOCl is the active chemical that provides sanitation as well as shows reactivity towards organic pollutants. When there is sufficient HOCl, the pollutants are easily oxidized. But in case of low level of HOCl compared to organic pollutants combined chlorine is formed. These combined chlorine compounds can be oxidized by increasing the level of HOCl level in water. The point at which all the organic impurities are oxidized is called the break point (Bahadori et al. 2013). To avoid the flaws (viz. corrosion, bleaching of hair and skin, foul smelling), sometimes superchlorination followed by dechlorination is necessary before the use of water. Superchlorination is practiced after the sunset as there are no possibilities to react with UV-rays from the sun.

28. CONCLUSIONS

The world is facing turbulent water future. With the growing economy and rising population, the theme of all nations is 'Save water'. Quantity and quality of water should be given equal importance. Awareness related to 'water conservation' and 'safe drinking water' is extremely important, and should be given a good thought to the people.

The technological solution depends on raw water characteristics, affordability and acceptability and level of application. Of course, sustainability depends on an awareness of the related issues. Since there are limitations in every individual treatment technologies and, thus, hybrid technologies are always beneficial; however, availability, selection, optimization, etc. are important for the best performances of the system. Lastly, it must be mentioned through the gambling of research that the future of the water treatment technology is highly prosperous and hope 1 day we will fulfill the demand 'fresh water for everyone'.

ACKNOWLEDGMENTS

The authors are grateful to SERB, Department of Science and Technology, India for research funding and Council of Scientific Industrial Research, New Delhi for the support. The authors also wish to thank Mr. Chirag Sharma and Mr. Mayank Saxena, CSMCRI, Bhavnagar for their help.

REFERENCES

1. Ahluwalia SS, Goyal D (2007) Microbial and plant derived biomass for removal of heavy metals from wastewater. Bioresour Technol 98:2243–2257
2. Akira F, Rao T, Tryk D (2000) Titanium dioxide photocatalysis. J Photochem Photobiol C 1:1–21
3. Alexandratos SD (2009) Ion-exchange resins: a retrospective from industrial and engineering chemistry research. Ind Eng Chem Res 48(1):388–398
4. Ali I, Gupta VK (2007) Advances in water treatment by adsorption technology. Nat Protoc 1:266–2667
5. Alireza B, Mohammadi S, Mowlavi A, Parvaresh P (2010) Measurement of heavy radioactive pollution: radon and radium in drinking water samples in Mashhad. Int J Curr Res 10:54–58
6. Andreozzi R, Caprio V, Insol A, Marotta R (1999) Advanced oxidation processes (AOP) for water purification and recovery. Catal Today 53(1):51–59
7. Anwar F (2003) Assessment and analysis of industrial liquid waste and sludge disposal at unlined landfill sites in arid climate. Waste Manag 23(9):817–824
8. Arino MD, Pinna F, Strukul G (2004) Nitrate and nitrite hydrogenation with Pd and Pt/SnO_2catalysts: the effect of the support porosity and the role of carbon dioxide in the control of selectivity. Appl Catal B 53:161–168
9. Ashbolt NJ (2004) Microbial contamination of drinking water and disease outcomes in developing regions. Toxicology 198:229–238
10. Bahadori A, Clark M, Boyd B (2013) Essentials of water systems design in the oil, gas, and chemical processing industries. Springer, Dordchrecht
11. Baudu M, Le Cloirec P, Martin G (1991) Pollutant adsorption onto activated carbon membranes. Water Sci Technol 23(7–9):1659–1666
12. Bekbolet M, Boyacioglu Z, Ozkaraova B (1998) The influence of solution matrix on the photocatalytic removal of color from natural waters. Water Sci Technol 38:155–162
13. Bergmann H, Iourtchouk T, Schops K, Bouzek K (2002) New UV irradiation and direct electrolysis—promising methods for water disinfection. Chem Eng J 85:111–117
14. Berndt H, Mönnicha I, Lücke B, Menzel M (2001) Tin promoted palladium catalysts for nitrate removal from drinking water. Appl Catal B 30:111–122
15. Beveridge TC, Doyle RJ (1989) Metal ions and bacteria. Wiley Interscience, New York
16. Bhattacharya A (2006) Remediation of pesticides polluted water through membranes. Sep Purif Rev 35:1–38
17. Bhattacharya A, Ghosh P (2004) Nanofiltration and reverse osmosis membranes: theory and application in separation of electrolytes. Rev Chem Eng 20(1–2):111–173

18. Bhattacharya A, Yogesh Popat KM, Ganguly B, Brahmbhatt H (2008) Studies on the separation performances of chlorophenol compounds from water by thin film composite membranes. Macromol Res 16(7):590–595
19. Blenkinsopp SA, Costerton JW (1991) Understanding bacterial biofilms. Trends Biotechnol 9:138–143
20. Bolognesi C (2003) Genotoxicity of pesticides: a review of human biomonitoring studies. Mutat Res 543:251–272
21. Bowe CA, Martin DF (2004) Extraction of heavy metals by 2-mercaptoethoxy group attached to silica gel. J Environ Sci Health A Tox Hazard Subst Environ Eng 39(6):1469–1485
22. Bowe CA, Poore DD, Benson RF, Martin DF (2003) Extraction of heavy metals by amine adsorbed onto silica gel. J Environ Sci Health A Tox Hazard Subst Environ Eng 38(11):2653–2660
23. Branda S, Vik A, Friedman L, Kolter R (2005) Biofilms: the matrix revisited. Trends Microbiol 13:20–26
24. Brenniman GR, Namekata T, Kojola WH, Carnow BW, Levyin PS (1979) Cardiovascular disease death rates in communities with elevated levels of barium in drinking water. Environ Res 20(2):318–324
25. Brostow W, Hagg Lobland HE, Pal S, Singh RP (2009) Polymeric flocculants for wastewater and industrial effluent treatment. J Mater Educ 31(3–4):157–166
26. Brown HS, Bishop DR, Rowan CA (1984) The role of skin absorption as a route of exposure for volatile organic compounds (VOCs) in drinking water. Am J Pub Health (AJPH) 74:5
27. Bull S (2007) Asbestos-toxicological overview. Health Protection Agency. In: Chemical hazards and poisons division, Version-1. HQ, UK, pp 1–15
28. Busch KW, Busch M (1997) Laboratory studies on magnetic water treatment and their relationship to a possible mechanism for scale reduction. Desalination 109(2):131–148
29. Cadotte JE, Peterson RJ (1981) in: A. F. Turbak (ED) Thin film composite reverse osmosis membranes: origin, development, and recent advances in Synthetic membranes, Desalination, vol 1, American Chemical Society, Washington, DC
30. Calmon C (1986) Recent developments in water treatment by ion exchange. React Polym Ion Exch Sorbents 4(2):131–146
31. Can-Bao W, Zhang W (1997) Synthesizing nanoscale iron particles for rapid and complete dechlorination of TCE and PCBs. Environ Sci Technol 31(7):2154–2156
32. Canizares P, Paz R, Saez C, Rodrigo MA (2008) Electrochemical oxidation of alcohols and carboxylic acids with diamond anodes: a comparison with other advanced oxidation processes. Electrochim Acta 53:2144–2153
33. Charcosset C (2009) A review of membrane processes and renewable energies for desalination. Desalination 245:214–231
34. Charles FH, Swartz CH, Badruzzaman ABM, Nicole KB, Yu W, Ali A, Jay J, Beckie R, Niedan V, Brabander D (2005) Groundwater arsenic contamination on the Ganges Delta: biogeochemistry, hydrology, human perturbations and human suffering on a large scale. C R Geosci 337(1/2):285–296
35. Chatterjee A, Das D, Mandal BK, Chowdhury TR, Samanta G, Chakraborti D (1995) Arsenic in ground water in six districts of West Bengal, India: the biggest arsenic calamity in the world. Part I. Arsenic species in drinking water and urine of the affected people. Analyst 120:643–650
36. Chen ASC, Snoeyink VL, Fiessinger F (1987) Activated alumina adsorption of dissolved organic compounds before and after ozonation. Environ Sci Technol 21(1):83–90
37. Chen CJ, Kuo TL, Wu MM (1988) Arsenic and cancers. Lancet 331:414–415
38. Chen W, Xu LP, Chen S (2009) Enhanced electrocatalytic oxidation of formic acid by platinum deposition on ruthenium nanoparticle surfaces. J Electroanal Chem 631:36–42
39. Chong MN, Jin B, Chow CWK, Saint C (2010) Recent developments in photocatalytic water treatment technology: a review. Water Res 44:2997–3027
40. Clarkson T (1992) Mercury: major issues in environmental health. Environ Health Perspect 100:31–38

41. Cole JT, Baird JH, Basta BT (1997) Influence of buffers on pesticide and nutrient run off from Bermudagrass turf. J Environ Qual 26:1589–1598
42. Comninellis C, Nerini A (1995) Anodic oxidation of phenol in the presence of NaCl for wastewater treatment. J Appl Electrochem 25:23–28
43. Cooper RG, Harrison AP (2009a) The exposure to and health effects of antimony. Indian J Occup Environ Med 13(1):3–10
44. Cooper RG, Harrison AP (2009b) The uses and adverse effects of beryllium on health. Ind J Occup Environ Med 13(2):65–76
45. Cooper C, Cummings A, Starostin M, Honsinger C (2007) Purification of fluids with nanomaterials. US Patent No 7,211,320, May, 2007
46. Counter SA, Buchanan LH (2004) Mercury exposure in children: a review. Tox Appl Pharmacol 198(2):209–230
47. Dagostino L, Goodman A, Marshall K (1991) Physiological responses induced in bacteria adhering to surfaces. Biofouling 4:113–119
48. Dahi E (1997) Development of the contact precipitation method for appropriate Defluoridation of water. 2nd international workshop on fluorosis prevention and defluoridation of water, pp 128–137
49. Damalas CA, Eleftherohorinos IG (2011) Pesticide exposure, safety issues, and risk assessment indicators. Int J Environ Res Public Health 8:1402–1419
50. Daneshvar N, Niaei A, Akbari S, Aber S, Kazemian N (2007) Photocatalytic disinfection of water polluted with *Pseudomonas aeruginosa*. Glob Nest J 9:132–136
51. Daschner FD, Rüden H, Simon R, Clotten J (1996) Microbiological contamination of drinking water in commercial household water filter system. Eur J Clin Microbiol Infect Dis 15(3):233–237
52. Deganello F, Liotta LF, Macaluso A, Venezia AM, Deganelloa G (2000) Catalytic reduction of nitrates and nitrites in water solution on pumice-supported Pd–Cu catalysts. Appl Catal B Environ 24:265–273
53. Delgado AN, Periago EL, Diaz-FierrosViqueira FDF (1995) Vegetated filter strips for waste water purification—a review. Bioresour Technol 51:13–22
54. Dierberg FE, DeBusk TA, Goulet NA Jr (1987) Removal of copper and lead using a thin film technique. In: Reddy KB, Smith WH (eds) Aquatic plants for water treatment and resource recovery. Magnolia Publishing Inc., Orlando, pp 497–504
55. Dillaha TA, Reneau RB, Mostaghimi S, Lee D (1989) Vegetative filter strips for agricultural nonpoint source pollution control. Trans ASAE 32(2):513–519
56. Ditsch A, Lindenmann S, Laibinis PE, Wang DIC, Hatton TA (2005) High-gradient magnetic separation of magnetic nanoclusters. Ind Eng Chem Res 44(17):6824–6836
57. Dolar SG, Kenney DR, Chesters G (1971) Mercury accumulation by *Myriophyllum spicatum*. Environ Lett 69:191–198
58. Dotzauer DM, Dai J, Sun L, Bruening ML (2006) Catalytic membranes prepared using layer by layer adsorption of polyelectrolyte/metal nanoparticle films in porous supports. Nano Lett 6:2268–2272
59. Eggins BR, Palmer FL, Bryne JA (1997) Photocatalytic treatment of humic substances in drinking water. Water Res 31:1223–1226
60. Eikebrokk B, Juhna T, Østerhus SW (2006) Water treatment by enhanced coagulation—operational status and optimization issues. Techneau D 5(3):1
61. Elimelech M (2006) The global challenge for adequate and safe water. J Water Supply Res Technol AQUA 55:3–10
62. Environmental Protection Agency US (2000) Wastewater technology sheet: chemical precipitation. United State Environmental Protection, 832-F-00-018
63. Environmental Protection Agency, US (2006) Inorganic Contaminant Accumulation in Potable Water Distribution Systems, Office of Groundwater and Drinking Water, USA
64. Environmental Protection Agency US (2009a) Health Effects Assessment for Asbestos. EPA/540/1-86/049 (NTIS PB86134608)
65. Environmental Protection Agency US (2009b) Pesticides: regulating pesticides. http://www.epa.gov/pesticides/regulating/index.htm

66. Environmental Protection Agency, US (2012) Disinfection by products: a reference resource
67. Esplugas S, Gime´nez J, Conteras S, Pascual E, Rodrı´guez M (2002) Comparison of different advanced oxidation processes for phenol degradation. Water Res 36:1034–1042
68. Eswar P, Devaraj CG (2011) Water defluoridation: field studies in India. Ind J Dent Adv 3:526–533
69. Fan AM, Kizer KW (1990) Selenium-nutritional, toxicologic and clinical aspects. West J Med 153:160–167
70. Fathia A, Mohamed T, Claude G, Maurin G, Mohamed BA (2006) Effect of a magnetic water treatment on homogeneous and heterogeneous precipitation of calcium carbonate. Water Res 40(10):1941–1950
71. Fawell J, Nieuwenhuijsen MJ (2003) Contaminants in drinking water. Br Med Bull 68:199–208
72. Fawell J, Bailey K, Chilton J, Dahi E, Fewtrell L, Magara Y (2006) Fluoride in drinking-water. World Health Organization, Published by IWA Publishing, London
73. Fields KA, Chen A, Wang L (2000) Arsenic removal from drinking water by coagulation/filtration and lime softening plants. EPA/600/R-00/063, USEPA, Cincinnati
74. Gabrielli C, Jaouhari R, Maurin G, Keddam M (2001) Magnetic water treatment for scale prevention. Water Res 35(13):3249–3259
75. Gala-Gorchev H (1996) Chlorine in water disinfection. Pure Appl Chem 68(9):1731–1735
76. Gao W, Guan N, Chen J, Guan X, Jin R, Zeng H, Liu Z, Zhang F (2003) Titania supported Pd–Cu bimetallic catalyst for the reduction of nitrate in drinking water. Appl Catal B 46:341–351
77. Gaya UI, Abdullah AH (2008) Heterogeneous photocatalytic degradation of organic contaminants over titanium dioxide: a review of fundamentals, progress and problems. J Photochem Photobiol C Photochem Rev 9:1–12
78. Ghosh U, Weber A, Jensen J, Smith J (1999) Granular activated carbon and biological active carbon treatment of dissolved and sorbed polychlorinated biphenyls. Water Environ Res 71(2):232–240
79. Ghrefat H, Nazzal Y, Batayneh A, Zumlot T, Zaman H, Elawadi E, Laboun A, Mogren S, Qaisy S (2014) Geochemical assessment of ground water contamination with special emphasizes on fluoride, a case study from Midyan Basin, north Western Saudi Arabia. Environ Earth Sci 71:1495–1505
80. Goodwin J, Forster F (1985) A further examination into the composition of activated sludge surfaces in relation to their settlement characteristics. Water Res 19:527–533
81. Gopal Ram, Ghosh PK (1985) Fluoride in drinking water its effects and removal. Def Sci J 35(1):71–88
82. Gray K, Yao C, O'Mella CR (1995) Polymeric inorganic coagulants. J Am Water Works Assoc 87:136–146
83. Gupta SK, Gupta RC, Seth AK, Gupta AB, Bassin JK, Guptathe A (2000) Methaemoglobinaemia in areas with high nitrate concentration in drinking water. Nat Med J Ind 13(2):58–61
84. Haki J, Hunyadi I, Varga K, Csige I (1995) Determination of radon and radium content of water samples by SSNTD technique. Radiat Meas 25(1–4):657–658
85. Harper TR, Kingham NW (1992) Removal of arsenic from wastewater using chemical precipitation methods. Water Environ Res 64(3):200–203
86. Harvey RW, Smith RL, George L (1984) Effect of organic contamination upon microbial distributions and heterotrophic uptake in a Cape Cod, Mass., aquifer. Appl Environ Microb 48(6):1197–1202
87. Heasman M, Mellentin J (2001) The functional foods revolution healthy people, healthy profits?. Earthscan Publisher, London, p 280. ISBN 978-1-85383-687-9
88. Heckel A, Seebach D (2000) Immobilization of TADDOL with a high degree of loading on porous silica gel and first applications in enantioselective catalysis. Angew Chem Int Ed 39(1):163–165
89. Hering JG, Chen P, Wilkie JA, Elimelech M (1997) Arsenic removal from drinking water during coagulation. J Environ Eng ASCE 123(8):800–807

90. Hijnen WAM, Beerendonk EF, Medema GJ (2006) Inactivation credit of UV radiation for viruses, bacteria and protozoan (oo)cysts in water: a review. Water Res 40:3–22
91. Hillis P (2000) Membrane technology in water and waste water treatment. The Royal Society of Chemistry, Cambridge **269**
92. Hitzfeld BC, Höger SJ, Dietrich DR (2000) Cyanobacterial toxins: removal during drinking water treatment, and human risk assessment. Environ Health Perspect 108:113–122
93. Hoffmann C, Franzreb M (2004a) A novel repulsive-mode high gradient magnetic separator—I. Design and experimental results Part 1. IEEE Trans Magn 40(2):456–461
94. Hoffmann C, Franzreb M (2004b) A novel repulsive-mode high gradient magnetic separator—II. Design and experimental results Part 2. IEEE Trans Magn 40(2):462–468
95. Hoffmann M, Martin S, Choi W, Bahnemann D (1995) Environmental applications of semiconductor photocatalysis. Chem Rev 95:69–96
96. Hollman AM, Bhattacharya D (2004) Pore assembled multilayers of charged polypeptides in microporous membranes for ion separation. Langmuir 20(13):5418–5424
97. Ibhadon AO, Fitzpatrick (2013) Hetergeneous photocatalysts; recent advances and applications. Catalysts 3:189–218
98. Iler RK (1979) The chemistry of silica. Wiley, New York
99. Inamori Y, Fujimoto N (2009) Water quality and standards—Vol. II, microbial/biological contamination of water. Encyclopaedia of Life support systems (EOLSS)
100. Iwashita Y, Ichikawa M, Tajima Y, Nakamura S, Kumada M, Spencer CM, Tauchi T, Kuroda S, Okugi T, Ino T, Muto S, Shimizu HM (2008) Strong variable permanent multipole magnets. IEEE Trans Appl Supercond 18(2):957–960
101. Jafarpour MM, Foolad A, Mansouri MK, Nikbakhsh Z, Saeedizade H (2010) Ammonia removal from nitrogenous industrial waste water using iranian natural zeolite of clinoptilolite type. World Acad Sci Eng Technol 46:939–945
102. Jiang JQ, Graham NJD (1998) Pre-polymerised inorganic coagulants and phosphorus removal by coagulation—a review. Water SA 24(3):237–244
103. Jianxin L, Jingxia L, Tao Y, Changfa X (2007) Quantitative study of the effect of electromagnetic field on scale deposition on nanofiltration membranes via UTDR. Water Res 41(20):4595–4610
104. Jiuhui QU (2008) Research progress of novel adsorption processes in water purification: a review. J Environ Sci 20(1):1–13
105. Kakihara Y, Fukunishi T, Takeda S, Nishijima S, Nakahira A (2004) Superconducting high gradient magnetic separation for purification of wastewater from paper factory. IEEE Trans Appl Supercond 14(2):1565–1567
106. Kalló D (2001) Applications of natural zeolites in water and wastewater treatment. Rev Min Geochem 45:519–550
107. Karapınar N (2009) Application of natural zeolite for phosphorus and ammonium removal from aqueous solutions. J Hazard Mat 170(2–3):1186–1191
108. Kass A, Yechieli Gavrieli Y, Vengosh A, Starinsky A (2005) The impact of freshwater and wastewater irrigation on the chemistry of shallow groundwater: a case study from the Israeli Coastal aquifer. J Hydrol 300(1–4):314–331
109. Kerwick M, Reddy S, Holt D, Chamberlain A (2005) A methodology for the evaluation of disinfection technologies. J Water Health 3(4):393–404
110. Khan MMH, Sakauchi F, Sonoda T, Washio M, Mori M (2003) Magnitude of arsenic toxicity in tube-well drinking water in Bangladesh and its adverse effects on human health including cancer: evidence from a review of the literature. Asian Pacific J Cancer Prev 4:7–14
111. Klavarioti M, Mantzavinos D, Kassinos D (2009) Removal of residual pharmaceuticals from aqueous systems by advanced oxidation processes. Environ Int 35:402–417
112. Koter S, Warszawski A (2000) Electromembrane processes in environment protection. Pol J Environ Stud 9(1):45–56
113. Lapworth DJ, Baran N, Stuart ME, Ward RS (2012) Emerging organic contaminants in groundwater: a review of sources, fate and occurrence. Environ Poll 163:287–303
114. Lawrence K, Tong D (2005) Feasibility of using biologically activated carbon for treatment of gaseous H_2S. J Inst Eng 45(4):15–23

115. Lawrence KW, David AV, Yan L, Nazih KS (2005) Chemical precipitation. In: Lawrence KW, Hung YT, Nazih KS (eds) Hand book of environmental engineering, physiochemical treatment processes, vol 3. The Humana Press Inc., Totowa, NJ, pp 141–197
116. Lazarova V, Manem J (1995) Biofilm characterization and activity analysis in water and wastewater treatment. Water Res 29(10):2227–2245
117. Li XL, Yao KL, Liu HR, Liu ZL (2007) The investigation of capture behaviors of different shape magnetic sources in the high-gradient magnetic field. J Magn Mater 311(2):481–488
118. Li Q, Mahendra S, Lyon DY, Brunet L, Liga MV, Li D, Alvarez PJJ (2008) Antimicrobial nanomaterials for water disinfection and microbial control: potential applications and implications. Water Res 42(18):4591–4602
119. Li M, Feng C, Zhang Z, Yang S, Sugiura N (2010) Treatment of nitrate contaminated water using an electrochemical method. Bioresour Technol 101:6553–6557
120. Ligor M, Buszewski B (2006) An Investigation of the formation of taste and odour contaminants in surface water using the headspace SPME-GC/MS method. Pol J Environ Stud 15(3):429–435
121. Liu A, Ming J, Ankumah RO (2005) Nitrate contamination in private wells in rural Alabama, United States. Sci Tot Environ 346(1–3):112–120
122. Louhichi B, Ahmadi MF, Bensalah N, Gadri A, Rodrigo MA (2008) Electrochemical degradation of an anionic surfactant on boron-doped diamond anodes. J Hazard Mat 158:430–437
123. Low KS, Lee CK, Lee KP (1993) Sorption of copper by dye treated oil palm fibers. Bioresour Technol 44:109–112
124. Maine MA, Maria VD, Noemi S (2001) Cadmium uptake by floating macrophytes. Water Res 35(11):2629–2634
125. Maine MA, Suñé NL, Lagger SC (2004) Chromium bioaccumulation: comparison of the capacity of two floating aquatic macrophytes. Water Res 38:1494–1501
126. Makwana BS, Desale GR, Thampy SK, Ghosh PK (2010) Attachment to domestic RO unit for high recovery of pure water from brakish water without compromising on quality. EDRO patent, US Patent 1418/DEL/
127. Manuel O, Fernando P, Hernán S, Bo L, Ricardo U (1998) Copper in infant nutrition: safety of World Health Organization provisional guideline value for copper content of drinking water. J Pediatr Gastroenterol Nutr 26(3):251–257
128. Margeta K, Logar NZ, Šiljeg M, Farkas A (2013) In: Elshorbagy W (ed) Natural zeolites in water treatment—how effective is their use, water treatment. InTech, Croatia. ISBN 978-953-51-0928-0
129. Martinez-Huitle CA, Ferro S (2006) Electrochemical oxidation of organic pollutants for the wastewater treatment: direct and indirect processes. Chem Soc Rev 35:1324–1340
130. Matlock MM, Howerton BS, Atwood DA (2002) Chemical precipitation of heavy metals from acid mine drainage. Water Res 36(19):4757–4764
131. Meenakshi Maheshwari RC (2006) Fluoride in drinking water and its removal. J Hazard Mater B137:456–463
132. Melvin A, Benarde W, Brewster S, Vincent P, Olivieri Burton D (1967) Kinetics and mechanism of bacterial disinfection by chlorine dioxide. Appl Microbiol 15(2):257–265
133. Michael N, Hughes JB, Wong MS (2005) Designing Pd-on Au bimetallic nanoparticles for trichloroethane hydrochlorination. Environ Sci Technol 39(5):1346–1353
134. Mikami I, Sakamoto Y, Yoshinaga Y, Okuhara T (2003) Kinetic and adsorption studies on the hydrogenation of nitrate and nitrite in water using Pd–Cu on active carbon support. Appl Catal B 44:79–86
135. Mikami I, Kitayama R, Okuhara T (2006) Hydrogenations of nitrate and nitrite in water over Pt-promoted Ni catalysts. Appl Catal A Gen 297:24–30
136. Mohana N, Balasubramanian N (2006) In situ electrocatalytic oxidation of acid violet 12 dye effluent. J Hazard Mat B136:239–243
137. Mok WM, Wai CM (1994) Mobilization of arsenic in contaminated river waters. In: Nriagu JO (ed) Arsenic in the environment, Part I: cycling and characterization. Wiley, John, pp 99–118

138. Moore MT, Bennett ER, Cooper CM, Smith S, Shields FD, Milam CD, Farris JL (2001) Transport and fate of atrazine and lambda cyhalothrin in an agricultural drainage ditch in the Missisipidelta, USA. Agric Ecosyst Environ 87:309–314
139. Morsia MS, Al-Sarawyb AA, Shehab El-Dein WA (2011) Electrochemical degradation of some organic dyes by electrochemical oxidation on a Pb/PbO_2 electrode. Desalin Water Treat 26:301–308
140. Mulligan CN, Yong RN, Gibbs BF (2001) Remediation technologies for metal contaminated soils and groundwater: an evaluation. Eng Geol 60(1–4):193–200
141. Mustafiz S, Basu A, Islam MR, Dewaidar A, Chaalal O (2002) A novel method for heavy metal removal using fish scales. Energy Sources 24:1043–1051
142. Nataraj SK, Hosamani KM, Aminabhavi TM (2009) Nanofiltration and reverse osmosis thin film composite membrane module for the removal of dye and salts from the simulated mixtures. Desalination 249(1):12–17
143. Needleman HL, Schell A, Bellinger D, Leviton A, Allred EN (1990) The long-term effects of exposure to low doses of lead in childhood—an 11-year follow-up report. N Eng J Med 322(2):83–88
144. Nishijima S, Takeda S (2007) Research and development of superconducting high gradient magnetic separation for purification of wastewater from paper factory. IEEE Trans Appl Supercond 17(2):2311–2314
145. Nriagu JO (1988) A silent epidemic of environmental metal poisoning? Environ Pollut 50:139–161
146. Nwachcuku N, Gerba CP (2004) Emerging waterborne pathogens: can we kill them all? Curr Opin Biotechnol 15:175–180
147. Okada H, Mitsuhashi K, Ohara T, Whitby ER, Wada H (2005) Computational fluid dynamics simulation of high gradient magnetic separation. Sep Sci Technol 40(7):1567–1584
148. Olson OE (1986) Selenium toxicity in animals with emphasis on man. Int J Toxicol 5:45–70
149. Ormerod J, Constantinides S (1997) Bonded permanent magnets: current status and future opportunities. J Appl Phys 81(8):4816–4820
150. Pagga U, Bruan D (1986) The degradation of dyestuffs: part II Behaviour of dyestuffs in aerobic biodegradation tests. Chemosphere 15:479–491
151. Pal A, Gin KY, Lin AYC, Reinhard M (2010) Impacts of emerging organic contaminants on freshwater resources: review of recent occurrences, sources, fate and effects. Sci Total Environ 408:6062–6069
152. Pal A, He Y, Jekel M, Reinhard M, Gin KY (2014) Emerging contaminants of public health significance as water quality indicator compounds in the urban water cycle. Environ Int 71:46–62
153. Park SK, Hu JY (2010) Assessment of the extent of bacterial growth in reverse osmosis system for improving water quality. J Environ Sci Health A Tox Hazard Subst Environ Eng 45(8):968–977
154. Pawlak Z, Zak S, Zablocki L (2005) Removal of hazardous metals from groundwater by reverse osmosis. Pollut J Environ Stud 15(4):579–583
155. Peel JW, Reddy KJ, Sullivan BP, Bowen JM (2003) Electrocatalytic reduction of nitrate in water. Water Res 37:2512–2519
156. Penlidis A, Vivaldo-Lima E, Wood PE, Hamielec AE (1997) An updated review on suspension polymerization. Ind Eng Chem Res 36:939–965
157. Pera-Titus M, Garcıa-Molina V, Baños MA, Giménez J, Esplugas S (2004) Degradation of chlorophenols by means of advanced oxidation processes: a general review. Appl Catal B Environ 47:219–256
158. Pintar A, Batista J, Muševi˘c I (2004) Palladium–copper and palladium–tin catalysts in the liquid phase nitrate hydrogenation in a batch-recycle reactor. Appl Catal B 52:49–60
159. Pontius FW, Brown KG, Chen CJ (1994) Health implications of arsenic in drinking water. J Am Water Work Assoc 86:52–63Google Scholar

160. Prevot AB, Baiocchi C, Brussino MC, Pramauro E, Savarino P, Augugliaro V, Marci G, Palmisano L (2001) Photocatalytic degradation of acid blue 80 in aqueous solution containing TiO_2 suspension. Environ Sci Technol 35:971
161. Public Health Service, US (1992) Toxicologocal profile for antimony and compounds. Agency for Toxic Substunces and Disease Registry
162. Punayani S, Narayana P, Sing H, Vasudevan P (2006) J Sci Ind Res 65:116–120
163. Qi G, Yang RT, Rinaldi FC (2006) Selective catalytic reduction of nitric oxide with hydrogen over Pd-based catalysts. J Catal 237:381–392
164. Qu J (2008) Research progress of novel adsorption processes in water purification: a review. J Environ Sci (China) 20(1):1–13
165. Radovici O, Banu A, Pirvu C (2009) Micro reactor for chlorophenols electrocatalytic oxidation. ECS Trans 16(27):1–9
166. Rahman MM, Awang MB, Yusof AM (2012) Preparation, characterization and application of zeolite-Y (Na-Y) for water filtration. Aus J Basic Appl Sci 6(1):50–54
167. Rai PK (2007a) Phytoremediation of Pb and Ni from industrial effluents using *Lemna minor*: an eco-sustainable approach. Bull Biosci 5(1):67–73
168. Rai PK (2007b) Wastewater management through biomass of *Azolla pinnata*: an ecosustainable approach. Ambio 36(5):426–428
169. Rai PK (2009) Heavy metal phytoremediation from aquatic ecosystems with special reference to macrophytes. Crit Rev Environ Sci Technol 39:697–753
170. Ram NM, Christman RF, Cantor KP (1990) Significance and treatment of volatile organic compounds in water supplies. Lewis Publishers, Chicago
171. Rao PV, Gupta N, Bhaskar AS, Jayraj R (2002) Toxins and bioactive compounds from cyanobacteria and their implication on human health. J Environ Biol 23(3):215–224
172. Ray(Arora) S, Ray MK (2009) Bioremediation of heavy metal toxicity-with special reference to chromium. Al Ameen J Med Sci 2(2):57–63
173. Reife A, Freeman H (1996) Environmental chemistry of dyes and pigments. Wiley, NewYork
174. Ronald E (1991) Cyanide hazards to fish, wildlife, and invertebrates: a synoptic review. US Fish Wildl Serv Biol Rep 85:1–23
175. Rukah A, Alsokhny K (2004) Geochemical assessment of groundwater contamination with special emphasis on fluoride concentration, North Jordan. Chem Erde Geochem 64(2):171–181
176. Rusin PA, Rose JB, Haas CN, Gerba CP (1997) Risk assessment of opportunistic bacterial pathogens in drinking water. Rev Environ Contam Toxicol 152:57–83
177. Saha NK, Bhattacharya A (2010) Urboniene Chapter 5: membrane desalination: methods, cost and technology. In: Irena A (ed) Desalination: methods, cost and technology. Nova Science Publishers, New York, pp 175–208
178. Savage N, Diallo MS (2005) Nanomaterials and water purification: opportunities and challenges. J Nano Res 7:331–342
179. Scholz M, Martin R (1997) Ecological equilibrium on biologicalactive carbon. Water Res 31(12):2959–2968
180. Selvaggi JA, Cottrell DL, Falconer TH, Daugherty MA, Daney DE, Hill DD, Prenger FC (1998) High gradient magnetic separation using a high temperature superconducting magnet. Appl Supercond 6(1):31–36
181. Semple AB, Parry WH, Phillips DE (1960) Acute copper poisoning: an outbreak traced to contaminated water from a corroded geyser. Lancet 2:700–701
182. Shaffer L, Mintz M (1980) Electrodialysis. In: Spiegler K, Laird A (eds) principles of desalination, 2nd edn, chapter 6. Academic Press, New York, pp 257–357
183. Shannon MA, Bohn PW, Elimelech M, Georgiadis JG, Marinas BJ, Mayes AM (2008) Science and technology for water purification in the coming decades. Nature 452:301–310
184. Smith AH, Lingas EO, Rahman M (2000) Contamination of drinking-water by arsenic in Bangladesh: a public health emergency. Bull World Health Org 78(9):1093–1103
185. Snoeyink V, Jenkins D (1980) Water chemistry. Wiley, New York, p 463
186. Soares OSGP, Orfao JJM, Pereira MFR (2008) Activated carbon supported metal catalysts for nitrate and nitrite reduction in water. Catal Lett 126:253–260

187. Soares OSGP, Orfao JJM, Ruiz-Martinez J, Silvestre-Albero J, Sepúlveda-Escribano A, Pereira MFR (2010) Pd–Cu/AC and Pt–Cu/AC catalysts for nitrate reduction with hydrogen: influence of calcinations and reduction temperatures. Chem Eng J 165:78–88
188. Stanley RA (1974) Toxicity of heavy metals and salts to Eurasian watermill foil (*Myriophyllum spicatum* L.). Arch Environ Contam Toxicol 2:331–340
189. Strathmann H (2010a) Electrodialysis, a mature technology with a multitude of new applications. Desalination 264:268–288
190. Strathmann H (2010b) Ion-exchange membrane processes in water treatment. Sustain Sci Eng 2(9):141–199
191. Stuart M, Lapworth D, Crane E, Hart A (2012) Review of risk from potential emerging contaminants in UK groundwater. Sci Total Environ 416:1–21
192. Susheela AK (1999) Fluorosis management programme in India. Curr Sci 77(10):1250–1256
193. Tee TW, Khan ARM (1988) Removal of lead, cadmium, and zinc by waste tea leaves. Environ Technol Lett 9:1223–1232
194. Teoh WY, Amal R, Scott J (2012) Progress in heterogenours photocatalysis: from classical radical chemistry to engineering nanomaterials and solar reactors. J Phys Chem Lett 3:629–639
195. Timoshenko EM, Ugarov GG (1994) Limiting efficiency of an electromagnet with a linear magnetic system—critical-review. J Min Sci 30(6):604–606
196. Tucker PM, Waite MJ, Hayden BE (2004) Electrocatalytic reduction of nitrate on activated rhodium electrode surfaces. J Appl Electrochem 34:781–796
197. Tzvetkova P, Nickolov R (2012) Modified and unmodified silica gel used for heavy metal ions removal from aqueous solutions. J Univ Chem Tech Metall 47(5):498–504
198. United Nations World Water Assessment Programme (UN WWAP) (2003) The World Water Development Report 1: water for people, water for life. UNESCO, Paris
199. University of Florida (1998) Institute of Food and Agricultural Sciences. Trihalomethanes and Our Water Supply
200. Valli F, Tijoriwala A, Mahapatra A (2010) Nanotechnology for water purification. Int Nuc Desalin 4:49–57
201. Veil J (2008) Thermal distillation technology for management of produced water and frac flowback water. Water Tech Brief #2008-1
202. Ventura A, Jacquet G, Bermond A, Camel V (2002) Electrochemical generation of the Fenton's reagent: application to atrazine degradation. Water Res 36:3517–3522
203. VonGunten U (2003a) Ozonation of drinking water: part I. Oxidation kinetics and product formation. Water Res 37:1443–1467
204. VonGunten U (2003b) Ozonation of drinking water: part II. Disinfection and by-product formation in presence of bromide, iodide or chlorine. Water Res 37:1469–1487
205. Wang Y, Qu J, Wu R, Lei P (2006) The electrocatalytic reduction of nitrate in water on Pd/Sn-modified activated carbon fiber electrode. Water Res 40:1224–1232
206. Watlungton K (2005) Emerging nanotechnologies for site remediation and waste water treatment. National Network for environmental Management Fellow North Carolina State University, Environmental Protection Agency, US
207. Weber W, Pribazari M, Melson G (1978) Biological growth on active carbon: an investigation by scanning electron microscopy. Environ Sci Technol 12:817R–819R
208. Wehrmann HA, Barcelona MJ, Varljen MD, Blinkiewicz G (1996) Ground-Water Contamination by Volatile Organic Compounds: Site Characterization, Spatial and Temporal Variability ISWS CR-591: Report 591, Prepared for the US Environmental Protection Agency Environmental Monitoring Systems Laboratory Advanced Monitoring Systems Division Aquatic and Subsurface Monitoring Branch
209. WHO (2008) Guidelines for drinking-water quality. Recommendations, vol 1, 3rd edn. World Health Organization, Geneva
210. WHO (2010) International Code of Conduct on the distribution and use of pesticides: guidelines for the registration of pesticides. World Health Organization, Rome
211. Wones RG, Stadler BL, Frohman LA (1990) Lack of effect of drinking water barium on cardiovascular risk factors. Environ Health Perspect 85:355–359

212. Xie W, Wang Q, Ma H, Ohsumi Y, Ogawa HI (2005) Study on phosphorus removal using a coagulation system. Process Biochem 40(8):2623–2627
213. Xu T (2005) Ion exchange membranes: state of their development and perspective. J Membr Sci 263:1–29
214. Comninellis C (1994) Electrocatalysis in the electrochemical conversion/combustion of organic pollutants for waste water treatment. Electrochim Acta 39(11/12):1857–1862
215. Yan LG, Nan HL, Yu YJ, Dai YM, Song SS, Ye ZX, Chen YL (1996) A fast ramp superconducting magnet for HGMS. IEEE Trans Magn 32(4):2707–2709
216. Yang RT (1997) Gas separation by adsorption process. Imperial College Press, London
217. Yang RT, Benton DF (2003) Adsorbents: fundamentals and applications, activated carbon, vol Chapter 5. Wiley, Oxford, p 80. doi:10.1002/047144409X.ch5
218. Yoshida T, Yamauchi H, Sun GF (2004) Chronic health effects in people exposed to arsenic via the drinking water: dose–response relationships in review. Tox Appl Pharmacol 198:243–252
219. Younes M, Galal-Gorchev H (2000) Pesticides in drinking water—a case study. Food Chem Toxicol 38(1):S87–S90
220. Zayed A, Gowthaman S, Terry N (1998) Phytoaccumulation of trace elements by wetland plants: duckweed. J Environ Qual 27:715–721
221. Zhang J, Dong MD, Li SK (1997) Cancer mortality in a Chinese population exposed to hexavalent chromium in water. J Occup Environ Med 39(4):315–319
222. Zhang X, Wnag J, Wang Z, Wang S (2005) Electrocatalytic reduction of nitrate at polypyrrole modified electrode. Synth Met 155:95–99
223. Zhang H, Quan X, Chen S, Zhao H, Zhao Y (2006) Fabrication of photocatalytic membrane and evaluation its efficiency in removal of organic pollutants from water. Sep Purif Technol 50:147–155
224. Zhu ZQ, Halbach HD (2001) Permanent magnet machines and applications: a review. IEEE Proc Electr Power Appl 148(4):299–308
225. Zhu YL, Zayed AM, Quian JH, Desouza M, Terry N (1999) Phytoaccumulation of trace elements by wetland plants: II. Water hyacinth. J Environ Qual 28:339–344
226. Zodrow K, Brunet L, Mahendra S, Li D, Zhang A, Li Q, Alvarez PJJ (2009) Polysulfone ultrafiltration membranes impreganated with silver nanoparticles show improved biofouling resistance and virus removal. Water Res 43(3):715–723

CHAPTER 7

Risk assessment of cyanobacteria-toxins for small drinking water treatment plants with lake water intakes

Uche A. Uduma, Edward A. McBean and Bahram Gharabaghi*

School of Engineering, University of Guelph, Guelph, Ontario N1G 2W1, Canada.

ABSTRACT

Recent increases in the frequency and severity of toxic algae blooms in freshwater lakes has been a major concern for small communities that rely on them for drinking water supply. A hazard quotient approach to risk characterization is employed to analyze the effectiveness of five conventional treatment methods for removal of cyanotoxins. The application of the method for risk assessment and mitigation is demonstrated for five case studies, including Lake Champlain (Quebec), Coal Lake (Alberta), Butte Lake (Alberta), Kubbani Lake (Nigeria) and Bomo Lake (Nigeria).

Keywords: Cyanotoxins, human health, hazard quotient, lake water intake, water quality.

1. INTRODUCTION

Agricultural non-point source pollution combined with global warming have caused major algal blooms in our freshwater rivers and lakes (Asnaashari et al., 2015; Stang et al., 2016; Gazendam et al., 2016). Cyanobacteria are photosynthetic prokaryotes that thrive well in various kinds of habitats ranging from freshwater and marine environments to hot springs and deserts (Duy et al., 2000; Ballot et al., 2003, 2010; Baxa et al., 2010; Bogialli et al., 2013; Chia et al., 2009, Chia and Kwaghe, 2015). They are popularly referred to as blue-green algae but their physiological, morphological and metabolic structures clearly identify them as bacteria. The photosynthetic origin of cyanobacteria is similar to that of algae but the pigments are located in the thylakoids, which is in the cytoplasm (Chorus et al., 2000; Codd et al., 2005; Drabkova and Marsalek, 2007; Echenique et al., 2014; Ostermaier and Kurmayer, 2010; Sayyad et al., 2015).One of the basic "metabolic processes of cyanobacteria" is the fixation of di-nitrogen in aerobic conditions using nitogenase, an enzyme that converts di-nitrogen to ammonium. This process enhances a bloom of cyanobacteria in surface waters (Ernst et al., 2006; Fastner et al., 2007; Fawell et al., 1999; Fischer et al., 2000; Graham et al., 2010; Gilroy et al., 2000; Harvey et al., 2015). Cyanobacteria have many properties which result in their relative success and predominance during the blooming season. The following factors are responsible for cyanobacteria blooms in aquatic habitat: aquatic temperatures above 25°C, low light intensity in water, and low nitrogen-to-phosphorous ratios (Hans and Timothy, 2013; Heisler et al., 2008; Hrudey et al., 1999; Griffiths and Saker, 2003; Kaushik and Balasubramanian, 2013; Paerl et al., 2011; Tencalla and Dietrich, 1997).

Presently,there about 3000 known species of cyanobacteria but not all produce toxins. Due to the increasing numbers of cases of cyanobacteria blooms, the occurrence of several harmful cyano-toxins in water supplies has also increased. There is growing concern about the potential for negative health effects on humans due to these toxins (Keijola et al., 1989; Baker et al., 2015; Makarewicz and Lewis, 2015; Merel et al., 2012; Mohamed et al., 2015).

Notwithstanding the recent scientific and technical advances in drinking water treatment plants around the globe, the concentrations of cyanobacterial-toxins have been reported to increase in treated drinking relative to raw water source for small water treatment plants. These toxins enter water supplies after lysis of cyano-bacterial cells, as induced by water collection and treatment activities resulting in subsequent release of toxins in finished treated waters. The toxins released by blue-green algae cannot be removed by conventional treatment methods (Merel et al., 2010;

Newcombe and Nicholson, 2002; Nicholson et al., 2003; Szlag et al., 2015; Westrick et al., 2010; Zamyadi et al., 2012).

Drinking water treatment processes may cause breakthrough of toxins into treated drinking water as demonstrated by Zamyadi et al. (2012) in a full-scale water treatment plant system. The passage of toxins, toxic cells and cell debris through filtration systems inhibited chlorination that resulted in the breakthrough of microcystins, resulting in exceedance of Canadian and WHO water quality standards for treated drinking water. Long-term consumption of water contaminated by cyanobacteria and algae is known to cause liver failure, cardiac arrhythmia, dysfunction of the nervous system and skin tumors (Ontario Health Unit, 2014; Farrer et al., 2015).

The objective of this study is to evaluate the use of conventional treatment options and riverbank filtration for managing cyano-toxins, to assess the vulnerability of existing municipal drinking water facilities using surface water sources.

2. STUDY AREAS

The application of the hazard index is demonstrated for five case studies, including Lake Champlain (Quebec), Coal Lake (Alberta), Butte Lake (Alberta), Kubbani Lake (Nigeria) and Bomo Lake (Nigeria).

2.1 Lake champlain's missisquoi bay (quebec)

The population is about 330,000. The Missisquoi River is 130 km long and a tributary of Lake Champlain. The Missisquoi River cuts across Vermont in the United States and southern Quebec in Canada located between Latitude 45° and 0′ E and longitude 72° and 35′ due west. The Missisquoi River catchment area covers the Green mountains along the US-Canada border (Lake Champlain and Eastern townships in Quebec before emptying into Lake Champlain's Missisquoi Bay at Richford (Zamyadi et al., 2012).

2.2 Coal Lake, City of Wetaskiwin (Alberta)

The population is about 12,525. Wetaskiwin is a city in the Province of Alberta Canada (coordinates 52°N and 113°W). Wetaskiwin is located 70 km south of Edmonton; it sits on what was "formerly a coast of the large sea that covered much of Alberta, millions of years ago". Wetaskiwin sits at an elevation of 760 m. Coal Lake, a reservoir developed on the Battle River, is on the east side of the City (Zurawell, 2002).

2.3 Butte lake, town of picture butte (alberta)

The population is about 1650. Picture Butte is a city in southern Alberta (coordinates 49° 52' N and 112° 46' W). Picture Butte has a landmass of 2.9 km^2 and sits at an elevation of 900 m, 27 km north of the City of Lethbridge. Old-man River and Picture Butte Lake is the sources of drinking water for this town (Zurawell, 2002).

2.4 Kubanni Lake (Nigeria)

The population of this community is about 560,000. The coordinates of this water body are 11° 08'N and 07° 43'E. It has two major tributaries: Kampagi River and Kubanni River. Kubanni Lake is the major source of water to Ahmadu Bello University community and its environs. Surrounding the Kubanni Lake catchment is agricultural land; the farmers in this location utilize agricultural practices that are not environmentally sustainable for the lake (Chia and Kwaghe, 2015).

2.5 Bomo Lake (Nigeria)

The population is about 300,000. This lake is 6 km north west of ABU Zaria; it is located 11° 12'N and 07° 38'E and about 671 m above sea level (Chia et al., 2009).

3. MATERIALS AND METHODS

Due to variations in the species, concentrations and locations of cyanotoxins, appropriate analytical methods are required in order to understand the abundance and occurrence and the potential toxicity of cyanobacterial populations, and to effectively assess the risk of cyanotoxins to humans in drinking water supplies (Kutovaya and Watson, 2014).

Data for small drinking water treatment plants in Eastern Township Quebec, Wetaskiwin Alberta, Picture Butte Alberta, Kubanni and Bomo Lake in Zaira Nigeria are adapted from different literature before, and after conventional treatment. Data are collected and monitored on a weekly basis for spring and summer seasons for the raw water samples from these plants; the changing characteristic of the toxins in the raw water and treated drinking water are the major foci for the drinking water data collected.

Table 1 lists the concentrations of cyanotoxins in raw and treated water for the selected study regions. The data in Table 1 are in agreement with the removal efficiencies of Cyanotoxins using the conventional treatment range from 25 to 99%. Apart from Picture Butte in Table 1, the concentrations of cyanotoxins in the raw water for all other locations are above the Canadian guidelines for microcystins in drinking water. Further, the concentrations of cyanotoxins in the treated waters, with the exception of Wetaskiwin and Picture Butte, remained above WHO and Canadian guideline.

Table 1. Concentrations of cyanotoxins in raw and treated water.

Location	Raw water (μg/L)	Treated water (μg/L)	References
Eastern Township (Quebec)	118.7	2.47	Zamyadi et al. (2012)
Wetaskiwin (Alberta)	14.8	0.155	Zurawell (2002)
Picture Butte (Alberta)	0.105	0.077	Zurawell (2002)
Kubanni (Nigeria)	3.8	2.8	Chia and Kwaghe (2015)
Bomo (Nigeria)	2.4	1.8	Chia et al. (2009)

3.1 Risk assessment of cyanotoxins

A hazard quotient is the ratio of the potential exposure to a substance and the level at which no adverse effects are expected. Risk assessment using the intake rate, hazard quotient and hazard indices using the equations presented below, are described to determine the risk associated with drinking water contaminated with cyano-bacterial toxins.

$$\text{Intake rate}\left(\frac{\text{mg}}{\text{kg-day}}\right) = \frac{\text{Cw} \times \text{IF} \times \text{EF} \times \text{ED}}{\text{BW} \times \text{AT}} \tag{1}$$

Where: CW = concentration in water (mg/l); IF= intake frequency (L/day); EF = exposure frequency (days); ED = exposure duration (years); BW = body weight (kg); AT = average time (days).

$$\text{Hazard quotient} = \frac{\text{Intake Rate}}{\text{Reference Dose}} \tag{2}$$

Where: HQ = hazard quotient.

The intake rate and hazard quotients for consuming raw water, water after conventional treatment in the two geographical locations are calculated using Equations 1 and 2 using the data presented in Table 2. The calculation conditions below are used in deriving Tables 3 and 4.

Table 2. Reference daily doses.

Food	Cyano-toxin	References
Reference daily dose for an adult	0.0001 mg/L	Ohio EPA, 2013
Reference daily dose for a baby	1.43×10^{-5} mg/L	Ohio EPA, 2013

Table 3. Hazard quotient summary for an adult and a baby consuming raw water.

Location	Raw water source	Hazard quotient for an adult	Hazard quotient for a baby
Eastern Township	Missisquoi River	33.9	197
Wetaskiwin	Coal Lake	4.2	24.5
Picture Butte	Butte Lake	0.03	0.18
Kubanni	Kubanni Lake	2.17	6.3
Bomo	Bomo Reservoir	1.3	3.9

Table 4. Hazard quotient summary for an adult and a baby consuming treated water.

Location	Raw water source	Water treatment technology	Hazard quotient for an adult	Hazard quotient for a baby
Eastern Township	Missisquoi River	Conventional Treatment (Coagulation, Powdered activated carbon, post clarification and chlorination)	1.6	4.1
Wetaskiwin	Coal Lake	Conventional Treatment (pre-oxidation, clarification and chlorination)	0.04	0.26
Picture Butte	Butte Lake	Conventional treatment, membrane filtration and chlorination	2.5×10^{-8}	0.02
Kubanni	Kubanni Lake	Riverbank filtration, sedimentation and chlorination	2.5	4.6
Bomo	Bomo Reservoir	Sand filtration, pre-oxidation and chlorination	1.86	2.9

Key assumptions

1. Body weight for adult, 70 kg.

2. Body weight for a baby, 10 kg.

3. Water ingestion rate for an adult is 2 L/day for Canada and 4 L/day for the hot climate in Nigeria.

4. Water ingestion rate for a ten-month-old baby is 0.24 L/day.

5. Unintentional ingestion is 1.5 L/day for Canada and Nigeria.

6. Exposure frequency is 182 days for an adult and a baby in Canada

7. Exposure frequency is 365 days for an adult and a baby in Nigeria

8. The reference dose for an adult is 0.0001 $\frac{\text{mg}}{\text{L}}$ per day for both water (Ohio State Environmental Protection Agency, 2013).

9. Reference dose for a baby is $\left(\frac{10\text{kg}}{70\text{ kg}} \times 0.0001\right) = 1.43 \times 10^{-5} \frac{\text{mg}}{\text{L}}$ per day.

10. The cyanobacteria toxins in the assumed intake are derived only from water.

4. RESULTS

Table 3 summarizes the hazard quotient (the ratio of the potential exposure to a substance and the level at which no adverse effects are expected) of raw water and cow milk for an adult and a baby drinking raw water. From the hazard quotient values presented in Table 3, the results indicate that the Butte Lake is the only safe drinking water in terms of microcystins.

Table 4 presents a summary that relates the hazard index of treated drinking water for an adult and a baby to the water treatment technology used drinking water treatment plants; from the hazard quotient values presented in Table 4, the results indicate that a combination of conventional treatment and membrane filtration which is used in Picture Butte is an efficient technique for removing cyanotoxins.

Table 1 was used together with Equations 1, 2, and 3 to calculate the non-carcinogenic risk for a person that weighs 70 kg and drinks 2 L (Canada) and 4 L (Nigeria) of water daily. The individual drank directly from the contaminated lake. Furthermore, the same approach is performed for a baby that weighs 10 kg and drinks 0.24 L of water per day.

A hazard quotient is the ratio of the potential exposure to a substance and the level at which no adverse effects are expected. The hazard quotients for an adult and a baby consuming raw water are summarized in Table 3. The hazard quotient for an adult consuming raw water in four communities (study areas in Canada and Nigeria) except Picture Butte exceeded unity. This was similar to the hazard quotient for a baby consuming raw untreated water for all the communities under study with the exception of Picture Butte.

The hazard quotients for an adult and a baby consuming treated drinking water is presented in Table 4; for an adult in Eastern Township, Kubanni and Bomo consuming treated drinking water the hazard quotient is greater than one, while for an adult in Wetaskiwin and Picture Butte consuming treated drinking water, the hazard quotient is less than one. Similarly, the hazard quotient result for a baby consuming treated drinking water in Eastern Township, Kubanni and Bomo exceeds 1, while for a baby consuming treated water in Alberta (Wetaskiwin and Picture Butte), the hazard quotient is less than one.

5. DISCUSSION

The presences of cyanotoxins in treated drinking water supply from small water treatment plants reaffirms the need to assess the risk of cyanotoxins from these small water treatment plants. In this study, the hazard quotients for an adult and a baby were calculated for a single exposure of treated water for the five treatment plants. The summary calculations are shown in Table 3. The results indicate that a combination of conventional treatment and membrane filtration used in Picture Butte (Table 4) is an efficient technique for removing cyanotoxins. The other Canadian water treatment plants that use combination of coagulation, activated carbon, oxidation, clarification and chlorination have shown a high percentage of removal of cyanotoxins (greater than 97%). This highlights the importance of adopting a more efficient water treatment technologies in affected areas to minimize the risk to a human receptor. On the other hand, in Nigeria, a simple sand filtration or riverbank filtration combined with only chlorination was found to be ineffective in reducing the cyanotoxins concentration to a safe and tolerable level. The efficiency of cyanotoxins removal for both treatment systems in Nigeria were found to around 25%. The people living in these two communities are exposed to elevated concentration of toxin in their drinking water supply.

The results in the different communities included in this study show clearly that hazard quotient for babies are much higher than that for adults. Excluding Picture Butte community, the increased risk for babies compared to adults is between 1.6 and 4.1 to all other four communities. Moreover, all the results above were calculated based on a single exposure (drinking water). Therefore, it is expected that the risk index, which is defined as the sum of hazard quotients due to different exposures, is also alarming. Other exposures that were not considered in this study are food crops irrigated with cyanotoxins-contaminated water, inhalation of water aerosols, and other water sports like swimming and surfing (Queensland Health, 2001).

6. CONCLUSIONS

The assessment of the cyanotoxin removal capacity for small water treatment plants shows that the concentrations of cyanobacterial toxins in treated drinking water in Quebec and northern Nigeria are of concern; the hazard quotient derived from a combination of treated or untreated water from drinking water treatment plants in Quebec and northern Nigeria exceeded unity which indicates the need for appropriate remedial action to be taken or an alternative water supply developed. Also, the intake rate exceeded the recommended World Health Organization intake guideline of 1 μg/L.

Applicable strategies to eliminate cyanobacteria blooms would be appropriate to adopt for water treatment facilities in Eastern Township (Quebec) and northern Nigeria in order to protect consumer's health. A more sustainable strategy that consists of reducing the introduction of nutrients in surface waters is ideal for a long-term solution. Advanced water treatment techniques such as membrane filtration should be adopted since they have been proven to be very effective in treating all kinds of organic contaminants. Small drinking water treatment plants can be upgraded to include membrane filtration and ozonation. This will help to achieve high removal efficiency.

REFERENCES

1. Asnaashari A, Gharabaghi B, McBean E, Mahboubi AA (2015). Reservoir Management Under Predictable Climate Variability and Change. J. Water Climate Change 6(3):472-485.
2. Baker LC, Manassaram-Baptiste D, LePrell R, Bolton B (2015). Cyanobacterial and algae blooms: Review of health and environmental data from the harmful algal bloom-related illness surveillance system (HABISS) 2007–2011. Toxins 7:1048-1064.
3. Ballot A, Fastern J, Wiedner C (2010). Paralytic shellfish toxin-producing Cyanobacterium. Alphanizomenon gracile in Noeast Germany. Appl. Environ. Microbiol. 76:1173-180.
4. Ballot A, Pflugmacher S, Wiegand C, Kotut K, Krienitz L (2003). Cyanobacterial toxins in Lake Baringo, Kenya. Limnologica. Tech. Sci. J. 33:2-9.
5. Baxa DV, Kurobe T, Ger KA, Lehman PW (2010). Estimating the abundance of toxic microcystins in San Francisco Esturay using qualitative real-time PCR. Harmful algae 9:342-349.
6. Bogialli S, Nigro Di Gregorio F, Lucentini L, Ferretti E, Ottaviani M, Ungaro N, Abis PP, Cannarozzi De Grazia M (2013). Management of toxic Cyanobacterium bloom (Planktothrix rubescens) affecting an Italian drinking water basin: A case study. Environ. Sci. Technol. 47:574-583.
7. Chia A, Abolude D, Laden Z, Akanbi O, Kalabonus A (2009). The presence of Microcyctins in Aquatic ecosystems in Northern Nigeria: Zaria as a Case Study. Resour. Environ. 3(4):170-178.
8. Chia MA, Kwaghe MJ (2015). Microcyctins Microcystins contamination of surface water supply sources in Zaria-Nigeria. Environ. Monitor. Assess. 187:10.1007/s10661-015-4829-3.
9. Chorus I, Falconer IR, Salas HJ, Bartram J (2000). Health Risks Caused By Freshwater Cyanobacteria In Recreational Waters. J. Toxicol. Environ. Health Part B 3:323-347.
10. Codd GA, Lindsay J, Young FM Morrison LF, Metcalf JS (2005). Harmful cyanobacteria: from mass mortalities to management measures. Springer: pp. 1-23.
11. Drabkova M, Marsalek B (2007). A review of in-lake methods of cyanobacterial blooms control and management. CyanoData – The Glogal Database of Methods for Cyanobacterial Blooms Management, Centre for Cyanobacteria and their Toxins.
12. Duy TN, Lam PKS, Shaw GR, Connell DW (2000). Toxicology and risk assessment of freshwater cyanobacterial (blue-green algal) toxins in water. Rev. Environ. Contam Toxicol. 163:113-186.
13. Echenique RO, Aguilera A, Giannuzzi L (2014). Problems on drinking water related to toxigenic Cyanobacteria: Some cases studied in Argentina. Adv. Limnol. (65):431-444.

14. Ernst B, Hoeger SJ, O' Brien E, Dietrich DR (2006). Oral toxicity of the microcystins-containing cyanobacterium planktothrix rubescens in European whitefish (Coregonus lavaretus). Aquat. Toxicol. 79:31-40.
15. Farrer D, Counter M, Hillwig R, Cude C (2015). Health-based cyanotoxin guideline values allow for cyanotoxin-based monitoring and efficient public health response to cyanobacterial blooms. Toxins 7:457-477.
16. Fastner J, Rucker J, Stuken A, Preubel K, Nixdorf B, Chorus I (2007). Occurrence of cyanobacterial toxin cylindrosperospin in northeast Germany. Environ. Toxicol. 22:26-32.
17. Fawell JK, Mitchell RE, Evertt DJ, Hill RE (1999). The toxicity of Cyanotoxins in the Mouse: microcystins exposure in human. Exp. Toxicol. 181:162-167.
18. Fischer WJ, Hiltzfield BC, Tencalla F, Eriksson JE, Mikhailov A, Deietrich DR (2000). Microcystin-LR Toxicodynamics, induced pathology and immunohistochemical localization in livers of blue-green algea exposed rainbow trout (Oncorhynchus mykiss). Toxicol. Sci. 54:365-373.
19. Gazendam E, Gharabaghi B, Ackerman J, Whiteley H (2016). Integrative Neural Networks Models for Stream Assessment in Restoration Projects. J. Hydrol. 536:339-350.
20. Gilroy DJ, Kauffman KW, Hall RA, Huang X, Chu FS (2000). Assessing Potential Health Risk from Microcystin Toxins in Blue-Green Algae Dietary Supplements. Environ. Health Perspective 108:435-439.
21. Graham J, Loftin K, Meyer M, Ziegler A (2010). Cyanotoxin mixtures and taste and odor compounds in cyanobacterial blooms from Midwest United States. Environ. Sci. Technol. 44:7361-7368.
22. Griffiths DJ, Saker ML (2003). The Palm Island mystery disease 20 years on: a review of research on the cyanotoxin cylindrospermopsin. Environ. Toxicol. 18:78-93.
23. Hans WP, Timothy GO (2013). Harmful Cyanobacterial Blooms: Causes, Consequences, and Controls. Environ. Microbiol. DOI 10.1007/s00248-012-0159.
24. Harvey R, McBean EA, Murphy HM, Gharabaghi B (2015). Using data mining to understand Drinking Water Advisories in small water systems: A case study of Ontario First Nations drinking water supplies. Water Resour. Manage. 29(14):5129-5139.
25. Heisler J, Glibert PM, Burkholder JM, Anderson DM, Cochlan W, Dennison WC, Dortch Q, Gobler CJ, Heil CA, Humphries E, Lewitus A, Magnien R, Marshall HG, Sellner K, Stockwell DA, Stoecker DK, Suddleson M (2008). Eutrophication and harmful algal blooms: a scientific consensus. Harmful Algae 8:3-13.
26. Hrudey S, Burch MD, Drikas M, Gregory R (1999). Remedial measures. In Toxic cyanobacteria in water: a guide to their public health consequences, monitoring, and management. Edited by Chorus I, Bartram JE, Spon FN New York. pp. 275-312.
27. Kaushik R, Balasubramanian R (2013). Methods and Approaches for Detection of Cyanotoxins in Environmental samples: A review: Critical Rev. Environ. Sci. Technol. 43:1349-1383.
28. Keijola A, Himberg K, Esala A, Sivonen K, Kiisvirata L (1989). Removal of cyanobacterial toxins in water treatment processes: laboratory and pilot plant experiments. Toxic. Assess. 3:643-656.
29. Kutovaya OA, Watson SB (2014). Development and application of a molecular assay to detect and monitor geosmin-producing cyanobacteria and actinomycetes in the Great Lakes. J. Great Lakes Res. 40:404-414.
30. Makarewicz JC, Lewis TW (2015). Exploring spatial trends and causes in Lake Ontario coastal chemistry: Nutrients and pigments. J. Great Lakes Res. 3:794-800.
31. Merel S, Clément M, Thomas O (2010). State of the art on Cyanotoxins in water and their behaviour towards chlorine. Toxicon 55:677-691.
32. Merel S, Walker D, Chicana R, Snyder S, Baures E, Thomas O (2012). State of Knowledge and concerns on Cyanobacterial blooms. Environ. Int. 59:303-327.
33. Mohamed ZA, Deyab MA, Abou-Dobara MI, El-Sayed AK, El-Raghi WM (2015). Occurrence of cyanobacteria and microcystin toxins in raw and treated waters of the Nile River, Egypt: implication for water treatment and human health. Environ. Sci. Pollution Res. 22:11716-11727.

34. Mouchet P, Bonnélye V (1998). Solving algae problems: French expertise and worldwide applications. AQUA 47(3):125–141.
35. Newcombe G, Nicholson BC (2002). Treatment options for the saxitoxin class of Cyanotoxins. Water Sci. Technol. Water Supply 2(5-6):271-275.
36. Nicholson BC, Shaw GR, Morrall J, Senogles PJ, Woods TA, Papageorgiou J, Kapralos C, Wickramasinghe W, Davis BC, Eaglesham GK, Moore MR (2003). Chlorination for degradingsaxitoxins (paralytic shellfish poisons) in water. Environ. Technol. 24(11):1341-1348.
37. Ohio State Environmental Protection Agency (2013). Public Water System Harmful Algal Bloom Response Strategy 1:1-47.
38. Ontario Health Unit (2014). Beach management guidance document; Ministry of health and long-term care pp. 1-16.
39. Ostermaier V, Kurmayer R (2010). Application of real-time PCR to estimate toxin production by the cyanobacterium Planktothrix sp. Appl. Environ. Microbiol. 76:3495-3502.
40. Paerl HW, Hall NS, Calandrino ES (2011). Controlling harmful cyanobacterial blooms in a world experiencing anthropogenic and climatic-induced change. Sci. Total Environ. 409:1739-1745.
41. Queensland Health (2001). Environmental health assessment guidelines: Cyanobacteria in recreational and drinking wa-ters. Environmental Health Unit Bris- bane, Australia.
42. Sayyad G, Vasel L, Besalatpour A, Gharabaghi B, Golmohammadi G (2015). Modeling Blue and Green Water Resources Availability in an Iranian Data Scare Watershed Using SWAT. J. Water Manage. Modeling doi: 10.14796/JWMM.C391.
43. Stang C, Gharabaghi B, Rudra R, Mahboubi A, Ahmed S (2016). Conservation management practices - success story of the Hog Creek and Sturgeon River Watersheds, Ontario, Canada. J. Soil Water Conserv. 71(3):237-248.
44. Szlag D, Sinclair J, Southwell B, Westrick J (2015). Cyanobacteria and Cyanotoxins Occurrence and Removal from five High-risk conventional Treatment drinking water plants. Toxins 7:2198-2220.
45. Tencalla F, Dietrich D (1997). Biochemical Characterization of Microcystin toxicity in trout (Oncorhyncnchus mykiss). Toxicon 35:583-595.
46. Westrick JA, Szlag DC, Southwell BJ, Sinclair J (2010). A review of cyanobacteria and Cyanotoxins removal /inactivation in drinking water treatment. Anal. Bioanal. Chem. 397:1705-1714.
47. Zamyadi A, Sherri L, Yan F, McQuaid N, Dorner S, Sauve S, Prevost M (2012). Toxic cyanobacterial breakthrough and accumulation in a drinking water treatment plant: A monitoring and treatment Challenge. Water Res. 46:1511-1523.
48. Zurawell R (2002). An initial Assessement of Microcystin in Raw and Treated Municipal drinking water derived from Eutrophic surface waters in Alberta. 1-15: http://environment.gov.ab.ca/info/library/6026.pdf

CHAPTER 8

Microbiological Evaluation of Water Quality from Urban Watersheds for Domestic Water Supply Improvement

A. Mark Ibekwe [1,*], Shelton E. Murinda [2] and Alexandria K. Graves [3]

[1]*U.S. Salinity Laboratory, USDA-ARS, 450 West Big Springs Road, Riverside, CA 92507, USA*

[2]*Department of Animal and Veterinary Sciences, California State Polytechnic University Pomona, 3801 West Temple Avenue, Pomona, CA 91768, USA*

[3]*Department of Soil Sciences, North Carolina State University, Williams Hall 3411E, P.O. Box 7619, Raleigh, NC 27695, USA*

ABSTRACT

Agricultural and urban runoffs may be major sources of pollution of water bodies and major sources of bacteria affecting the quality of drinking water. Of the different pathways by which bacterial pathogens can enter drinking water, this one has received little attention to date; that is, because soils are often considered to be near perfect filters for the transport of bacterial pathogens through the subsoil to groundwater. The goals of this study were to determine the distribution, diversity, and antimicrobial resistance of pathogenic *Escherichia coli* isolates from low flowing river water and sediment with inputs from different sources before water is discharged into ground water and to compare microbial contamination in water and sediment at different sampling sites. Water and sediment samples were collected from 19 locations throughout the watershed for the isolation of pathogenic *E. coli*. Heterotrophic plate counts and *E. coli* were also determined after running tertiary treated water through two tanks containing aquifer sand material. Presumptive pathogenic *E. coli* isolates were obtained and characterized for virulent factors and antimicrobial resistance. None of the isolates was confirmed as Shiga toxin *E. coli* (STEC), but as others, such as enterotoxigenic *E. coli* (ETEC). Pulsed field gel electrophoresis (PFGE) was used to show the diversity *E. coli* populations from different sources throughout the watershed. Seventy six percent of the isolates from urban sources exhibited resistance to more than one antimicrobial agent. A subsequent filtration experiment after water has gone through filtration tanks containing aquifer sand material showed that there was a 1 to 2 log reduction in *E. coli* in aquifer sand tank. Our data showed multiple strains of *E. coli* without virulence attributes, but with high distribution of resistant phenotypes. Therefore, the occurrence of *E. coli* with multiple resistances in the environment is a matter of great concern due to possible transfer of resistant genes from nonpathogenic to pathogenic strains that may result in increased duration and severity of morbidity.

Keywords: pathogenic *Escherichia coli*; indicator bacteria; surface water; sediment; contamination; watershed

1. INTRODUCTION

The Santa Ana River (SAR) is the largest river in the Santa Ana Region of southern California and it is a major source of domestic water supply for over 2 million people that live in Orange County, California. The Santa Ana River is critical for replenishment of Orange County's Groundwater Basin since over 2 million residents in Orange County depend on groundwater for 75% of their water supply. Any factor in the watershed which degrades the river affects the drinking water supply. The river extends from its headwaters in the San Bernardino Mountains into the Prado Basin and Santa Ana Canyon. Below Prado Dam, there are extensive facilities to recharge much of the flows in the River into the underlying groundwater basin. Sources of non-point contaminants into the river may be from municipal wastewater, agricultural waste discharges, urban runoffs, and a combination of the above factors. Currently, the Santa Ana River in southern Californian is impacted by one of the highest concentrations of cattle in the United States. The watershed is undergoing drastic changes. In general, the varying land uses in the middle SAR (Chino Basin) watershed include agriculture, open space, and rapidly growing urban areas [1]. In 1995, approximately 340 animal-confinement facilities having over 386,000 animals, mostly dairy cows, operated within the area that is mostly drained by Chino, Cypress, and Cucamonga Creeks. Pollutants in the watershed mainly consist of pathogens and nutrients due to the densely populated areas, agricultural activities, and urban and storm-water runoff in the region. Different federal, state, and private agencies have monitored fecal bacterial composition in the surface water [1,2], but little has been done to determine the different *Escherichia coli* strains within the water bodies.

E. coli are very diverse in the environment with about 173O, 103K, and 56H antigen and the numbers of newly discovered antigens is increasing [3]. Most *E. coli* are nonpathogenic, but there are some such as *E. coli* O157:H7 that cause human diseases such as hemorrhagic colitis (HC) and hemolytic uremic syndrome (HUS). In addition to *E. coli* O157:H7, there are other *E. coli* pathogroups that causes diseases in human such as enteropathogenic *E. coli* which causes diarrhea in children especially in developing countries, enterotoxigenic *E. coli* which causes traveler's diarrhea and others [3]. There is an extensive review of sources of pathogenic *E. coli* in the environment [4], but their distributions in urban water has been limited to very few studies [5,6]. Due to the increasing urbanization and the large number of cattle in the studied watersheds, the health risk from pathogenic *E. coli* is a major concern to drinking water quality. There is virtually no information on the occurrence of pathogenic *E. coli* in the middle Santa Ana River watershed despite the high concentration of cattle in the watershed.

Most pathogenic *E. coli* are commonly carried by healthy cattle in their feces. The fecal excretion of these organisms by cattle appears to be seasonal, with excretion rates highest in spring and late summer [7,8]. This studies sought to characterize pathogenic *E. coli* isolates obtained in terms of their virulence profiles, and pulsed field gel electrophoresis (PFGE) genomic profiles. PFGE DNA banding patterns, in conjunction with virulent factors, may assist in the epidemiologic tracing of pathogenic *E. coli* isolates of medical concern. The goals of this study were to determine the distribution, diversity, and antimicrobial activities of pathogenic *E. coli* isolates from low flowing river water and sediment with inputs from different sources before water is discharged into ground water and to compare microbial contamination in water and sediment at different sampling sites. We also incorporated the evaluation of fecal bacterial contamination of drinking water aquifer sand material at a specific site that receives water from the above sources before discharge into ground water.

2. MATERIALS AND METHODS

2.1. Study Area and Sample Collection

The study area is as previously described [9], which outlines the sampling points and their relative locations (Figure 1; Table 1). Surface water samples and sediment were collected from a natural/open-space location (S1, M1) to evaluate bacterial contributions from natural or background source. The S1 sampling point is located in National Forest land. Effluent from three wastewater treatment plants (S11WW, S13 and S14) were also analyzed (Table 1). All sampling locations and their land use types are listed in Table 1. Reference samples were taken quarterly for 12 months. All samples were collected at the water surface in duplicate in sterile recipients, and sediments from the bank of the river, stored at 4 °C, and analyzed within 6–8 hours. Sediment samples were collected in duplicate from the river banks with a stainless steel instrument and analyzed within 24 hours.

The topography in the Santa Ana River watershed ranges from steep, rugged mountains with peaks as high as 3,261 m above sea level, to a broad alluvial-filled valley, bordered by the San Gabriel/San Bernardino Mountains to the northeast and the elevated Perris Block/San Jacinto Mountains to the south (Figure 1). The Santa Ana River is the main tributary draining the valley. The Chino Basin study area, located in the northwestern part of this watershed, was formed as a result of tectonic activity along major fault zones [10]. The bottom of the basin—the effective base of the freshwater aquifer—consists of relatively impermeable sedimentary and igneous bedrock formations that are exposed at the surface in the surrounding mountains and hills. Sediments eroded from the surrounding mountains have filled the Chino Basin, providing reservoirs for ground water. In the deepest portions of the basin, these sediments are greater than 304 m thick. The sediments consist of geologically old and young alluvium. The thickness of the older alluvium varies from about 60 m near the southwestern end of the Chino Basin to over 334 m in the eastern parts of the valley and averages about 152 m throughout the basin.

The surface outcrop is commonly distinguishable by its red-brown or brick-red color and is generally more weathered than the overlying younger alluvium. The younger alluvium occupies streambeds, washes, and other areas having recent sedimentation. The thickness of the younger alluvium varies from over 30 m near the mountains to just a few meters in the center of the valley. The younger alluvium generally covers most of the northern half of the Chino Basin in undisturbed areas [10].

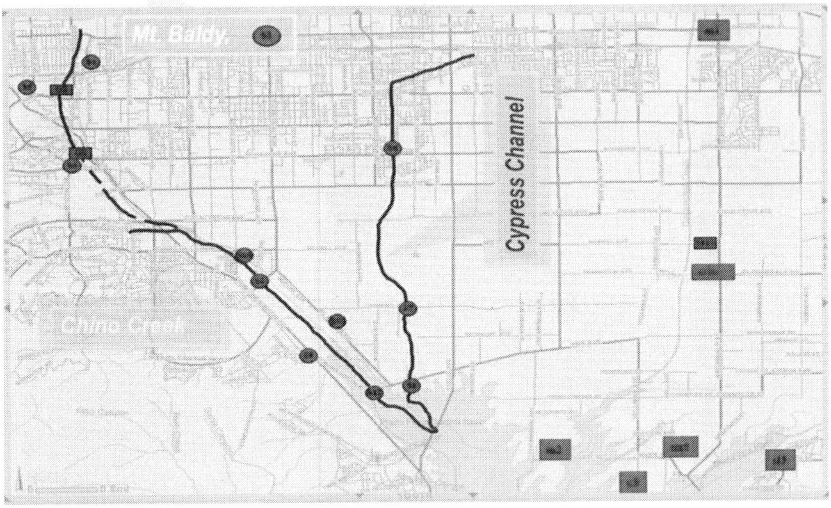

Figure 1. Sites used for the study along Chino Basin also known as the middle Santa Ana River Watershed (MSAR). Chino Creek and Cypress channels are the two main channels in the basin with inputs from urban and agricultural activities, respectively. Both Creeks empty into the Santa Ana River.

The stratigraphy of the Chino Basin can be described by two natural divisions: (1) the pervious formations that comprise the ground-water reservoirs are termed the water-bearing sediments and (2) the less pervious formations that enclose the ground-water reservoirs are termed the consolidated bedrock. The consolidated bedrock is further differentiated as metamorphic and igneous rocks of the basement complex partially overlain by consolidated sedimentary rocks. The water-bearing sediments overlie the consolidated bedrock, with the bedrock formations coming to the surface in the surrounding hills and highlands.

Most recharge to the ground-water reservoirs of the Chino Basin is from percolation of direct precipitation and infiltration of stream flow within tributaries exiting the surrounding mountains and hills and within the Santa Ana River. Potential sources of recharge in the Chino Basin include the following: (1) infiltration of flow (and, locally, imported water) within unlined stream channels overlying the basin, (2) infiltration of storm water flow and municipal wastewater discharges within the channel of the Santa Ana River, (3) underflow from the saturated sediments and fractures within the bounding mountains and hills, (4) artificial recharge at spreading grounds of storm water, imported water, and recycled water, (5) underflow from seepage across the bounding faults, (6) intermittent underflow from adjacent basins, and (7) deep percolation of precipitation and returns from use [10].

Table 1. Sampling locations for MSAR pathogen source evaluation study. *

Site #	Site locations	Land use	Geographic positioning system (GPS)
S1	Ice House Canyon	Open Space	N34° 15.057 min.; W117° 37.977 min; 1,447 m elevation
M1	Cucamonga Creek. at OCWD Ponds	Open Space	San Bernardino County Flood Control District (SBCFCD)
S2	Chino Creek at Central Ave.	Urban runoff	N33° 58.420 min.; W117° 41.302 min; 174 m elevation
S3	Chino Creek at Schaefer Ave.	Urban runoff	N34° 0.246 min.; W117° 43.628 min; 207 m elevation
S4	San Antonio Wash at County Drive	Urban runoff + Commercial wash out	N30° 1.543 min.; W117° 43.652 min; 222 m elevation;
S5	Chino Creek. at Riverside Drive	Urban runoff	N34° 1.144 min.; W117° 44.204 min; 207 m elevation;
S6	Cypress Channel at Schaefer Ave.	Agricultural Runoff	N34° 0.262 min.; W117° 39.766 min 208 m elevation;
S7	Cypress Channel at Kimball Ave.	Agricultural Runoff	N33° 58.113 min.; W117° 39.624 min 177 m elevation;
S8	Cypress Channel at Golf Course	Agricultural Runoff	N33° 57.057 min.; W117° 39.555 min; 160 m elevation;
S9	Big League Dreams at storm drain	Urban runoff	N33° 57.364 min.; W117 °40.788 min; 163 m elevation;
S11ww	Cucamonga Creek at Regional Water Recycling Plant #1	Effluent from wastewater treatment plant	N34°; 1.853 min; W117° 35.946 min; Altitude:246 m
S11ur	Cucamonga Creek at Regional Water Recycling Plant #1	Urban runoff+ wastewater	N34°; 1.853 min; W117° 35.946min; Altitude:246 m.
S12	Chino Creek at Pine Ave.	Urban runoff+ wastewater	N33° 56.941 min.; W117° 39.986 min; 155 m elevation;
S13	Inland Empire Utilities Agency (IEUA) Regional Water Recycling Plant #5	Effluent from wastewater treatment plant	N33° 57.840 min.; W117 °40.826 min; 180 m elevation;

S14	IEUA Carbon Canyon Waste Reclamation Facility (CCWRF)	Effluent from wastewater treatment plant	N33 °58.799 min.; W117° 41.655 min; 184 m elevation;
ST2	Santa Ana River at Prado Dam	Urban Runoff	N33°; 54.737 min; W117° 38.711 min Altitude: 141 m.
C3	Prado Park outlet	Urban Runoff+ waste water discharge	N33°; 56.402 min; W117° 38.763min, 166 m
ST5	Santa Ana River at River road	Urban Runoff	N33°; 55.405 min; W117° 35.894 min Altitude:155 m.
M5	OCWD (Prado) Wetlands Effluent	Wetland treated (bacteria loaded) Orange County Water District (OCWD	N33°; 54.737 min; W117° 38.711 min Altitude: 141 m.

* From Ibekwe *et al.* [9] with slight changes.

Warm, dry summers and cool, moist winters characterize the climate of the study area. Average annual precipitation ranges from about 450 mm in the lower part of the Chino Basin to about 1,000 mm in the San Gabriel Mountains. Most precipitation occurs during the winter rainy season between November and March. Average precipitation for January ranges from about 76 mm in the southern valley to more than 254 mm in the mountains, whereas average precipitation for July is less than 2.54 mm across the whole basin [11]. The spatial distribution of average monthly precipitation is similar for most months and is characterized by the topographic effect of the San Gabriel Mountains. For example, although average November precipitation has a spatial distribution that is similar to the January pattern, the average January precipitation is over 30% greater than that for all other months. Air temperatures across the basin in the winter are cool, with average daily temperatures in January ranging from about 2 °C in the north, resulting in persistent mountain snowpack at the higher altitudes, to as high as 15 °C in the southern valley. Average daily air temperatures for July can be quite warm, ranging from 18 to 27 °C.

2.2. Enumeration of TC, FC, and E. coli from Chino Basin

Water samples were processed in the laboratory within six hours of sample collection. All water samples were transported on ice to the laboratory and analyzed by adding 100 mL of water sample to a Colilert vessel and processing following the manufacturer's protocol. TC, FC, and *E coli* populations were enumerated and expressed as Most Probable Number (MPN/100mL). For isolation of *E. coli* colonies from Colilert vessels, 100 μL liquid sample was removed from positive wells, then spread plated onto Chromagar ECC agar (CHROMagar Microbiology, Paris, France), and was incubated at 37 °C for 24 h. Individual colonies of pure cultures that were isolated were stored at –80 °C for further characterization. Moist sediment samples (10 g) were diluted with 90 mL of phosphate buffered saline (PBS) water (0·0425 g/L KH_2PO_4 and 0·4055 g/L $MgCl_2$) and shaken for 15 minutes. Ten mL of the suspension was added to Colilert vessel, diluted 1:10 and mixed. One mL from the 1:10 dilution was transferred to another vessel and was further diluted 1:1,000, and an aliquot was added to the Colilert media, mixed, then sealed in QuantiTrays and incubated at 37 °C for 24 h. Samples were processed following the manufacturer's protocol in accordance with method 9223 [12].

2.3. Isolation of Pathogenic E. coli from Chino Basin

One gram or 1 mL of environmental samples was added to 9 mL of PBS, vortexed briefly, serially diluted and plated for the enumeration of *E. coli* O157 on Harlequin cefixime-tellurite sorbitol MacConkey (CT-SMAC) agar with BCIG (5-bromo-4-chloro-3-indoxyl-β-D-glucuronide) containing 0.05 mg of cefixime L^{-1} and 2.5 mg of tellurite L^{-1} (LAB M: IDG–Lancashire, UK). The plates were incubated at 37 °C for *E. coli* O157 for 16 h. Six sorbitol-negative, translucent colonies per sample were tested by multiplex PCR to determine the presence of the three genes. Additionally, isolates that were sorbital positive or β-glucuronidase positive (red/pink colonies with a purple center or green colonies) were enumerated as other *E. coli* or non O157 or presumptive pathogenic *E. coli.* The presumptive pathogenic *E. coli* isolates were probed for the following genes: heat labile toxin (LT), heat stable toxins a & b (STa/STb), shiga-like toxins 1 & 2 (*stx*1/*stx*2), cytotoxin necrotizing factors 1 & 2 (*cnf*1/*cnf*2, intimin (*eaa*), O and H types [13]. These analyses were performed at the *E. coli* Reference Center (The Pennsylvania State University, University Park, PA, USA).

2.4. Pulsed-Field Gel Electrophoresis

XbaI PFGE analysis was carried out as previously described [14]. Comparison of digested profiles to identify restriction enzyme digestion pattern clusters (REPCs) was performed with the BioNumerics software, version 5.0 (Applied Maths, Austin, TX, USA). Fingerprints were clustered by using the Jaccard coefficient evaluated by the unweighted-pair group method (UPGMA). A tolerance and optimization of 1% was allowed to account for gel-to-gel differences. Isolates were considered to be related and to belong to the same PFGE cluster if their similarity index was >85%, according to Tenover's criteria (≤6 bands of difference) [15].

2.5. Antimicrobial Resistance

Antimicrobial susceptibility tests (phenotypes) of pathogenic *E. coli* isolates were done using a disk diffusion assay following CLSI standards [16]. Mueller-Hinton II agar (Difco) was used and cells were harvested from the surface of the medium with a cotton swab after 24 h growth at 37 °C. Cells were suspended in sterile saline (0.85% NaCl) and cell density was adjusted to a 0.5 McFarland turbidity standard. The diluted cells were plated, and then antibiotic disks were placed on them. Following incubation (24 h at 37 °C), zone sizes (diameter) were measured to two decimal points and were used for quantitative analysis. Isolates resistant to two or more antimicrobials were defined as multiply drug resistant. *E. coli* ATCC 25922 (American Type Culture Collection, Manassas, VA, USA) was included in each assay as a negative control strain. Antimicrobial agents were tested with BD BBL Sensi-Disc antimicrobial susceptibility test discs (Becton Dickinson & Co., Sparks, MD, USA) with the breakpoints (μg mL^{-1}) indicated as follows: amoxocillin/clavulanic acid, 20/10 μg, ampicillin, 10 μg, cephalothin, 30 μg, erythromycin, 15 μg, rifampin, 5 μg, streptomycin, 10 μg, and tetracycline, 30 μg.

Multiplex PCR screens were performed on the isolates using genes encoding for ampicillin resistance (bla_{TEM}), tetracycline resistance (*tet* A, *tet* B, and *tet* C), and streptomycin resistance (*aadA*I). Details of primers, annealing temperatures, and amplicon sizes are as previously provided [9,17,18].

2.6. Sampling Collection During Sand Filtration Experiment

Water samples were collected from the Santa Ana River at Orange County Water District Field Station. The water consists of source water (water from the river) and filtrate water (water from aquifer sand material after passing through a sand filtration system).This process was repeated three times to determine reliability of data. The experiment was conducted in a 1.2 × 1.2 × 1.8 m filtration tank built with stainless steel outside the field station. Aquifer sand material was heterogeneous native lake sediment that had been processed through a sand washing plant to remove the majority of silt and clay particles [19,20]. The material was trucked to the station and packed into the tanks. Samples from aquifer material were collected on day 15 at the end of each experiment. Source water was obtained from the Santa Ana River water stored in an artificial lake. The water ran through a 1 inch PVC pipe into the sand filtration unit. The water runs another 1 m through the filtration tank containing aquifer materials and collected at the outlet for analysis of fecal bacteria.

2.7. Enumeration of Heterotrophic Bacteria and E. coli in Water Before and After Sand Filtration

Water sample was collected in 1-L sterile bottles, transported on ice to the laboratory, and processed within 6 h using standard procedure [21]. Various dilutions and volumes were filtered with the goal of achieving 30–300 colonies per dilution. Surface water samples was vortexed and volumes of 100, 10 and 1 mL were filtered in phosphate buffered saline (PBS) water (0.0425 g/L KH_2PO_4 and 0.4055 g/L $MgCl_2$) to obtain the best sample conditions. Tenfold and 100-fold dilutions were also prepared in PBS, vortexed, and 1 mL of each dilution was filtered in duplicate. Volumes of 1 mL, 10 mL and 100 mL (via membrane filtration) were plated onto tryptic soy agar (TSA) (for heterotrophic plate counts [HPC]) and sorbitol-MacConkey agar (SMAC-BCIG without cefixime-tellurite) for *E. coli*, and incubated at 37 °C for 24 h and colonies were enumerated. *E. coli* strains are β-glucuronidase positive and/or sorbitol positive, so produce pink/red colonies with a purple center, or green colonies (some may be translucent with a green center).

2.8. Statistical Analysis

All analyses (PFGE, culture isolation and PCR) were performed in triplicate, and the data shown in the graphs are the average of three separate measurements conducted. Thus, an analysis of variance (ANOVA) was conducted with $\log_{10}$-transformed density of *E. coli* bacteria using SAS version 9.1 [22] to determine statistical significant differences using Tukey's studentized range (HSD) test for mean separation.

3. RESULTS

3.1. Fecal Indicator Bacterial Concentrations in Chino Basin

Indicator bacteria in sediment and surface water were determined on 447 water and sediment samples collected from 19 sites over a 12-month period. Counts ranged from undetectable (detection limit 1 MPN $100mL^{-1}$) in the surface water to 2.5×10^4 MPN $100g^{-1}$ in the sediment. Basic univariate summary statistics for TC, FC, *E. coli* and pathogenic *E. coli* counts included in Figure 2 are from sediment samples because most of the potentially pathogenic *E. coli* were recovered from sediment samples. The statistics summarized the $\log_{10}$ transformed counts for each indicator bacterial group. Total coliform counts were the highest, and with the greatest variability in concentrations. Presumptive pathogenic *E. coli* were small in numbers and most of the times below 10 cfu/g.

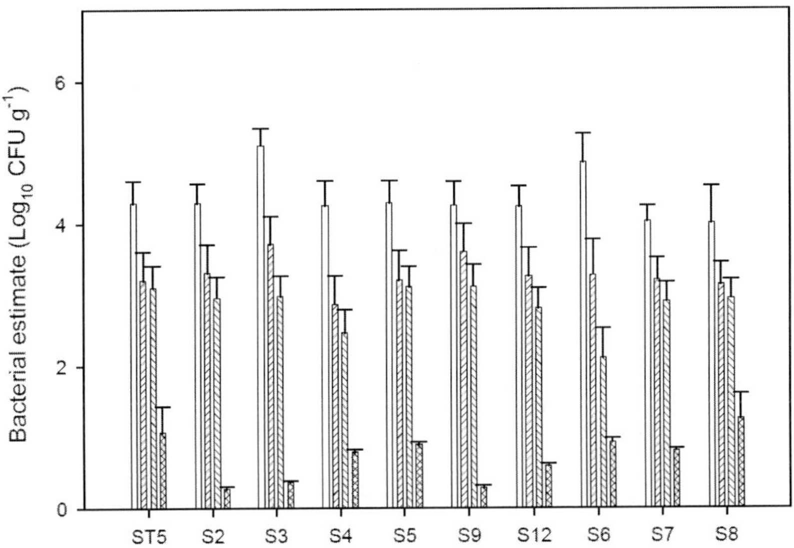

Figure 2. Concentrations of indicator bacteria in sediment on various sampling points along the major sources. Samples ST5, S2, S3, S4, S5, S9, and S12 are from urban runoff and samples S6, S7, and S8 are from agricultural inputs. Symbols □ total coliform (TC), ▨ fecal coliform (FC), ▧ *E. coli*, ▩ pathogenic *E. coli*. Error bars represent standard errors of three replicate samples. Only samples with potential pathogenic *E. coli* are shown.

3.2. Characterization Pathogenic E. coli

A total of about 389 isolates from the selected sites in Figure 2 were screened on CT-SMAC agar and these isolates were rescreened by PCR for various virulence factors. Only 17 isolates showed the presence of a combination of these factors (Table 2). These isolates were sent to Pennsylvania State University *E. coli* Reference Center for complete typing (Table 2). All strains were classified as non-O157. Two isolates were classified as *E. coli* O11:H40 and in 15 the O serotype could not be determined, but were identified as H11.

Table 2. Antimicrobial susceptibility, genetic variations, and resistant genotypes of selected isolates of presumptive pathogenic *E. coli*. *

Sample Name	Amox	Amp	Ceph	Eryth	Rif	Strep	Tet	LT	STa	STb	CNF1	CNF2	O	H	stx2	stx1	eae	bla_{TEM}	*aadA*I	*tetA*	*tetB*	*tetC*
S8-1	S	S	S	I	R	S	S	-	-	-	-	-	11	40	-	-	-	-	-	-	-	-
S8-2	S	S	S	I	R	S	S	-	-	-	-	-	11	40	-	-	-	-	-	-	-	-
S8-3	S	R	I	I	R	S	R	-	-	-	-	-	-	11	-	-	+	-	-	-	-	-
S8-4	S	S	S	I	R	S	S	-	-	-	-	-	-	11	-	-	+	-	-	-	-	-
S8-5	S	S	S	I	R	S	R	-	-	-	-	-	-	11	-	-	+	-	-	-	-	-
S8-6	S	S	S	R	R	S	S	-	-	-	-	-	-	11	-	-	+	-	-	-	-	+
ST5-1	S	R	S	R	R	S	S	-	-	-	-	-	-	11	-	-	+	-	-	-	-	+
ST5-2	S	R	S	R	R	S	S	-	-	-	-	-	-	11	-	-	+	-	-	-	-	+
ST5-3	S	S	R	R	R	S	S	-	-	-	-	-	-	11	-	-	+	-	-	-	-	-
ST5-4	S	R	S	R	R	S	S	-	-	-	-	-	-	11	-	-	+	-	-	-	-	+
ST5-5	S	R	S	R	R	S	S	-	-	-	-	-	-	11	-	-	+	-	-	-	-	+
ST 5-6	S	S	S	R	R	S	S	-	-	-	-	-		11	-	-	+	-	-	-	-	+
ST5-7	S	S	S	R	R	S	S	-	-	-	-	-	-	11	-	-	+	-	-	-	-	+
ST5-8	S	S	S	R	R	S	S	-	-	-	-	-	-	11	-	-	+	-	-	+	-	+
ST 5-9	S	S	S	R	R	S	R	-	-	-	-	-	-	11	-	-	+	-	-	-	-	+
ST5-10	S	S	S	I	R	S	S	-	-	-	-	-	-	11	-	-	+	-	-	-	-	-
S12	S	R	R	I	I	I	S	-	-	-	-	-	-	11	-	-	+	-	-	-	-	-

* S = antimicrobial susceptibility, R = resistance, I = intermediate; - undetectable and + detectable genes by PCR. The six different isolates from site S8 are sediment samples from April 2005 and the ST5 from three sampling days.

Seven antibiotics were used for susceptibility tests of the 17 isolates. Resistant phenotypes were determined for the 17 isolates from CT-SMAC plates. All isolates were resistant to rifampacin except S12, followed by erythromycin and ampicillin. Only isolates from ST5 sites showed multiples resistances to rifampacin, erythromycin, and ampicillin. Site S8 also showed multiple resistances to ampicillin, rifampicin, and tetracycline. Site S8 is located along Cypress channel and this site is highly impacted by agricultural (dairy) activities.

Sites ST5 receives water from mostly urban sources from the watershed before it flows through the Prado wetland. One isolate from S8 and seven from ST5 were resistant to *tet*C genes based on positive PCR signal. Genes for ampicillin resistance (bla_{TEM}) and streptomycin resistance (*ant3")-Ia* (also called *aadA1*) were not detected in any of the samples that showed resistant phenotype, however, *tet*C gene was detected in most of the samples that were susceptible to tetracycline.

3.3. Genetic Variations among E. coli Isolates from Watershed

The 17 isolates were characterized by pulsed field gel electrophoresis using XbaI restriction endonucleases (Figure 3).

Most of the restriction endonuclease digestion profiles clustered within specific sources. Most of the isolates from S8 site along Cypress channel clustered together with one exception. This isolates were collected on April 2004. This site is mainly agricultural source. The other cluster was isolates from urban runoff that flows through Prado wetland. This shows that *E. coli* isolates from the different sources have different PFGE profiles.

3.4. Fecal Indicator Bacterial Levels in Source Water and Aquifer Sand Material During Sand Filtration

There were no differences in the levels of heterotrophic bacteria as determined by plate count in the source water (influent) and the filtration (water that has gone through sand filtration (Figure 4). Significantly higher levels ($P < 0.001$) of HPC were found in water samples in late May and June than in April and early May. There were significant ($P = 0.05$) higher numbers of *E. coli* in source water in May and early June than in April. After water has gone through filtration tanks containing aquifer sand material, there was a 1 to 2 log reduction in *E. coli* in aquifer sand tank. This showed that the filtration unit with aquifer material had limited impact on *E. coli* population.

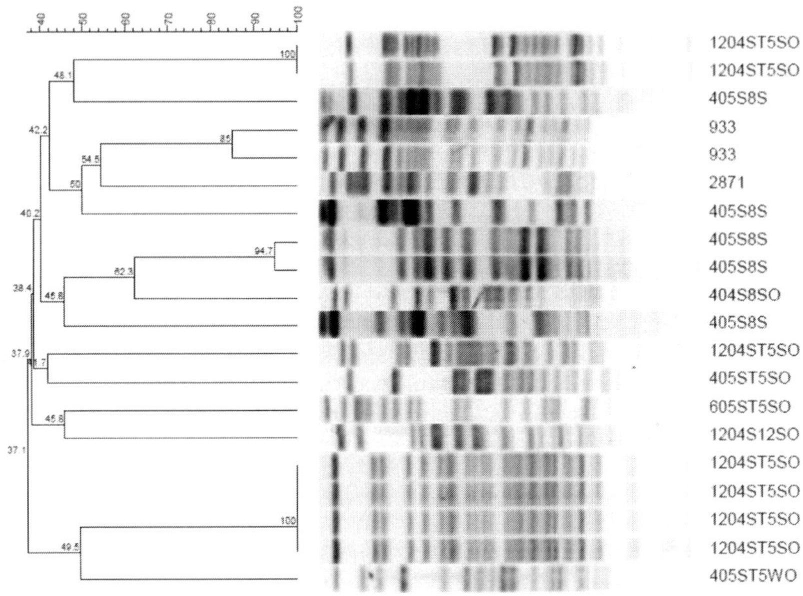

Figure 3. Representative PFGE fragment patterns and dendrogram analysis of potentially pathogenic antimicrobial-resistant *E. coli* stains from urban watershed impacted by different sources of pollutants. Most of the *E. coli* isolates were obtained from sediment except one that was obtained from surface water (405ST5 WO). Sample identifications are as follow; e.g., 405S8S and 1204ST5SO indicates month (4 or 12) followed by year (05 or 04), site description (S8 (1 to 6) or ST5-1 to 10 from Table 2), and sample source SO or S (sediment) or water (WO). 933 and 2,871 are pathogenic and nonpathogenic *E. coli* controls.

4. DISCUSSION

Examination of each site throughout the watershed indicated that indicator bacterial concentrations along Chino Creek and Cypress channel routinely exceeded the applicable water quality objectives. The exception was TC in the control sites (S1, M1) and WWTPs. The same trends were observed for fecal coliform [23]. Therefore, Figure 2 presents data mainly from the sediment of these two channels (Chino and Cypress) to illustrate the major sources of pollution to the watershed. From our previous studies, peak concentrations of *E. coli* depended more on larger storms and on pervious-area bacteria sources and loading rates [24]. Only the larger storms generated runoff, and thus bacteria wash off, on areas in the Chino Basin. Land uses that were assigned the highest bacteria-loading

(Chino Creek and Cypress channel) values affected the time-averaged bacteria loads and the frequency of concentrations exceeding 235 cfu/100mL [24]. Therefore land use was the major factor affecting the concentration of *E. coli* in the Chino Basin area of the watershed. In contrast, Prado Park and open space land use areas had a significant decrease in the frequency that bacteria concentrations in waterways exceeded 235 cfu/100mL.

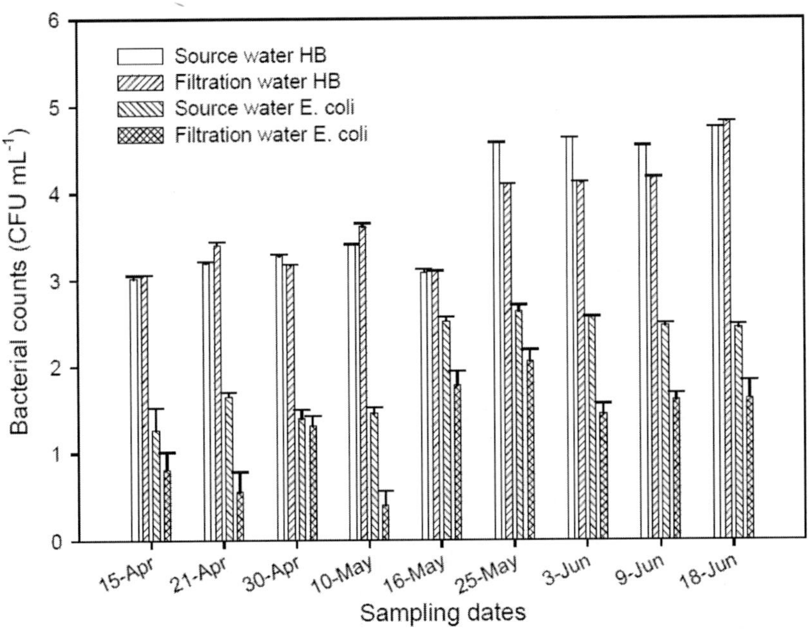

Figure 4. Levels of heterotrophic bacteria (HB) and *E. coli* as determined by plate count in the source water and after filtration through aquifer sand material. All samples were taken in April through June 2004.

The continuous and increasing use of antibiotics has led to the emergence of pathogenic bacteria that are resistant to many antibiotics [25]. In this study, more that 50% of our isolates were resistant to *tet*C gene and three isolates carry resistant phenotype. Since tetracycline resistance genes are located on the mobile genetic elements, they are transmissible between bacteria. Two of these isolates are from agricultural sources and one is from urban source. Also, eight isolates from urban sources carry *tet*C resistance gene while one from agricultural sources carried this resistant gene. Therefore, most fecal bacteria from human or agricultural sources released into the environment may carry antibiotic resistance genes [26].

Their fate and the transfer of antibiotic resistances by gene transfer to other bacteria are of great concern to human health [27]. The great threat to drinking water of antibiotic resistant bacteria may be the high concentrations of such bacteria in the source water that may result in the transfer of genetic elements from nonpathogenic to pathogenic strains. This was confirmed by our recent study with 600 isolates of generic *E. coli* from the same watershed [9]. Resistance genes are often associated with integrons or mobile DNA elements such as plasmids and transposons that facilitate the spread of resistance genes [28,29,30]. More often, there is a linkage between many of these resistance genes on mobile elements and the distribution of antibiotic resistant bacteria in the environment [17,18,31]. We did not study the exact mechanisms of resistance in the current work; however previous molecular studies have shown strong statistical associations between resistance genes [32,33]. No isolate showed resistance to amoxicillin-clavulanic acid and this was in agreement with our previous study using 600 generic *E. coli* [9] that showed less that 2% were resistant to this antibiotic from urban runoff and none from agricultural sources. The correlation between antimicrobial resistance and the presence of antibiotic resistant genes was better for Streptomycin than than tetracycline and ampicillin. *aad*AI gene for streptomycin was not detected in any of our samples and this corresponded to all our isolates being susceptible to streptomycin (Table 2). However, such a correlation was not observed with ampicillin and tetracycline.

Pathogens with increased resistances may be transported from the animal or human via feces or other mechanism into rivers and groundwater [11,34] where the water is use as a source for domestic water supply. In this watershed, there are networks of channels and creeks that surface water are transported to the Santa Ana River. Through this process antibiotic resistant bacteria may be transported from human or animal sources to the river that is subsequently used to recharge ground water for domestic water use. In a study to determine the impact of nontherapeutic use of antibiotics on swine manure-impacted water sources, surface water and groundwater situated up and down gradient from a swine facility were assessed for antibiotic-resistant enterococci and other fecal indicators. As expected, the median concentrations of enterococci, fecal coliforms, and *Escherichia coli* were 4 to 33 fold higher in down-gradient *versus* up-gradient surface water and groundwater [5]. Higher amounts of erythromycin- and tetracycline-resistant enterococci were detected in down-gradient surface waters. These findings demonstrated that water contaminated with swine manure could contribute to the spread of antibiotic resistance in the environment. Recently, Ibekwe *et al.* [9] found in the watershed used for this study that more *E. coli* with higher multiple resistant phenotypes were present in water samples from urban sources that from agriculture sources. These isolates were also more diverse genetically that isolates from agricultural sources. Therefore, there is no doubt whether there are fecal bacterial is drinking water sources, but the question is how we manage such systems that these bacteria are not in the actual drinking water. This is achievable in developed countries but this is a serious problem in developing countries because of inadequate water treatment plants.

The microbiological data provided in this study can help water utility companies in their understanding of source water quality and help them in the processing of tertiary treated water that may be subsequently available for domestic use. After water has gone through the filtration tanks containing aquifer material, there were reductions in *E. coli* population. The data presented here demonstrates high level of indicator bacteria in the river as it continues to the ocean. Our study also suggest that when surface water is diverted through aquifer sand material significant reduction in fecal bacterial population occurs as the water passes through aquifer sand material by a natural filtration process. Part of the source water (Santa Ana River) used for this study has gone through tertiary treatment and wetlands before flowing into the artificial lakes at the Orange County Water District field station. Water from these lakes is subsequently used for ground water recharge and discharge. After this process the water is treated further for domestic use by over 2 million residence and businesses. Part of the Santa Ana River continues to flow and empties into the Pacific Ocean near Huntington Beach. In southern California, it is well recognized that a major cause of bacterial pollution of coastal waters is urban runoff in rivers/channels and storm drains that discharge into the ocean [35,36]. In a recent paper enumerating enterococci in marine and intertidal sediments [37], high densities of fecal indicator bacteria were reported in Santa Ana River near Huntington Beach. These authors indicated that shoreline waters at Huntington State Beach may be recipients of fecal indicator bacteria originating from intertidal sediments in the Santa Ana River that contain high levels of bacteria.

5. CONCLUSIONS

In this study no *E. coli* O157 was found in the surface water or sediment despite the fact that the study area has one of the highest concentrations of cattle in the United States and cattle is the main reservoir of pathogenic *E. coli.* The concern is that the presence of antibiotic resistant non pathogenic *E. coli* in this watershed is a matter of great concern because of possible horizontal gene transfer from generic to pathogenic *E. coli* which may lead to increased duration and severity of morbidity.

ACKNOWLEDGEMENTS

This research was supported by the 206 Manure and Byproduct Utilization Project of the United States Department of Agriculture-Agriculture research Service (USDA-ARS). Mention of trademark or propriety products in this manuscript does not constitute a guarantee or warranty of the property by the USDA and does not imply its approval to the exclusion of other products that may also be suitable.

REFERENCES

1. Izbicki, J.A.; Pimentel, M.I.; Leddy, M.B. Microbial and Dissolved Organic Carbon Characterization of Storm Flow in the Santa Ana River at Imperial Highway, Southern California, 1999–2002; Scientific Investigation Report 2004-5116; U.S. Geological Survey: Reston, VA, USA, 2004.
2. Rice, B.W. Staff Report on Bacterial Indicator Total Maximum Daily Loads in the Middle Santa Ana River Watershed; California Regional Water Quality Control Board: Santa Ana, CA, USA, 2005.
3. Ishii, S.; Sadowsky, M.J. Escherichia coli in the environment: Implication for water quality and human health. Microbes Environ. **2008**, 23, 101–108.
4. Kaper, J.B.; Nataro, J.P.; Mobley, H.L.T. Pathogenic Escherichia coli. Nature Rev. **2004**, 2, 123–140.
5. Abraham, W.-R. Megacities as sources for pathogenic bacteria in rivers and their fate downstream. Int. J. Microbiol. **2011**.
6. Shelton, D.R.; Karns, J.S.; Higgins, J.A.; Van Kessel, J.S.; Perdue, M.L.; Belt, K.T.; Russell-Anelli, J.; DebRoy, C. Impact of microbial diversity on rapid detection of enterohemorrhagic Escherichia coli in surface waters. FEMS Microbiol. Lett. **2006**, 261, 95–101.
7. Hancock, D.D.; Besser, T.E.; Rice, D.H.; Herriott, D.E.; Tarr, P.I. Longitudinal study of Escherichia coli O157 in fourteen cattle herds. Epidemiol. Infect. **1997**, 118, 193–195.
8. Sherman, P.; Soni, R.; Karmali, M. Attaching and effacing adherence of Vero cytotoxin-producing Escherichia coli to rabbit intestinal epithelium in vivo. Infect. Immun. **1988**, 56, 756–761.
9. Ibekwe, A.M.; Murinda, S.E.; Graves, A.K. Genetic diversity and antimicrobial resistance of Escherichia coli from human and animal sources uncovers multiple resistances from human sources. PLoS One **2011**, 6.
10. Wildermuth Environmental, Inc, Optimum Basin Management Program—Chino Basin Dry-Year Yield Program; Prepared for Chino Basin Watermaster, Inland Empire Utilities Agency; Wildermuth Environmental, Inc.: Chino, CA, USA, 2003; pp. 1–141.
11. Daly, C.; Gibson, W.P.; Doggett, M.; Smith, J.; Taylor, G. Up-to-date monthly climate maps for the conterminous United States. In Proceedings of the 14th Conference on Applied Climatology, 84th American Meteorological Society Annual Meeting, Seattle, WA, USA, 13–16 January 2004.

12. Eaton, A.D.; Clesceri, L.S.; Greenberg, A.E.; Franson, M.A.H. Standard Method for Examination of Water and Wastewater; American Public Health Association: Washington, DC, USA, 1998.
13. DebRoy, C.; Maddox, C.W. Identification of virulence attributes of gastrointestinal Escherichia coli isolates of veterinary significance. Anim. Health Res. Rev. **2001**, 1, 129–140.
14. Mora, A.; López, C.; Dabhi, G.; Blanco, M.; Blanco, J.E.; Alonso, M.P.; Herrera, A.; Mamani, R.; Bonacorsi, S.; Moulin-Schouleur, M.; Blanco, J. Extraintestinal pathogenic Escherichia coli O1:K1:H7/NM from human and avian origin: Detection of clonal groups B2 ST95 and D ST59 with different host distribution. BMC Microbiol. **2009**, 9.
15. Tenover, F.C.; Arbeit, R.D.; Goering, R.V.; Mickelsen, P.A.; Murray, B.E.; Persing, D.H.; Swaminathan, B. Interpreting chromosomal DNA restriction patterns produced by pulsed-field gel electrophoresis: criteria for bacterial strain typing. J. Clin. Microbiol. **1995**, 33, 2233–2239.
16. Clinical and Laboratory Standard Institute, Performance Standards for Antimicrobial Susceptibility Tests—Ninth Edition: Approved Standard M02-A9; Clinical and Laboratory Standards Institute: Wayne, PA, USA, 2006.
17. Gow, S.P.; Waldner, C.L.; Harel, J.; Boerlin, P. Associations between antimicrobial resistance genes in fecal generic Escherichia coli isolates from cow-calf herds in Western Canada. Appl. Environ. Microbiol. **2008**, 74, 3658–3666.
18. Sharma, R.; Munns, K.; Alexander, T.; Entz, T.; Mirzaagha, P.; Yanke, L.J.; Mulvey, M.; Topp, E.; McAllister, T. Diversity and distribution of commensal fecal Escherichia coli bacteria in beef cattle administered selected subtherapeutic antimicrobials in a feedlot setting. Appl. Environ. Microbiol.**2008**, 74, 6178–6186.
19. Ibekwe, A.M.; Lyon, S.R. Microbial characteristics through drinking water aquifer sand material. Eng. Life Sci. **2007**, 7, 81–89.
20. Ibekwe, A.M.; Lyon, S.R. Microbiological evaluation of fecal bacterial composition from surface water through aquifer sand material. J. Water Health **2008**, 6, 411–421.
21. American Public Health Association, Standard Methods for the Examination of Water and Wastewater, 19th ed; American Public Health Association: Washington, DC, USA, 1995.
22. SAS Institute Inc, 2009 SAS/STAT User's Guide; SAS Institute Inc.: Cary, NC, USA, 2009; Release 9.1.
23. EPA, Ambient Water Quality Criteria for Bacteria—1986; EPA 440/5-84-002; EPA Office of Water: Washington, DC, USA, 1986.
24. Ibekwe, A.M.; Lesch, S.M.; Bold, R.M.; Leddy, M.B.; Graves, A.K. Variations of indicator bacteria in a large urban watershed. In Proceedings of 21st Century Watershed Technology: Improving Water Quality and Environment Conference, Universidad EARTH, Costa Rica, 21–24 February 2010.
25. Call, D.R.; Borucki, M.K.; Loge, F.J. Detection of bacterial pathogens in environmental samples using DNA microarrays. J. Microbiol. Methods **2003**, 53, 235–243.
26. Allen, H.K.; Donato, J.; Wang, H.H.; Cloud-Hansen, K.A.; Davies, J.; Handelsman, J. Call of the wild: Antibiotic resistance genes in natural environments. Nature Rev. Microbiol. **2010**, 8, 251–259.
27. Normark, B.H.; Normark, S. Evolution and spread of antibiotic resistance. J. Intern. Med. **2002**, 252, 91–106.
28. Jacoby, G.A. Extrachromosomal resistance in gram-negative organisms: The evolution of beta-lactamase. Trends Microbiol. **1994**, 2, 357–360.
29. Murinda, S.E.; Ebner, P.D.; Nguyen, L.T.; Mathew, A.G.; Oliver, S.P. Anitimicrobial resistance and class 1 integrons in pathogenic Escherichia coli from dairy farms. Foodborne Pathog. Dis. **2005**, 2, 348–352.
30. Tenover, F.C.; Rasheed, J.K. Genetic methods for detecting antimicrobial and antiviral resistance genes. In Manual of Clinical Microbiology, 7th, Murray, P.R., Baron, E.J., Pfaller, M.A., Tenover, F.C., Yolken, R.H., Eds.; American Society for Microbiology: Washington, DC, USA, 1998; pp. 1578–1592.

31. Smith, D.L.; Harris, A.D.; Johnson, J.A.; Silbergeld, E.K.; Morris, J.G. Animal antibiotic use has an early but important impact on the emergence of antibiotic resistance in human commensal bacteria. Proc. Natl. Acad. Sci. USA **2002**, 99, 6434–6439. [Google Scholar]
32. Boerlin, P.; Travis, R.; Gyles, G.L.; Reid-Smith, R.; Janecko, N.; Lim, H.; Nicholson, V.; McEwen, S.A.; Friendship, R.; Archambault, M. Antimicrobial resistance and virulence genes of Escherichia coli from swine in Ontario. Appl. Environ. Microbiol. **2005**, 71, 6753–6761.
33. Travis, R.M.; Gyles, C.L.; Reid-Smith, R.; Poppe, C.; McEwen, S.A.; Friendship, R.; Janecko, N.; Boerlin, P. Chloramphenicol and kanamycin resistance among porcine Escherichia coli in Ontario. J. Antimicrob. Chemother. **2006**, 58, 173–177.
34. Aarestrup, F.M.; Kruse, H.; Tast, E.; Hammerum, A.M.; Jensen, L.B. Associations between the use of antimicrobial agents for growth promotion and the occurrence of resistance among Enterococcus faecium from broilers and pigs in Denmark, Finland, and Norway. Microbial. Drug Resist. **2000**, 6, 63–70.
35. Dwight, R.H.; Semenza, J.C.; Baker, D.B.; Olson, B.H. Association of urban runoff with coastal water quality in Orange County, California. Water Environ. Res. **2002**, 74, 82–90.
36. Reeves, R.L.; Grant, S.B.; Mrse, R.D.; Copil Oancea, C.M.; Sanders, B.F.; Boehm, A.B. Scaling and management of fecal indicator bacteria in runoff from a coastal urban watershed in southern California. Environ. Sci. Technol. **2004**, 38, 2637–2648.
37. Ferguson, D.M.; Moore, D.F.; Getrich, M.A.; Zhowandai, M.H. Enumeration and speciation of enterococci found in marine and intertidal sediments and coastal water in southern California. J. Appl. Microbiol. **2005**, 99, 598–608.

CHAPTER 9

Non-Fermentative Gram-Negative Bacteria in Drinking Water

Daiva Staradumskyte, Algimantas Paulauskas

Vytautus Magnus University, Kaunas, Lithuania

ABSTRACT

The description of a microbiological community of drinking water distribution systems is intended for a safe and proper use of drinking water. These studies were aimed at characterising the microbial condition of the drinking water supply system in Lithuania by means of culture-based methods and biochemical techniques. Samples of drinking water (in cities supplied in centralised way from taps, wells, and boreholes) were collected in different locations in Lithuania. Seeking to grow microflora present in water the membrane filtration method was applied to study the samples. Most often water samples were studied to identify coliform bacteria (*Citrobacter* spp., *Enterobacter* spp., *Klebsiella* spp., Serratia spp.), *Escherichia coli* (or faecal coliform), Enterococci (or fecal Streptococcus), Pseudomonas aeruginosa (other non-fermentative gram-negative bacteria) according to the international ISO standards. Large amounts of non-fermenting gram-negative bacteria were found in centralised urban water. The investigation showed that more than 50% of the wells under study had increased microbial contamination with faecal coliform bacteria and faecal *Streptococcus*. Bacteria of the *Pseudomonas* group, CDC group and other non-fermentative gram-negative bacteria were detected in more than 30% samples of centralised urban water studied.

Keywords :Drinking Water; Microbiology; CDC; *Pseudomonas*

1. INTRODUCTION

Drinking water is suitable for drinking if it is clean and does not contain any micro-organisms, parasites and other substances, which can pose danger to human health. To ensure the quality of drinking water, microbiological investigations of each tap are carried out. The greatest risk from water contaminated with microbes is related to consumption of drinking-water that is contaminated with human and animal excreta, although other sources and routes of contamination can also be of significance [1,2].

Groundwater is a source of drinking water in Lithuania. Publicly supplied water usually comes from deep-water aquifers which are not affected by environmental pollution very much. Microbiological properties of this water are usually good too. There are, however, exceptions, when surface water is artificially added to groundwater [3].

Almost one million people use well water in Lithuania [3]. This is the only source of drinking water for many people in Lithuania. Unfortunately, groundwater is not actually not safe to use. Investigations show that more than 50% of well water samples studied are noted for an increased level of microbial contamination, approximately 30% of all the wells and 90% of the wells located in towns or villages are contaminated with nitrates.

According to the international ISO standards, water samples are most often investigated to detect coliform bacteria (*Citrobacter* spp., *Enterobacter* spp., *Klebsiella* spp. *Serratia* spp., *Enterobacter* spp.), *Escherichia coli* (or faecal coliform) and faecal *Enterococcus*, but it was noticed that water samples contained large amounts of non-fermentative gram-negative bacteria.

Non fermentative gram-negative bacilli (non-fermenters) are ubiquitous in the environment. Non fermenters can cause a vast variety of infections [4-6]. Characteristically, these organisms fail to acidify the butt of Kligler or triple sugar iron agar or oxidative-fermentative media and grow significantly better under aerobic conditions than they do under aerobic conditions; many strains even fail to grow anaerobically [7].

Biochemically inactive bacteria of group NO-1 reproduce in drinking water and *Pseudomonas* spp. bacteria with the positive citochrom oxidase activity. Group NO-1 bacteria are oxidase negative, asaccharolytic, immobile, medium-sized gram-negative rods forming small colonies on sheep blood agar (SBA) [7]. In humans, P. *aeruginosa* is the second most frequent gram-negative nosocomial pathogen in hospitals and has the highest case-fatality rate of all hospital acquired bacteremia's [8]. *Pseudomonas* genera represents opportunistic pathogens, which disperse and adhere easily to surfaces forming a biofilm, which interferes with the cleaning and disinfection procedures in hospitals and industrial environments [9].

The main objective of these investigations was to assess the total microbial community in the samples of different drinking water (centralised tap water, well water, borehole water).

2. MATERIALS AND METHODS

Samples were taken from the following sources: tap water, well water, borehole water. All the samples were collected in Kaunas and Marijampolė regions. After collection the samples were transported to the laboratory as soon as possible. Different dilutions and volumes were filtered through membrane filter (0.45 μm pore size), which is used on a vacuum pump. The filter is placed on a petri dish containing selective agar. Membrane filters are transferred onto Tergitol 7 agar (T7A), which is used to detect coliform bacteria, when incubated at elevated temperatures (36°C ± 2°C, 44°C ± 0.5°C, 24 h). To confirm the identity of coliform bacteria, all red, yellow, orange and other colonies are cultivated in Tryptone water for the indole test [10]. Phenol Red Lactose Peptone Broth (PRLPB) for lactose degradation and gas formation test and Tryptone soy agar for oxidase test. Bacteria colonies whose oxidase test is positive and whole indole is negative are confirmed to be coliform bacteria. If Triple Sugar Iron agar and the column are not decomposed and no gas is produced, these colonies or cultivars are gram-negative rods, non-fermenters. Bacteria whose oxidase test is positive and indole is negative, Triple Sugar Iron agar and the column are not decomposed and no gas is produced, are non-fermenting gram-negative rods with positive citochrom oxidase activity. To identify the species biochemical tests (Triple Sugar Iron Agar, Lysine Iron Agar, Urease Test Broth, RapID ONE or RapID NF systems were used (Figure 1).

To detect and count *Pseudomonas aeruginosa* and other bacteria of *Pseudomonas* group membrane filters were placed on Pseudomonas detection agar (PIA). Agar is incubated at elevated temperatures (36°C ± 2°C, 24 - 48 h). All brown, rose-coloured or white colonies are counted. The colonies are lit in darkness with ultraviolet light to check if they fluorescent blue/green pigment. All fluorescent colonies are confirmed as *Pseudomonas aeruginosa*. The investigation is carried out according to LST EN ISO 16266:2008, the detection scheme is represented in Figure 2 [11].

The organism's phenotypic characteristics were similar to the asaccharolytic strains of *Acinetobacter* (Table 1) but differed in their cellular morphology and the cellular fatty acid profile.

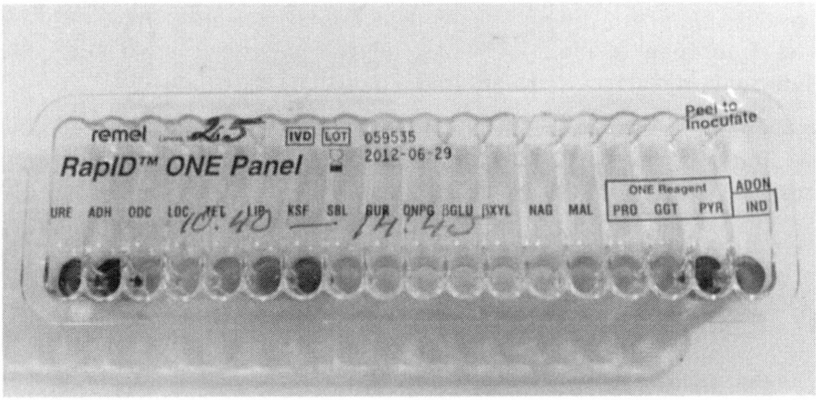

Figure 1. Rapid One System.

Pseudomonas (Figure 3) are aerobic, non-sporeforming, gram-negative rods that are straight or slightly curved. These bacteria are catalase positive and most of them are oxidase positive. With the exception of certain strains of *P. vesicularis* these organisms grow on Mac Conkey agar and appear as non-fermenters. Most strains fail to decompose glucose and most species degrade nitrate to either nitrite or nitrogen gas. Certain species have distinctive colony morphologies or pigmentation [7].

Pseudomonas aeruginosa when grown in the suitable media produces non-fluorescent bluish pigment pyocianin. Many strains also produce fluorescent green pigment pyoverdin. *Pseudomonas aeruginosa,* like other fluorescent pseudomonas, produces catalase, oxidase and amnion from arginine and can grow on citrate as the sole source of carbon [11].

3. RESULTS AND DISCUSSION

From 2010 to 2012 drinking water from different water groups—tap water, well water, borehole water—was analysed (Figure 4).

In 2010, a total of 2607 water samples were analysed to identify microbial water contamination. The heaviest bacterial contamination was detected in well water, 55.8% of intestinal *enterococci* in all the samples tested. In 2011, a total 2174 samples were analysed to identify microbial water contamination. The highest bacterial contamination was detected in well water: 56.1% of *Escherichia coli* in all the samples. In 2012, as many as 1678 samples were analysed to identify microbial water contamination.

The highest bacterial contamination was detected in well water 93.3% of coliform bacteria in all the samples tested.

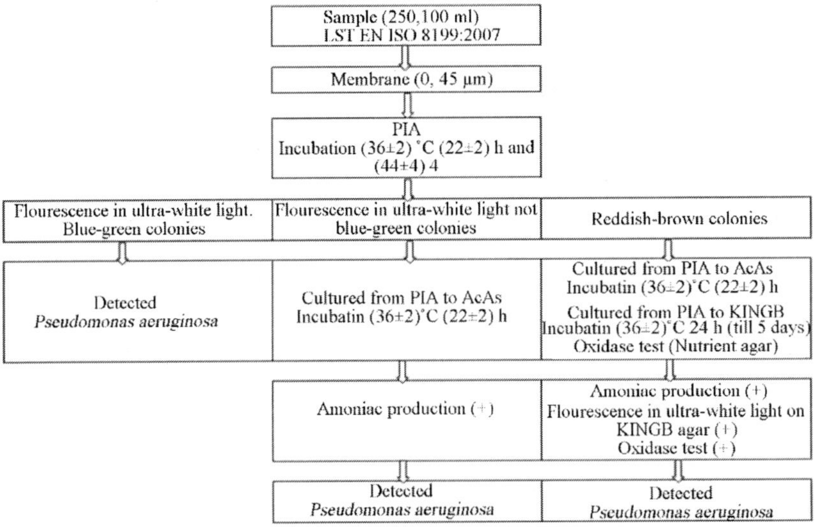

Figure 2. Detection and enumeration of Pseudomonas aeruginosa according to LST EN ISO 16266:2008 [11].

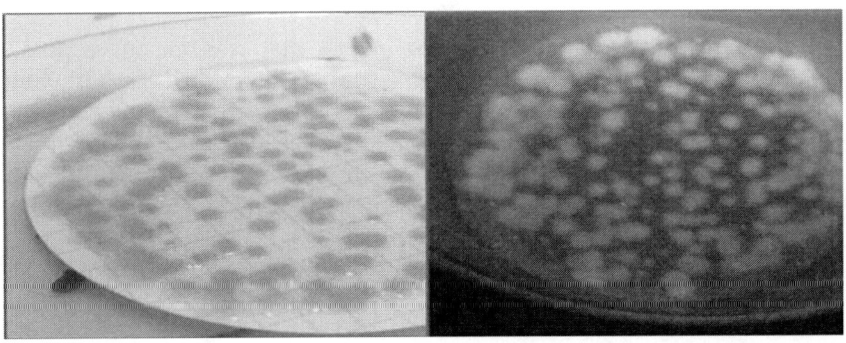

Figure 3. a) Pseudomonas aeruginosa on PIA agar from drinking water b) Pseudomonas aeruginosa on PIA agar fluorescence in ultra-violet light.

The results of the investigated water samples showed that water contained large amounts of non-fermentative gram-negative bacteria. Our results show (Figure 4) that the majority of the inhabitants think that the quality of drinking water is good because of the use of water filters alone, therefore the number of people who wish the microbiological quality of their drinking water to be tested is on the decrease.

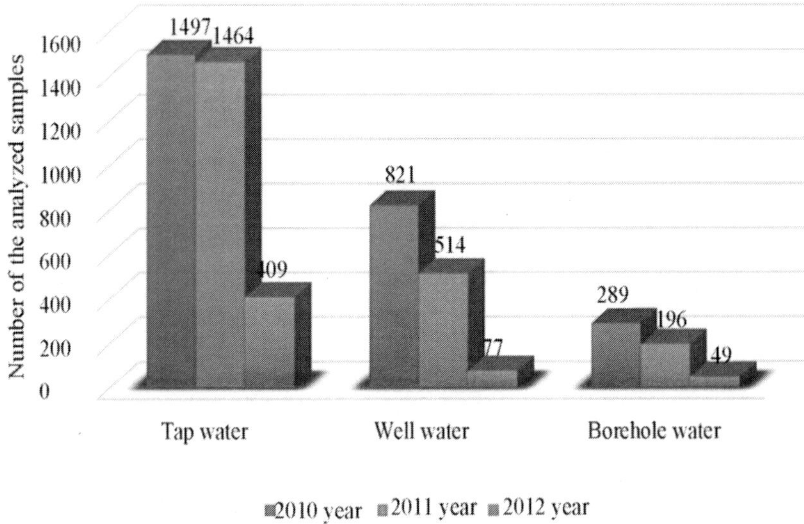

Figure 4. Number of the analyzed samples by water groups, from 2010 to 2012.

The investigation results showed (Figure 5) that centralised urban tap water is most often analysed by water control laboratories in Lithuania. This is because it is obligatory for public institutions to have the quality of water tested once a year, but, unfortunately, the tests are carried out to detect only those micro-organisms, which are indicated in the hygiene standards (coliform bacteria and faecal enterococci) [12].

The microbiological analysis of drinking water from individual boreholes showed (Table 2, Figure 6) that in 2012 non-fermentative gram-negative bacteria were found in eight water samples.

The results showed (Figure 5) that in water samples people look for micro-organisms, which are indicated in the hygiene standards. Our investigations, however, showed that their water systems contain non-fermentative gram-negative bacteria *Acinetobacter, Pseudomonas* or CDC (Table 2, Figure 6); these bacteria are not indicated in the hygiene standard as an indicated organism to assess the quality of drinking water therefore people do not ask samples of their drinking water to be tested for the said micro-organisms.

Table 1. Oxidase-negative, indole negative, non-fermentative gram–negative rods [7].

Test	*Acinetobacter* spp.	CDC
Urease	V	-
Arginine dehy-	-	-
Drolyse		
DNase	-	ND
ONPG	-	-
Starch hydrolysis	-	ND
Acid from:		
Glucose	V	-
Maltose	V	-
Sucrose	-	-
Mannitol	-	-
Xylose	V	-
Molitarity	-	-
Pigmentation	-	-
Growth:		
On MacConkey	V	V
Agar		
At 42°C	V	V
On MBM +	+	ND
Acetate		
Nitrate to nitrate	-	+
Esculin hydrolysis	-	-
Gelatin hydrolysis	V	-

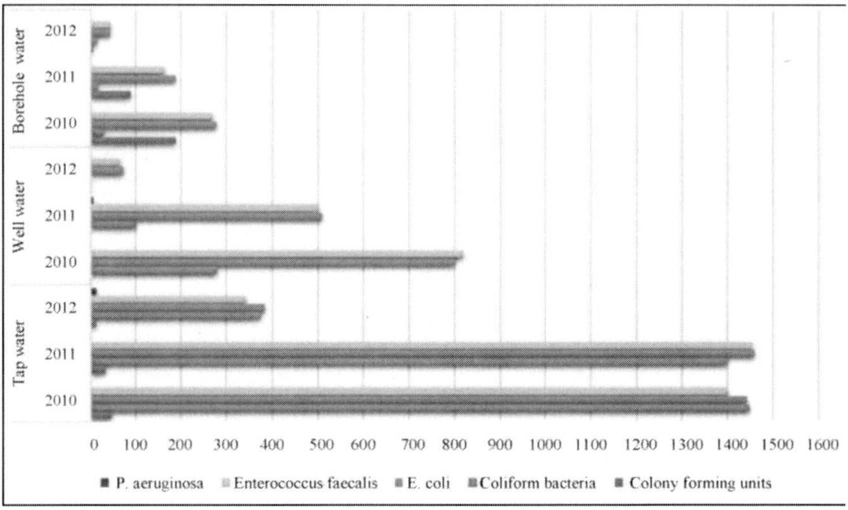

Figure 5. The number of samples analyzed by microbiological indicators between 2010 and 2012 by water source groups.

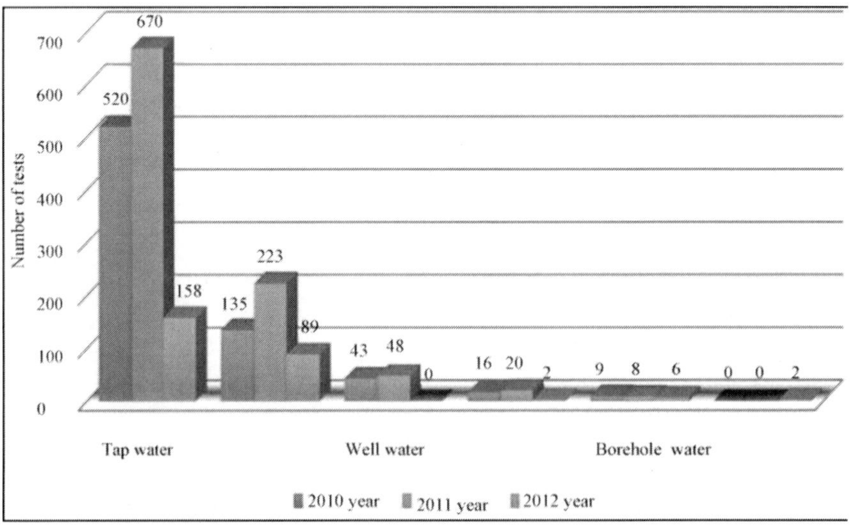

Figure 6. Non-fermentative bacteria in drinking water by group, from 2010 to 2012.

Table 2. Detected non-fermentative gram-negative bacteria in drinking water (*NF-Non-fermentative).

	Tap water		Well water		Borehole water	
Year	*NF**	*P. aeruginosa*	*NF*	*P. aeruginosa*	*NF*	*P. aeruginosa*
2010	520	135	43	16	9	0
2011	670	223	48	20	8	0
2012	158	89	0	2	6	2

P. aeruginosa is a bacterium that does not require any nutrients. This is indicated by the ability of these bacteria to grow even in distilled water [8]. *P. aeruginosa* tends to grow in moist environment and this is most likely a reflection of their natural existence in soil and water. They can grow on the inside surface areas of polyvinyl chloride pipes and garden watering hoses. *P. aeruginosa* is a bacterium, which causes skin rash and infections acquired in swimming pools and saunas. In water these bacteria stick to the skin and penetrate into the hair follicles where they reproduce, and products of their metabolism cause inflammation. The affected spot can look like red itching rash or like a burn. The affected spot reminds of chemical rash and since its incubation period is long some cases might be wrongly diagnosed as chemical rash rather than bacterial infections. *P. aeruginosa* is the main cause of the so-called "swimmer's air" (*otitas externa*). This bacterium can also cause infection of the cornea in people who wear contact lenses. Besides, it is one of the main causes of urinary tract infections, surgical infections and burn wound infections. *P. aeruginosa* is an opportunistic pathogen. This means that it is able of making use of the problems of the host's immune system to penetrate into the organism and develop infections.

P. aeruginosa is the species most commonly associated with human diseases, ear and skin infections acquired in contaminated pools, hot tubs and whirlpools. It causes serious nosocomial (hospital-acquired) infections in cystic fibrosis and burn patients, and immunocompromised individuals [13]. *P. aeruginosa* is a common environmental organism and it can be found in faeces, soil, water and sewage. It can multiply in water environment and also on organic surfaces that are in contact with water [1,2]. The cells of such ubiquitous bacterial species as *P. aeruginosa* when responding to favourable nutrient conditions and being stuck to the present surfaces at the time of binary fission and exopolymer production form strong biofilms [14,15]. *P. aeruginosa* biofilms cause persistent infections in the individuals with health problems. For example, most people with the genetic cystic fibrosis disease suffer from chronic P. aeruginosa biofilm infections of the lungs [9]. *P. aeruginosa* biofilms are developed communities with individual bacterial cells embedded in an extracellular polysaccharide matrix and are inherently resistant to antimicrobial treatment [15,16].

4. CONCLUSIONS

The investigations carried out show that the quality of centralised urban drinking water is good; however, some cases of increased microbial contamination were determined. Lithuanian users are supplied with drinking water from deep aquifers in a centralised way. The environment exerts no considerable impact on water from deep aquifers. Unfortunately, groundwater is not safe for consumption.

The majority of violations of the requirements set forth in the Lithuanian Hygiene Standard HN 24:2003 were discovered in individual water supply systems, especially in dug wells. Contamination with coliform bacteria was discovered in 72.9% of all the wells under investigation. Contamination of water with *Escherichia coli* 54.8% and contamination with intestinal enterococci 56.2% were found in all samples of well water studied.

The reason for micro-biological contamination of wells is that polluted surface water can get into the well water. It is the owners of a homestead themselves that most often contaminate drinking water. Traditionally a well is dug in the centre of the homestead, close to dwelling houses and outside buildings, kitchen gardens. Groundwater can be contaminated if nearby soils is fertilized or if pollutants get directly into a well.

Violations of the requirements laid down in the Lithuanian Hygiene Standard HN 24:2003 were discovered in centralised urban drinking water where contamination with coliform bacteria was detected in 11.5% of the samples. All the samples were contaminated with intestinal rods (*Escherichia coli*) 0.2% and intestinal enterococci 0.1%.

The Lithuanian Hygiene Standard 24:2003 *Pseudomonas aeruginosa* contamination is controlled only in pre-packed drinking water (they cannot be found in 250 ml of water) [12]. Since *P. aeruginosa* do not require any nutrients, they can survive and reproduce in water prepacked drinking water containing very small amounts of organic substances). *P. aeruginosa* can be discovered in drinking water, especially in pre-packed water, they can be related to complaints about the taste, odour and turbidity. The bacteria release poisonous endotoxin and enzymes therefore *Pseudomonas* is constantly monitored in bottled drinking water. Publicly supplied drinking water is neither regulated nor controlled with respect to *P. aeruginosa*.

ACKNOWLEDGEMENTS

Daiva Staradumskyte, Algimantas Paulauskas (NPHSL), Kaunas Department, Microbiology subunit, for assistance provided in our investigations.

REFERENCES

1. World Health Organization, "Guidelines for DrinkingWater Quality," Incorporating First Addendum, Vol. 1, Recommendations, 3rd Edition, 2006, pp. 121-144, 221- 247.
2. World Health Organization, "Guidelines for DrinkingWater Quality," 4th Edition, 2011.
3. V. Kiguoliene, "Ką Reikia Žinoti Apie Geiamąjį Vandenį," 2013.
4. L. Dijkshoorn, A. Nemec and H. Seifert, "An Increasing Threat in Hospitals: Multidrug-Resistant Acinetobacter baumannii," Nature Reviews Microbiology, Vol. 5, No. 12, 2007, pp. 939-951. http://dx.doi.org/10.1038/nrmicro1789
5. J. J. LiPuma, B. J. Currie, G. D. Lum and P. A. R. Vandamme, "Burkholderia, Stenotrophomonas, Ralstonia, Cupriavidus, Pandoraea, Brevundimonas, Comamonas, Delftia, and Acidovorax," In: P. R. Murray, E. J. Baron, J. H. Jorgensen, M. L. Landry and M. A. Pfaller, Eds., Manual of Clinical Microbiology, 9th Edition, ASM Press, Washington DC, 2007, pp. 749-769.
6. S. C. Su, M. Vaneechoutte, L. Dijkshoorn, Y. F. Wei, Y. L. Chen and T. C. Chang, "Identification of Non-Fermenting Gram-Negative Bacteria of Clinical Importance by an Oligonucleotide Array," Journal of Medical Microbiology, Vol. 58, No. 5, 2009, pp. 596-605. http://dx.doi.org/10.1099/jmm.0.004606-0
7. P. R. Murray, E. J. Boran, M. A. Pfallek, F. C. Tenover and R. H. Yolken, "Manual of Clinical Microbiology," 6th Edition, ASM Press, Washington DC, 2005, pp. 509- 530.
8. M. David and D. V. M. Moore, "Reference Paper: Pseudomonas and the Laboratory Animal," 1997.
9. E. Banin, M. L. Vasil and E. P. Greenberg, "Iron and Pseudomonas aeruginosa biofilm formation," Proceedings of the National Academy of Sciences of the United states of America, Vol. 102, No. 31, 2005, pp. 11076- 11081. http://dx.doi.org/10.1073/pnas.0504266102.
10. LST EN ISO 9308-1:2001, "Water Quality—Detection and Enumeration of Escherichia coli and Coliform Bacteria—Part 1: Membrane Filtration Method," ISO 9308- 1:2000.
11. LST EN ISO 16266:2008, "Water Quality—Detection and Enumeration of Pseudomonas aeruginosa—Method by Membrane Filtration," ISO 16266:2006.
12. Lithuanian Hygiene Norm HN 24:2003, Drinking Water Safety and Quality Requirements.

13. V. T. C. Penna, S. A. M. Martins and P. G. Mazzola, "Identification of Bacteria in Drinking and Purified Water during the Monitoring of a Typical Purification System," BMC Public Health, Vol. 2, 2002, p. 13. http://dx.doi.org/10.1186/1471-2458-2-13
14. K. D. Mena and C. P. Gerba, "Risk Assessment of Pseudomonas aeruginosa in Water," Reviews of Environmental Contamination and Toxicology, Vol. 201, 2009, pp. 71-115http://dx.doi.org/10.1007/978-1-4419-0032-6_3
15. J. W. Costerton, Z. Lewandowski, D. E. Caldwell, D. R. Korber and H. M. Lappin-Scott, "Microbial Biofilms," Annual Review of Microbiology, Vol. 49, 1995, pp. 711-745.http://dx.doi.org/10.1146/annurev.mi.49.100195.003431
16. J. W. Costerton, Philip S. Stewart and E. P. Greenberg, "Bacterial Biofilms: A Common Cause of Persistent Infections," Science, New Series, Vol. 284, No. 5418, 1999, pp. 1318-1322.

CHAPTER 10

Microfungi in Drinking Water: The Role of the Frog *Litoria caerulea*

Noel B. Sammon [1,*], Keith M. Harrower [1], Larelle D. Fabbro [2] and Rob H. Reed [1]

[1]*Centre for Plant and Water Science, CQUniversity Australia, Bruce Highway, Rockhampton, Queensland, Australia 4702*

[2]*Centre for Environmental Management, CQUniversity Australia, Bruce Highway, Rockhampton, Queensland, Australia 4702*

ABSTRACT

Microfungi were recovered from all parts of a municipal water distribution system in sub-tropical Australia even though virtually no colony-forming units were recovered from the treated water as it left the treatment plant. A study was then undertaken to determine the potential sources of the microfungal population in the distribution system. Observation of frogs (*Litoria caerulea*) using the internal infrastructure of a reservoir as diurnal sleeping places, together with observation of visible microfungal growth on their faecal pellets, led to an investigation of the possible involvement of this animal. Old faecal pellets were collected and sporulating fungal colonies growing on their surfaces were identified. Fresh faecal pellets were collected and analysed for microfungal content, and skin swabs were analysed for yeasts. It was found that the faeces and skin of *L. caerulea* carried large numbers of yeasts as well as spores of various filamentous fungal genera. While there are many possible sources of microfungal contamination of municipal drinking water supplies, this study has revealed that the Australian green tree frog *L. caerulea* is one of the important sources of filamentous microfungi and yeasts in water storage reservoirs in sub-tropical Australia where the animal is endemic.

Keywords: filamentous fungi; yeasts/yeast-like; Australia

1. INTRODUCTION

An extensive identification and enumeration study of microfungi in a treated municipal water supply system in sub-tropical Australia over an 18 month period showed that microfungi are present throughout the system [1]. Analysis of the treated water leaving the single water treatment plant indicated that the treatment processes are very effective in removing or inactivating microfungal propagules since only occasional microfungal colony-forming units (CFU) were recovered from samples taken at the treatment plant outlet. However, microfungal CFU were routinely recovered from treated water storage reservoirs and distribution mains throughout the distribution system.

This finding led to a further investigation to attempt to identify the source(s) of such post-treatment microfungal population. During the collection of monthly water samples from one of the roofed storage reservoirs, designated R3, over an 18 month period it was noticed that numbers of the large Australian green tree frog, *Litoria caerulea*, were always resident within the reservoir, occupying the internal infrastructure of stairway, supporting columns and roofing beams as diurnal sleeping places. They particularly favoured the sloping C-section side rails of the stairway (Figure 1).

Frog excrement was also evident on these structures but it was clear that most faecal pellets are deposited directly into the water or are dislodged into the water by movement of the animals. During the study period a number of unexplained cases of low *Escherichia coli* counts were reported by a number of water authorities throughout Queensland. The regular observation of *L. caerulea* within water storage reservoir R3 in the water supply system under study led to the hypothesis that the source of those *E. coli* counts was the excrement of *L. caerulea.* Subsequent investigation demonstrated that the frog *L. caerulea* carries very large *E. coli* loads [2]. The water authority concerned has since taken steps to exclude *L. caerulea* from the reservoirs in its water distribution system.

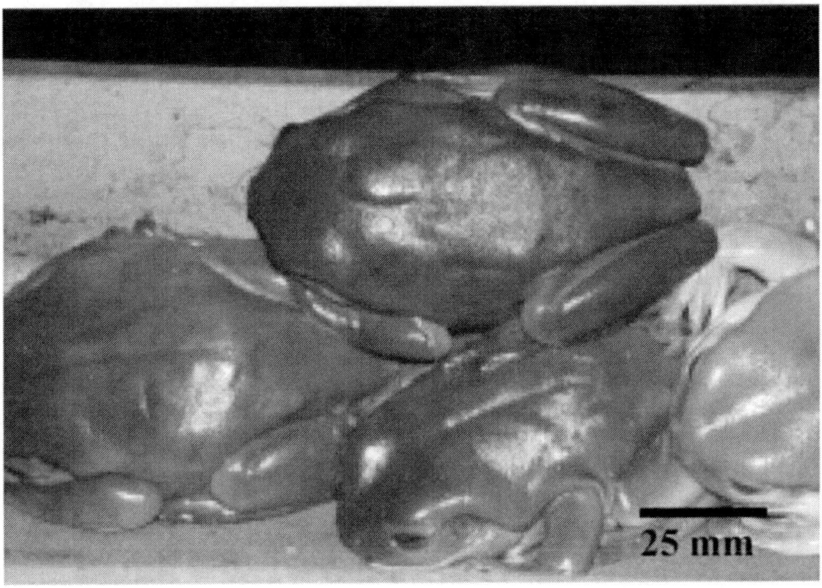

Figure 1. Group of *Litorea caerulea* specimens asleep on a side rail of an internal stairway in a reservoir.

Concurrently, it was noticed that old faecal pellets of *L. caerulea* which remained on the internal structures of reservoir R3 were showing evidence of extensive microfungal growth on their surfaces. This observation suggested three possibilities; (i) that frog excrement contained viable microfungal propagules which germinated post-voidance, (ii) that frog excrement acted as a suitable substrate for the germination and growth of airborne spores which landed on it, or (iii) both. This led to an investigation which aimed to determine whether or not the Australian green tree frog *L. caerulea* may be one of the sources of the microfungal propagules recovered from the mains and reservoir water during the identification and enumeration study mentioned earlier.

2. MATERIALS AND METHODS

Samples were collected from the interior stairway and accessible structures of a roofed water service reservoir (R3) in a city in Central Queensland, Australia.

2.1. The Reservoir

Reservoir R3, which was the site for this study, is eminently suitable for that purpose. Its walls and floor are cast concrete and it is roofed with galvanised sheeting on galvanised beams supported by concrete columns. It has a maximum capacity of *ca.* 10.2 megalitres and is filled solely by an underground trunk main fed directly from another reservoir, designated R2, located *ca.* 2 km away. Reservoir R2 is automatically re-chlorinated to *ca.* 0.60 ppm. During the 18 month period over which the identification and enumeration study was conducted the average recovery of yeasts from reservoir R2 was <1 cell per litre and filamentous microfungi was only 3 cells per litre. Consequently the most likely explanation for the existence of the microfungal population in reservoir R3 was through introduction of propagules at the reservoir itself. Reservoir R3 was not re-chlorinated at the time of this study, and free chlorine readings during the identification and enumeration study period averaged less than 0.10 ppm.

2.2. Surface Growth of Microfungi

Twenty aged faecal pellets of the Australian green tree frog *L. caerulea* were collected from the internal stairway and accessible infrastructure of reservoir R3. These were placed in individual sterile 30 mL polycarbonate screw-top tubes (Sarstedt) which were capped immediately after each sample was collected. At the laboratory, the surfaces of the pellets were examined for the presence of microfungal colonies under a Wild M3Z stereomicroscope. Fragments of sporulating colonies were then picked with a dissecting needle from the surfaces of five randomly selected pellets. These fragments were placed in drops of fuchsin acid/lactic acid stain (0.1 g acid fuchsin in 100 mL lactic acid) on glass microscope slides and examined under a Leica DMLB light microscope. Sporulating microfungal colonies were identified by their reproductive structures and spores to genus level by reference to Carmichael *et al.* [3], Ellis [4,5], Gilman [6], and Kendrick and Carmichael [7].

2.3. Microfungal Analysis of Frog Excrement

Ten freshly deposited *L. caerulea* faecal pellets were collected for further microfungal analysis. Each pellet was weighed then carefully broken apart and a needle-point sample was taken from the interior of the pellet, placed in a drop of acid fuchsin/lactic acid stain on a glass microscope slide, and examined under a Leica DMLB light microscope for visual evidence of the presence of microfungal spores. Each pellet was then placed in a sterile 30 mL polycarbonate screw-top tube (Sarstedt), 20 mL of sterile reverse osmosis water (SROW) was added, and a faecal suspension was produced by mixing the contents thoroughly for 1 minute on a CSV90 Auto vortex mixer at maximum setting No. 8. Each suspension was filtered through sterile muslin and the filtrate was brought up to 100 mL with SROW. Streptomycin was added to the filtrate at the rate of 100 mg L^{-1} to inhibit bacteria, and decimal dilution series to 10^{-3} were prepared from the filtrate using SROW. An aliquot of 100 µL of each dilution was spread onto malt extract agar amended with chloramphenicol (MEAC) (malt extract 10 g, glucose 10 g, bacteriological agar 10 g, peptone 0.5 g, reverse osmosis water 500 mL, chloramphenicol 50 mg) in standard plastic Petri dishes. The culture plates were surface-dried in a biological safety cabinet for 1 h immediately before use. The spread plates were incubated at 25 °C in the dark. Yeast and filamentous fungal colonies were counted over 7 days and the estimated number of CFU g^{-1} of frog excrement was calculated. Filamentous colonies were identified to genus level and the estimated percentage of each genus recovered was calculated.

This procedure was repeated on another five fresh faecal pellets of *L. caerulea* collected in September/October 2009 after several months of extremely dry, low-humidity weather. Ten yeast colonies, which appeared to be different to each other in terms of gross colony morphologies and which were represented in all of the samples, were picked from the cultures and sub-cultured. These were sent to Sullivan Nicolaides Pathology Laboratories in Rockhampton, Queensland for identification.

2.4. Skin Swabs—Yeast

An area of 9 cm^2 of the dorsal skin of each of 10 sleeping adult specimens of *L. caerulea* (CQUniversity Animal Ethics permit No. A10/04-258) was gently swabbed to determine if yeasts were endemic on the skin of these animals. A sterile cotton bud, with long wooden handle, dipped in SROW was used for this purpose. The cotton bud was then placed in a sterile 30 mL polycarbonate screw-top tube (Sarstedt) containing 10 mL of SROW, the wooden handle of the cotton bud was broken off, and the tube was capped. Sterile rubber gloves were worn to prevent accidental cross-contamination between operator and frog. The tubes were transported back to the laboratory in pre-cooled containers and were processed immediately on arrival. Each tube was vortexed on a CSV90 Auto vortex mixer at setting 8 for 30 sec. and the swab was discarded after expressing excess solution by pressing it against the side of the tube. A decimal dilution series to 10^{-3} was prepared from the contents of each of the ten tubes using SROW. Spread plates were prepared using MEAC culture plates which had been surface-dried in a biological safety cabinet for 1 hr. immediately before use. An aliquot of 200 µL of each of the dilution series was pipetted onto these plates and spread with sterile plastic spreaders. The culture plates were incubated in the dark at 25 °C and yeast colonies were counted daily for 3 days. Ten yeast colonies, which appeared to be different to each other in terms of gross colony morphologies and which were represented in all of the samples, were picked from the cultures and sub-cultured. These were sent to Sullivan Nicolaides Pathology Laboratories in Rockhampton, Queensland for identification.

2.5. Photographs and Photomicrographs

Photographs were taken with a Nikon Coolpix 995 digital camera. Photomicrographs were taken with an Olympus Colorview III digital camera mounted on a Leica DMLB light microscope.

3. RESULTS

3.1. Surface Fungal Growth

All 20 aged faecal pellets collected and examined for surface fungal growth supported extensive, visible, sporulating microfungal colonies (Figure 2a,b). The genera identified on the five pellets selected at random for colony characterisation included *Aspergillus* (three species), *Penicillium* (two species), *Absidia, Memnoniella, Stilbella, Fusarium, Acremonium, Syncephalastrum* and *Geotrichum.*

3.2. Microfungi Recovered from Excrement

Most of the faecal pellets from which needle-point samples were taken and examined microscopically in May 2009 contained visible microfungal spores (Figure 2d). Many of the faecal pellets contained visible grass fragments and soil particles (Figure 2a,c).

The estimated CFU per gram of excrement and characterisation of filamentous microfungi and yeasts, based on the colony recoveries from the decimal dilution series made from the 10 pellets collected in May 2009 and the five collected in September, 2009 are in Table 1.

3.3. Yeast Identification—Excrement

The 10 yeast isolates sub-cultured from colonies recovered from the five frog pellets collected in Sept/Oct 2009 were identified as 1 *Rhodotorula* sp.,

6 *Trichosporon* spp., 1 *Candida* sp., 1 *Candida famata*, and 1 *Candida glabrata*.

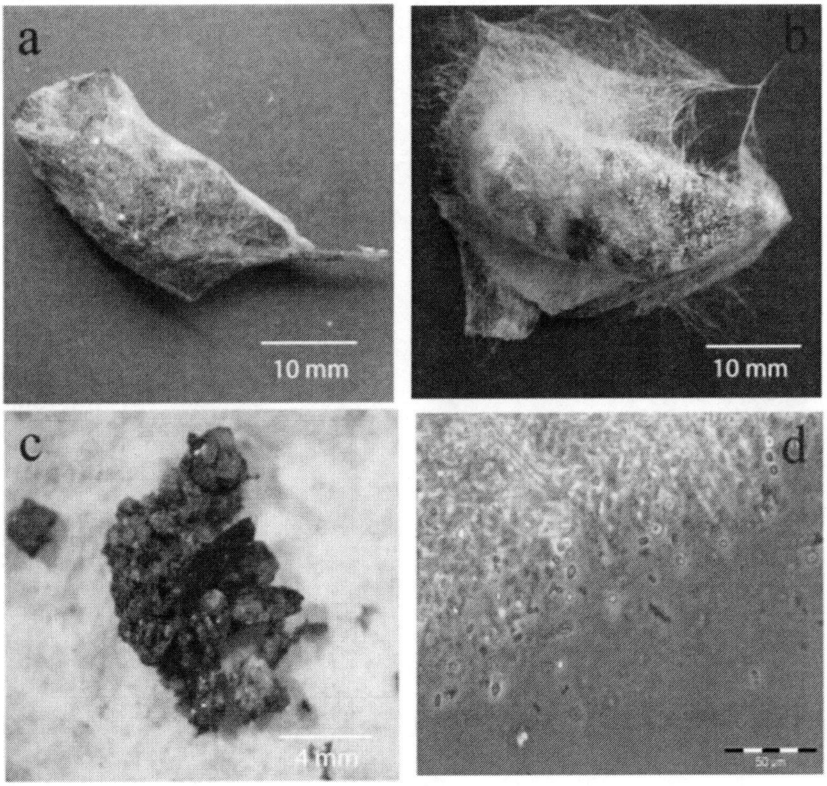

Figure 2. (a,b) Two aged faecal pellets of *Litoria caerulea* covered with extensive growth of sporulating microfungi. (c) Soil particles from a faecal pellet of *Litoria caerulea*. (d) Microfungal spores from a needle-point fragment of the interior of a fresh faecal pellet of *Litoria caerulea* showing visible microfungal spores.

3.4. Yeasts Recovered from Skin Swabs

The skin swabs from all 10 specimens of *L. caerulea* tested positive for yeasts. The estimated CFU per cm^2 of dorsal skin, based on the CFU recovered from the decimal dilution series made from the 10 swab samples collected in May 2010 are shown in Table 2. The 10 yeast isolates sub-cultured from colonies recovered from the skin swabs were identified as 3

Trichosporon asahii, 3 *Trichosporon* spp., 1 *Rhodotorula* sp., 1 *Aureobasidium pullulans*, 1 *Candida* sp. and 1 unknown species.

Table 1. Composition of microfungal CFU recovered from *Litoria caerulea* excrement.

Sample No.	Total CFU g^{-1} of excretum	*Penicillium*	*Paecilomyces*	*Aspergillus*	*Cladosporium*	Other f/fungi incl. asporous	Yeasts and yeast-like
		Percentage of total					
May 2009							
1	15.2×10^6	1.1					98.9
2	4.3×10^6	2.3	97.7				0.0
3	30.6×10^6						100.0
4	9.5×10^6	46.3	9.3			24.0	20.4
5	11.4×10^6	3.9					96.1
6	4.4×10^4	9.9		3.9		5.0	81.2
7	7.8×10^6						100.0
8	1.2×10^5			45.1	0.6	0.5	53.8
9	3.2×10^6	3.2		9.8	2.2		84.8
10	22.1×10^6						100.0
Sept/Oct 2009							
11	6.5×10^5	20.3					79.7
12	4.0×10^6	1.0				3.8	95.2
13	44.0×10^6						100.0
14	1.3×10^5	6.5	3.3	6.6		8.2	75.4
15	4.2×10^6	4.2	4.2				91.6
Average	10.5×10^6	6.6	7.6	4.4	0.2	2.8	78.5

Table 2. Yeast CFU recovered from *Litoria caerulea* skin swabs.

Sample No.	1	2	3	4	5	6	7	8	9	10	Average
Yeast CFU per cm^2 of skin	956	211	617	228	978	133	300	2,167	639	1,333	623

It was notable that the yeast populations recovered from several of the frogs showed a dominance of one species and the dominant species varied from frog to frog.

4. DISCUSSION

The large green tree frog *L. caerulea* is endemic across the northern half of Australia and parts of Papua New Guinea [8,9]. In its natural habitat this nocturnal, arboreal species hides during the day in hollow tree limbs, in dense foliage, and in similar protected places. It also has an affinity for human dwellings and other man-made structures including roofed water storage reservoirs as daytime hiding places, and seems to prefer these to its natural habitat. These structures provide a dark, warm, humid atmosphere, close proximity to water, and protection from predators, conditions which are ideal for this animal.

On every occasion that reservoir R3 was visited for water sampling purposes, mature specimens of *L. caerulea* were always found sleeping on the internal stairway, adjacent roof superstructure, and supporting columns within the reservoir. The suction pads on the toes of these animals enable them to easily climb vertical surfaces. At night this ambush predator comes to ground to prey on coleopterans and other insects as well as small vertebrates. The prey is engulfed, secured with its sticky tongue and vomerine teeth, and swallowed whole. As a consequence of this ground-feeding method the frog frequently ingests dried grass fragments and soil particles along with its prey, and these were commonly observed in the faecal pellets (Figure 2 a,c). These materials are natural substrates for many of the saprobic microfungi which were recovered from frog faecal pellets and also from water samples taken from the water body in the reservoirs [1]. Consequently it would be expected that such debris would introduce microfungal spores into the frog's digestive system. Furthermore, many of the frog's prey species are known to carry entomopathogenic fungi such as *Paecilomyces* spp. [10] which are also found in soil and on plants. The large number of fungal spores observed in needle-point samples of fresh frog faecal pellets also supports the theory that fungal spores are ingested with prey and extraneous debris, and pass through the animal's digestive system unharmed.

It is clear that in areas of Australia where *L. caerulea* is endemic, the species is a potential source of the microfungal population of drinking water systems, and this could occur in two ways. Firstly by introduction of fungal spores from outside the reservoir in the frog's faeces which are, in most cases, voided directly into the water body of the reservoir. The frog's ground-feeding method ensures ingestion of extraneous material along with spore-laden prey and it is clear from the data that many of those spores survive the frog's digestive system. This is supported by the variety of filamentous microfungal genera recovered from the frog pellets and the occasional large percentages of individual genera. For example, Table 1 shows that 97.7% of microfungal CFU recovered from frog pellet #2 was *Paecilomyces*, a known entomopathogenic genus. Samples #4 and #11 carried high percentages of *Penicillium* spp. while 45.1% of the filamentous fungi CFU recovered from sample #8 were *Aspergillus* spp. These genera are saprobes of plant debris and are also found in soil. The data in Table 3 also suggests that the frog's digestive system naturally supports very high yeast populations. With the sole exception of sample #2, high numbers of yeasts were recovered from all of the pellets tested. It is notable that yeast and yeast-like colonies were frequently recovered from water samples collected from reservoir R3 during the separate identification and enumeration investigation [1]. Secondly, when excrement is deposited onto the internal superstructure above the maximum water level it provides a substrate where microfungi grow and sporulate thus adding further to the airborne spore load within the reservoir and subsequently in the water body.

The high prevalence and number of yeasts and yeast-like microfungi recovered from frog excrement and from skin swabs suggests that these microorganisms are resident commensals of the gut and skin of these animals. These high numbers, together with the knowledge that the animal carries high *E. coli* loads, suggests that *L. caerulea* should be excluded from reservoirs wherever possible and that adequate residual free chlorine levels should be maintained at all times. It is noteworthy that during the identification and enumeration study, 18 monthly water samples from reservoir R3, which was not rechlorinated during that study, produced an average of 147 yeast cells per litre of water. The water utility involved in this study has already taken appropriate action, and is constructing a rechlorination facility at reservoir R3 to manage the risk of recontamination.

The yeast genera *Candida*, *Rhodotorula* and *Trichosporon* which were recovered from excrement of *L. caerulea* collected from reservoir R3 contain species which can be opportunistic human pathogens [11]. Three of the ten representative yeast isolates from frog excrement were *Candida* species. It is known that metabolites of *Candida* are the aetiological agents of candidids, an allergic skin reaction [12]. The recovery of *Candida glabrata* is of some concern since it is a major cause of candidiasis in humans. This yeast, like *C. albicans*, is a commensal organism and is also an opportunistic pathogen causing superficial and systemic infections such as endocarditis, cystitis, osteomylitis, vaginitis and fungemia [13]. *C. albicans* was not found in this study but has been recovered from overseas water systems [14]. Thuraisingam and Denning [15] reported increasing frequency of systemic and vaginal infections with *C. glabrata* and stated that such infections are difficult to treat. Vaginitis is caused by both *C. albicans* and *C. glabrata*. However, increased prevalence of *C. glabrata* in diabetic women was reported by Ray *et al.* [16] who isolated *C. glabrata* from 61.3% and *C. albicans* from only 28.8% of 111 consecutive diabetic patients suffering from vulvovaginal candidiasis during their study into the prevalence of *C. glabrata* and its response to certain treatments. According to Gugic *et al.* [17], *C. glabrata* is the second most common species of *Candida* found in hospital patients in the U.S.A.

The other yeasts recovered from frog excrement and skin, *Rhodotorula* spp. and *Trichosporon* spp., are commonly occurring genera and while they have, on rare occasions, been associated with opportunistic human infections their involvement is of little importance [12].

It is not known if infections and allergies caused by *C. glabrata* or other yeasts can be acquired through contact with water containing them, but it is reasonable to expect that this may be so. In recent years it has been increasingly recognised that opportunistic, pathogenic microfungi, particularly filamentous genera such as *Aspergillus*, may be the waterborne aetiological agents of nosocomial mycoses but only a limited number of studies have been undertaken into this potential mode of infection.

5. CONCLUSIONS

The evidence presented in this study suggests that the Australian green tree frog *L. caerulea*, a diurnal resident of covered water reservoirs, is a source of yeast/yeast-like and filamentous microfungi found in water supply systems in northern Australia.

The excrement of this species introduces fungal spores directly into the water body of water storage reservoirs and, when deposited on internal reservoir architecture, it also acts as a substrate on which microfungi grow and sporulate thus increasing the spore load within the reservoir air and consequently in the water body. The results presented here suggest that the frog species *L. caerulea* is an important source of yeasts in water storage reservoirs in areas where the animal is endemic.

While this study was focused on a municipal water supply system where the water is treated and water quality is monitored closely, the findings would also be applicable to household water supply derived from rainwater tanks. In fact, if *L. caerulea* can gain access to domestic rainwater tanks to which exclusion devices are not fitted then a potential health problem could very well exist. Domestic rainwater tanks in Australia are rarely, if ever, chlorinated and the frog to water volume ratio is likely to be high because of the relatively small water-holding capacity of such tanks.

ACKNOWLEDGMENTS

Funding for this project, in-kind support, and access to water storage and distribution facilities were provided by Rockhampton Regional Council through Fitzroy River Water. This funding and the co-operation and assistance of Fitzroy River Water staff are gratefully acknowledged and appreciated.

We sincerely thank Jason Plumb (Fitzroy River Water) for his valued comments and support.

We are grateful for the assistance freely given by Patrick Applegarth, Anthony Jennings and Sullivan Nicolaides Pathology in identifying yeast samples.

REFERENCES

1. Sammon, NB; Harrower, KM; Fabbro, LD; Reed, RH. Incidence and Distribution of Microfungi in a Treated Municipal Water Supply System in Sub-tropical Australia. *Int. J. Environ. Res. Public. Health* 2010, *7*, 1597–1611.
2. Sammon, NB; Harrower, KM; Fabbro, LD; Reed, RH. Green tree frogs: Contamination of covered reservoirs in Northern Australia. *Water* 2009, *36*, 97–101.
3. Carmichael, JW; Kendrick, WB; Connors, IL; Sigler, L. *Genera of Hyphomycetes*; The University of Alberta Press: Edmonton, Canada, 1980.
4. Ellis, MB. *Dematiaceous hyphomycetes*; The Commonwealth Mycological Institute: Kew, UK, 1971.
5. Ellis, MB. *More Dematiaceous Hyphomycetes*; Commonwealth Mycological Institute: Kew, UK, 1976.
6. Gilman, JC. *A Manual of Soil Fungi*; Constable and Company: London, UK, 1959.
7. Kendrick, WB; Carmichael, JW. Hyphomycetes. In *The Fungi*; Ainsworth, GC, Sparrow, AS, Sussman, AS, Eds.; Academic Press: New York, NY, USA, 1973; Volume IV A, pp. 323–509.
8. Cogger, HG. *Reptiles and Amphibians of Australia*; Reed Books: Chatswood, Australia, 1992.
9. Cronin, L. *Australian Reptiles and Amphibians*; Envirobook: Annandale, Australia, 2001.
10. Inglis, PW; Tigano, MS. Identification and taxonomy of some entomopathogenic *Paecilomyces* spp. (Ascomycota) isolates using rDNA-ITS Sequences. *Genet. Mol. Biol* 2006, *29*, 132–136.
11. Kocková-Kratochvílová, A. *Yeasts and Yeast-like Organisms*; VCH Publishers: New York, NY, USA, 1990.
12. Rippon, JW. *Medical Mycology—The Pathogenic Fungi and The Pathogenic Actinomycetes*, 3rd ed; W.B. Saunders Company: Philadelphia, PA, USA, 1988.
13. Al-Hedaithy, SSA; Fotedar, R. Incidence of *Candida glabrata* in Patient Specimens. *Saudi. J. Bio. Sci* 1996, *6*, 76–80.
14. Alvarez, CIV. Caracterización micológica de aquas "crudas" y filtradas en la Planta de Tratamiento de Tres Ríos, Costa Rica. *Rev. Biol. Trop* 1993, *41*, 417–422.
15. Thuraisingam, AI; Denning, DW. *Candida glabrata* Oesophagitis in a Patient Without HIV Infection. *Eur. J. Clin. Microbiol. Infect. Dis* 2000, *19*, 561–569.
16. Ray, D; Goswami, R; Banerjee, U; Dadhwal, V; Goswami, D; Mandal, P; Sreenivas, V; Kochupillai, N. Prevalence of *Candida glabrata* and its Response to Boric Acid Vaginal Suppositories in Comparison with Oral Fluconazole in Patients with Diabetes and Vulvovaginal Candidiasis. *Diabetes Care* 2007, *30*, 312–317.
17. Gugic, D; Cleary, T; Vincek, V. *Candida glabrata* infection in gastric carcinoma patient mimicking cutaneous histoplasmosis. *Dermatol Online J* 2008, *14*(2).

CHAPTER 11

Drinking Water from Hand-Pumps in Mali, Niger, and Ghana, West Africa: Review of Health Effects

Alexandra Lutz[1*], Samuel Diarra[2], W. Braimah Apambire[1], James M. Thomas[1], Jarvis Ayamsegna[2]

[1]*Division of Hydrologic Sciences, Desert Research Institute, Reno, USA*

[2]*World Vision West Africa WASH, Accra, Ghana*

ABSTRACT

The purpose of this study was to evaluate parameters of drinking water quality from hand-pumps in Ghana, Mali, and Niger, evaluate possible sources of parameters, and provide an overview of potential health effects in the population. Concentrations of 22 parameters in 3337 groundwater samples were analyzed and compared with World Health Organization drinking water guidelines. In general, F, Mn, and Al had relatively larger and more common rates of occurrence, though there was by country. For F, there were reports of skeletal fluorosis in Niger and dental fluorosis in Ghana. For Mn and Al it was difficult to assess health effects due to scarce information.

Keywords: Drinking water; Groundwater; Water chemistry; Health

1. INTRODUCTION

This paper presents data on concentrations of 22 parameters in 3337 groundwater samples from hand pumps in Ghana, Mali, and Niger in West Africa. Samples were grouped by three-country "region" or by individual country. All samples were analyzed for concentrations of Ag, Al, As, Ba, Be, Cd, Co, Cr, Cu, Fe, Mn, Mo, Ni, Pb, Sb, Se, Sr, Tl, U, V, Zn, and F with respect to World Health Organization (WHO) drinking water guidelines. The purpose of this study was to evaluate drinking water quality from hand-pumps, assess risk as percent of population exposed, provide overview of human health effects from hand-pump water, and evaluate evidence of these health effects in the population, if available.

Collection and analyses of samples was done in collaboration by World Vision International (WVI) and the Desert Research Institute (DRI) and is part of an ongoing effort to provide rural populations with potable water. Samples reported in this study were collected and analyzed between 2002 and 2012. The distribution of sampling points was driven by hand-pump location, which, in turn, was dictated by local need for potable water. Thus, there existed a compromise in sampling for the specific purpose of hypothesis testing, such as analysis of elevated concentrations and geographic trends, or elevated concentrations and specific population. Approximate sampling areas are shown in **Figure 1**.

To-date, several thousand hand-pumps have been commissioned, providing approximately one million rural community members with potable water via this particular project. Across sub-Saharan Africa, however, approximately 880 million people still lack access to safe drinking water [1]. Groundwater is increasingly relied on as a source of safe drinking water for rural communities in the countries of Ghana, Mali, and Niger in West Africa and elsewhere. Reasons for developing groundwater include ease of developing hand-pumps in remote locations, availability during drought, relative low price compared with methods of treating surface water, and superior chemical and biological quality as compared with surface water sources [2,3].

Though generally considered to be a safer source of drinking water than surface water, groundwater may contain parameters that have either adverse or beneficial effects on human health. Some parameters have typically low concentrations and, in the case of Co, Cr, Cu, Fe, I, Mn, Mo, Se, V and Zn, are essential to human life, but at higher concentrations, there may be potential health risks for humans including diseases, disorders, cancers, reduced growth, and increased mortality and mutagenic effects [4]. The effects depend on concentration and a different range of acceptable concentrations exist for each parameter.

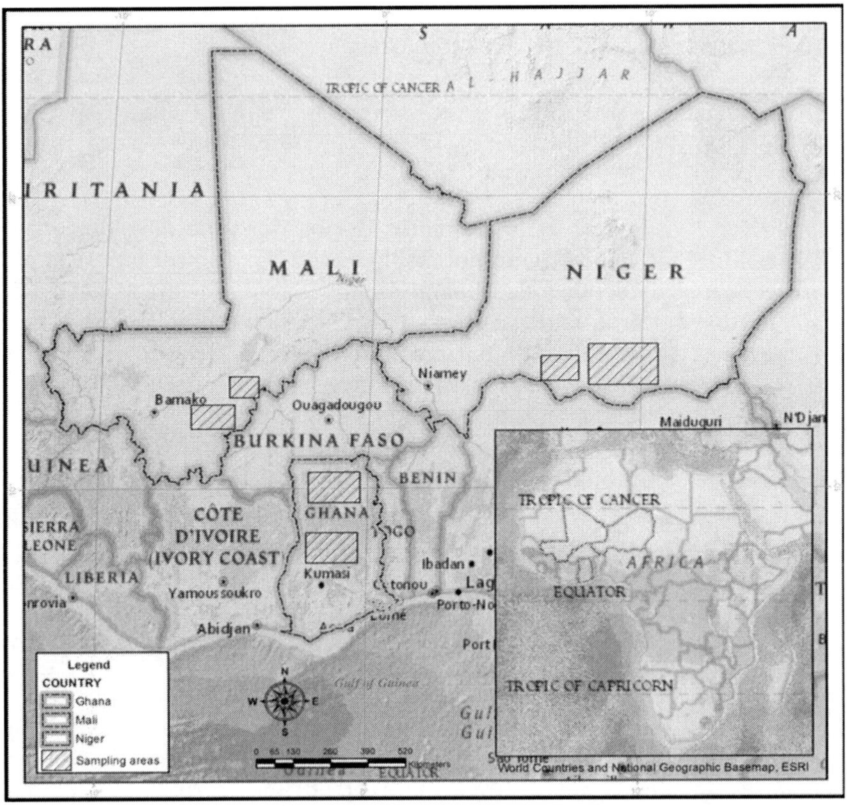

Figure 1. Location of sampling sites.

To evaluate risk and avoid adverse health effects, WHO drinking water guidelines may be used. These guidelines are derived on the basis of health and internationally agreed procedures for risk assessment [5]. These guidelines are neither regulations nor standards, though they may be used to develop local guidelines or standards on water quality. On the one hand there are guidelines for those considered toxic and for which a relative amount of research has been conducted (As, Fl), while there are no guidelines for those whose effects on human health are not know or unclear (Ag, Be, Co, Sr, and V).

In each country, the water projects seek to supply potable water in rural areas. These areas are characterized by small holder subsistence agriculture, and are generally devoid of industrial activity. While some hot spots may be associated with highly localized cottage industries, parameters in excess of WHO guidelines are largely geologic in origin, naturally occurring, and not likely to be from anthropogenic sources. The distribution of parameters found in the samples, thus, reflects the diverse geological terrains from the three countries.

2. METHODS

Samples were collected at hand-pumps during borehole development or after drilling but before hand-pump placement. In some cases, samples were collected at actively-used hand-pumps. Samples were collected by staff at Ghana Rural Water Project (GRWP), Mali Rural Water Project (MRWP), and Niger Rural Water Project (NRWP). All water projects are overseen by WVI. After collection, samples were shipped to DRI for analysis. As much as possible, standard methods were applied, however, samples were not field-filtered. Timing was random as sample collection was hampered by monsoonal rains and extreme heat.

Each sample was analyzed for concentrations of 22 parameters. Fluoride was analyzed according to EPA standard methods 4500F [6]. The parameters Ag, Al, As, Ba, Be, Cd, Co, Cr, Cu, Fe, Mn, Mo, Ni, Pb, Sb, Se, Sr, Tl, U, V, and Zn were analyzed on an inductively coupled plasma-mass spectrometry (ICPMS) according to EPA methods standard methods for drinking water 200.8 [7].

Between 2002 and 2011, detection limits of several parameters analyzed on the ICPMS changed, ultimately becoming lower. Thus, on some figures, previous detection limits will appear as solid lines of data, rather than points. In this study, all data were reported in their originnal format and no efforts were made to adjust, infill, or perform additional measures for values below the detection limit. The results provided are thus reproducible.

3. RESULTS

Table 1 shows WHO guidelines for each parameter and percentage of samples exceeding the guidelines. **Table 2** shows the WHO guideline, mean, median, and range of each parameter. In most cases, minimum values were the detection limits. Both tables show samples grouped by region and individual country.

When viewed by region, F had the largest occurrence, with 9.7% of samples exceeding the guideline. The next largest numbers exceeding guidelines were Mn, Al, Ba, U, and As (6.8, 5.9, 5.8, 5.6, and 5.1%, respectively). Guidelines were exceeded for Cd, Mo, Ni, Pb, Se, and Zn for 3.9% or less of all samples. For these elements and F, mean values were larger than median values, which indicated some unusually high values in the data set. For instance, the mean and median values for Ba were 252 and 122 μg/L, with a maximum value of 28.071 μg/L.

Table 1. WHO guidelines and samples exceeding.

Parameter	WHO Guideline (μg/L)	% Exceeding Guideline			
		Ghana	Mali	Niger	Region
Ag	NG	NG	NG	NG	NG
Al	200	5.6	8.6	5	5.9
As	10	6.6	*	0.7	5.1
Ba	700	6.6	2.1	4.2	5.8
Be	NG	NG	NG	NG	NG
Cd	3	*	0.2	0.5	0.1
Co	NG	NG	NG	NG	NG
Cr	50	*	*	*	*
Cu	2000	*	*	*	*
Fe	NG	NG	NG	NG	NG
Mn	400	7.5	5.6	4	6.8
Mo	70	2	*	*	1.6
Ni	70	<0.01	*	*	<0.01
Pb	10	1.2	14.2	9.7	3.9
Sb	20	*	*	*	*
Se	10	3.7	1.6	1.7	3.2
Sr	NG	NG	NG	NG	NG
Tl	NG	NG	NG	NG	NG
U	15	4.3	2.6	17.1	5.6
V	NG	NG	NG	NG	NG
Zn	5000	*	*	0.01	<0.01
F (mg/L)	1.5	11.4	0.7	8.4	9.7

NG = No guideline. *No samples exceeding.

Table 2. WHO guideline, mean, median, and range of each parameter.

	WHO guideline (µg/L)	Ghana			Mali			Niger			Region		
		Mean	Med.	Range	Mean	Med.	Range	Mean	Med.	Range	Mean	Med.	Range
Ag	NG	0.61	1	<0.10 - 14.1	0.9	1	<0.10 - 31	0.9	1	<0.10 - 1.13	0.7	1	<0.10 - 31.5
Al	200	67	3.06	<0.10 - 7046	68	18	<0.2 - 1159	76	6	0.96 - 6804	68	4.4	<0.10 - 7046
As	10	2.99	1	<0.01 - 122	1.3	1	<0.01 - 5	2.4	1	<0.02 - 295	2.7	1	<0.01 - 295
Ba	700	275	128	<1.0 - 28,071	174	108	6.3 - 3107	185	109	<1.0 - 1582	252	122	<1.0 - 28,071
Be	NG	0.6	1	<0.06 - 22	0.9	1	<0.06 - 2.11	1	1	<0.06 - 2.39	0.7	1	<0.06 - 22
Cd	3	0.57	1	<0.01 - 2	0.8	1	<0.01 - 3.03	0.9	1	<0.01 - 5.63	0.6	1	<0.01 - 5.63
Co	NG	1.13	1	<0.10 - 61	1.6	1	<0.10 - 21	1.1	1	<0.10 - 15.4	1.2	1	<0.10 - 61
Cr	50	1.31	1	<0.10 - 46	1.1	1	<0.10 - 10.9	1.3	1	<0.10 - 25	1.3	1	<0.10 - 45.8
Cu	2000	6.8	1	<0.10 - 728	2.1	1	<0.10 - 21	3.2	1	<0.10 - 101	5.8	1	<0.10 - 728
Fe	NG	215	17.2	<0.10 - 19,244	206	28	0.61 - 4943	286	20	0.64 - 14,537	223	19	<0.10 - 19,244
Mn	400	118	36	<0.10 - 3683	127	37	0.68 - 4998	79	12	0.41 - 4023	114	33	<0.10 - 4998
Mo	70	7.39	1	<0.05 - 837	1.2	1	<0.05 - 28	6.7	1	<0.05 - 1546	6.5	1	<0.05 - 1546
Ni	70	3.32	1.36	<0.10 - 388	2.4	1.2	<0.10 - 34	1.5	1	0.11 - 30	3	1.1	<0.10 - 388
Pb	10	1.09	1	<0.10 - 51.8	14	1.9	<0.10 - 649	5.9	1	<0.10 - 206	3.3	1	<0.10 - 649
Sb	20	0.56	0.15	<0.02 - 6.74	0.9	1	<0.02 - 14.1	0.9	1	<0.02 - 1.57	0.7	1	<0.02 - 14.1
Se	10	7.43	1	<0.02 - 11,200	1.9	1	<0.02 - 27	2.5	1	<0.02 - 29	6.1	1	<0.02 - 11,200
Sr	NG	808	421	<1.0 - 53,493	304	80	3.26 - 11,002	480	265	<1.0 - 9339	704	326	<1.0 - 53,493
Tl	NG	0.58	1	<0.03 - 1.04	0.8	1	<0.03 - 10	0.9	1	<0.03 - 2.18	0.7	1	<0.03 - 10
U	15	3.2	1	<0.10 - 146	2.4	1	<0.10 - 32	12	1.1	<0.10 - 355	4.2	1	<0.10 - 35.8
V	NG	5.27	1	<0.03 - 159	1.4	1	<0.03 - 28	6	2.1	<0.03 - 116	4.9	1	<0.03 - 159.2
Zn	5000	14.7	2	<0.10 - 1822	40	7.8	0.52 - 3675	244	23	0.97 - 22,042	45	3	<0.10 - 22,042
F (mg/L)	1.5	0.73	0.32	<0.10 - 14.9	0.2	0.1	<0.10 - 3.6	0.5	0.2	<0.10 - 10.8	0.6	0.3	<0.10 - 14.9

When viewed by country, F exceeded the guideline for 11.4% of samples in Ghana. The next largest numbers exceeding guidelines were Mn, As, Ba, and Al (7.5, 6.6, 6.6, and 5.6%, respectively). Guidelines were also exceeded for Mo, Ni, Pb, Se, and U, though at rates 4.3% or less. For these parameters, mean values were larger than median values.

In Mali, Pb exceeded the guideline for 14.2% of samples. The next largest numbers were Mn and Al (8.6 and 5.6%, respectively). Guidelines were also exceeded for Ba, Cd, Se, U, and F at rates 2.6% or less. With the exception of Cd, mean values were larger than median values.

In Niger, U exceeded the guideline for 17.1% of samples. The next largest numbers were Pb, F, and Al (9.7%, 8.4%, and 5.0%, respectively). Guidelines were also exceeded for As, Ba, Mn, Se, and Zn at rates 4.2% or less. With the exception of Cd, mean values were larger than median values.

4. DISCUSSION

4.1. Potential Sources

Project emphasis was on safe supply of potable water in rural areas, which were characterized by smallholder subsistence agriculture and generally devoid of industrial activity. While some "hot spots" of water chemistry may be associated with highly localized cottage industries, parameters in excess of WHO guidelines were largely geologic in origin and unlikely to have originated from anthropogenic sources.

In Ghana and Niger, for instance, occurrence of F was related to weathering of granitic rocks. In Ghana, As was likely released from iron oxides under reducing conditions, which may also explain the occurrence of Mn [8]. Under reducing conditions, release of As from iron hydroxides may also be bacterially mitigated [9]. Oxidizing, alkaline conditions in Niger may be associated with release of Pb and U from urano-carbonate compounds [10]. Values of pH both higher and lower than neutral (pH 6 to 8) solubilize Al into groundwater.

Hydraulic conductivity values may impact on groundwater chemistry [9]. For instance, groundwater residence times and subsurface geological formations may suggest high value parameters to have been released from one zone and transported to another. Site characteristics, such as values of pH, redox, hydraulic conductivity, and local geology, may further identify potential sources of high value parameters were not available for the samples presented in this study. The distribution of parameters found in the samples, thus, reflects the diverse geological terrains from the three countries.

4.2. Potential Health Effects

The following seven parameters exceeded guidelines for at least 4% of samples: Al, As, Ba, Mn, Pb, U, and F. Human health effects are discussed below. Discussion begins with F, Mn, and Al, which had relatively larger and more common rates of occurrences. Discussion includes evidence for human health effects and comparisons with other surveys done in similar areas of this study, if available.

There is epidemiological evidence that F in excess of the WHO guideline carries increasing risks of fluorosis of the enamel (mottling of the teeth) and, at progressively higher concentrations, leads to increasing risks skeletal of skeletal fluorosis [11,12]. Guidelines were exceeded in 11.4% of samples from Ghana, 0.7% from Mali, 8.4% from Niger, and the highest maximum value of 14.9 mg/L was observed in Ghana. **Figure 2** shows cumulative distribution as percentage of F in Ghana and Niger. Mitigation measures for high F levels are being taken by GRWP in Ghana and NRWP in Niger.

There is relatively more information available as to health effects and occurrence of F in drinking water as compared with other parameters evaluated in this study, especially in Ghana. Rossiter et al. [1] reports 6.7% of samples exceeded the guideline, with rates increasing to 17% in particular regions of Ghana. Apambire et al. [13] reports 62% of the total population of school children in the Bongo area of Ghana showing dental fluorosis. Both studies suggest high concentrations of F to be localized.

Ayoob and Gupta [14] describe both Niger and Ghana as places where fluorosis is endemic. More specifically, 425 cases of children with skeletal fluorosis are reported in Tibiri, Niger; drinking water samples from the area showed 4.8 to 6.6 mg/L of F [15]. Relatively less information on exposure is available from Mali. One study reports F ranging from <0.2 to 1.7 mg/L (n = 19), with only one sample above the guideline [16]. In a geologically-based model of F occurrence above WHO guidelines, Amini et al. [17] predict probability increasing from close to 0% to 80% in Mali and Niger along southwest to northeast axes in both countries.

Epidemiological studies show exposure to Al and Mn to be risk factors associated with mental impairment. Specifically, Al is associated with the development or acceleration of onset of Alzheimer's disease [18]. Exposure to Mn is associated with intellectual impairment in children. Studies link levels of Mn greater than 1000 μg/L in drinking water with: child attention and memory impairments, impaired manual dexterity and speed, and neurological symptoms including a repetitive stuttered, speech, poor balance, coordination, and fine motor skills [19].

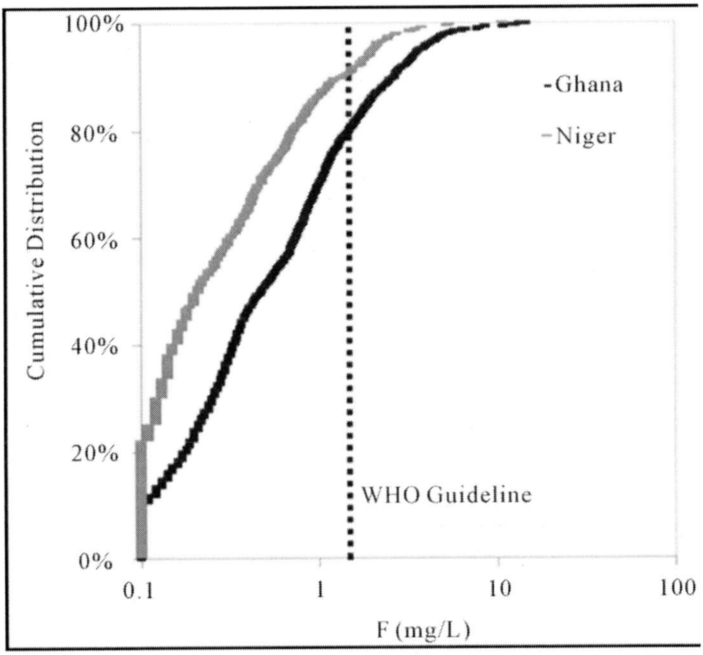

Figure 2. Cumulative distribution as percenatge of fluoride samples from Ghana and Niger.

The Al guideline was exceeded in 5.6% of samples from Ghana, 8.6% from Mali, and 5.0% from Niger (**Figure 3**). In Ghana, high concentrations of Al (mean value 4455 μg/L, n = 60) in groundwaters were reported to be localized [20]. A review of chemical water quality from 230 boreholes, wells, and standpipes of different regions in Ghana reported 95% of samples had Al above the guideline value, in particular in the Volta Region [1]. The localized nature of Al occurrence may explain why relatively fewer samples were in excess of the WHO guideline in this study, which evaluated only hand pumps and sampled a larger geographical area. No similar studies as to Al concentrations in Mali or Niger were found.

The guideline for Mn was exceeded in 7.5% of samples from Ghana, 5.6% from Mali, and 1.0% from niger (**Figure 1**). Rossiter et al. [1] report Mn to be one of the most widespread parameters in Ghana, exceeding the WHO guideline in 11% of samples from that study. No studies as to Mn concentrations in boreholes from Malior Niger were found.

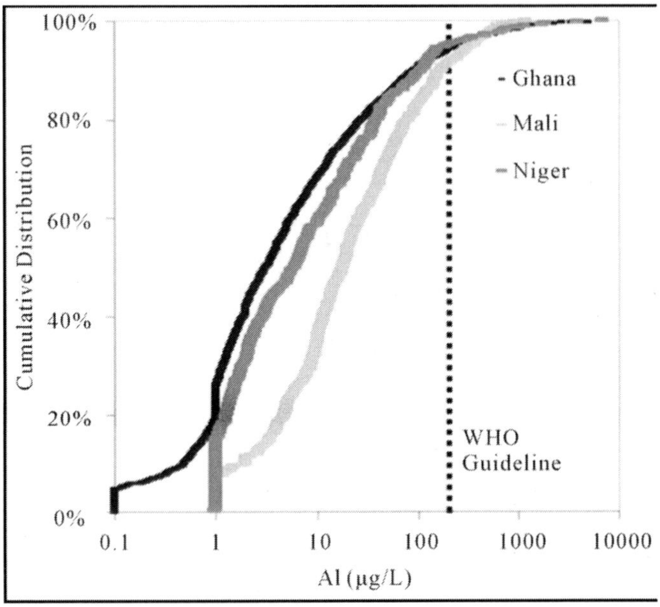

Figure 3. Cumulative distribution as percenatge of aluminum samples from Ghana, Mali, and Niger.

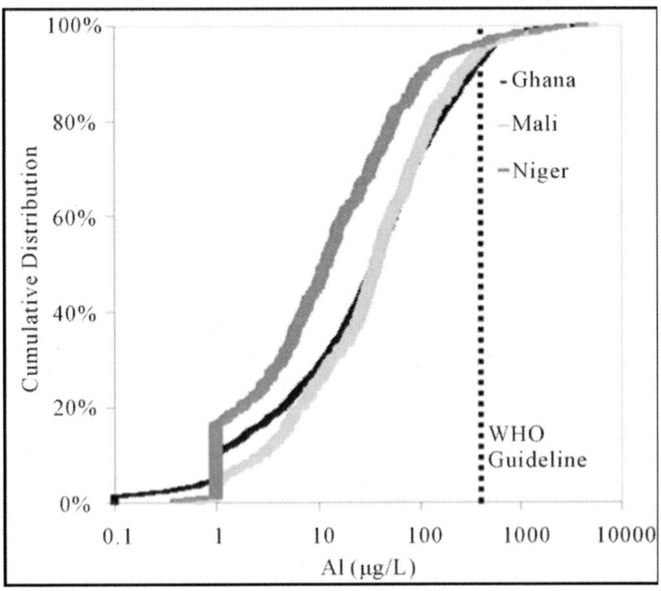

Figure 4. Cumulative distribution as percenatge of aluminum samples from Ghana, Mali, and Niger.

It is difficult to evaluate rates of mental impairment resulting from exposure from drinking water with few drinking water data and without a baseline of mental health. In general, there is little information as to mental health disorders and rates of diagnoses in developing countries. A general survey of mental health services in 185 developing countries (including Ghana) reports a scarcity of available resources relative to the populationlevel need, chronic shortages of trained personnel, shortfalls between the total health expenditure and those for mental health, and incomplete mental health policy frameworks [21,22].

Even with a baseline of mental health, the underlying cause may be difficult to determine. Severe malnutrition during childhood can lead to impaired psychological, intellectual, and motor skill development, ultimately leading to functional impairment [23]. Percentages of the under-five population characterized as moderately and severely underweight are: 14% in Ghana, 27% in Mali, and 40% in Niger [24]. Rates of wasting, which is defined as being two standard deviations below the median height-for-weight of the WHO child growth standards, are: 9% in Ghana, 15% in Mali, and 16% in Niger [24].

Heavy metals such as As, Pb, and U are associated with DNA damage, cancers, and damage to the central nervous system [2,5]. Relatively more information is available for health effects of As in drinking water than for some of the other parameters discussed in this study. Drinking water contaminated with arsenic is a major public health problem reported in many countries of the world [11]. Epidemiological studies show that long-term consumption of elevated levels of As in drinking water is directly related to increased cancers, skin ailments, neurological disorders, and type 2 diabetes [2,5]. In children, exposure to As from drinking water is associated with intellectual impairment [19].

Arsenic exceeded the guideline in 6.6% of all samples from Ghana, none from Mali, and 0.7% from Niger. Mitigation measures for high As levels are being taken by GRWP in Ghana. Other studies in Ghana report guidelines exceeded for less than 2% of samples and occurrence to be localized [1,25]. Pelig-Ba et al. [26] reports no arsenic from boreholes (n = 33). Few data are available for Mali and Niger, but a global model predicting As in groundwater shows probabilities of occurrence as: none for Niger, up to 50% for localized areas of Mali and Ghana, and up to 100% for localized areas of Ghana [27].

Though U is deposited at bone surfaces, where alpha radiation is emitted, risks arising from the biochemical toxicity of U as a heavy metal are considered to be about six orders of magnitude higher than those deriving from its radioactivity [28]. The guideline for Pb was exceeded for 1.3% of samples from Ghana, 14.2% from Mali, and 9.7% from Niger; the guideline for U was exceeded for 4.3% of samples from Ghana, 2.6% from Mali, and 17.1% from Niger.

In Ghana, Pb and U exceeding WHO guidelines are attributed to mining activities [1]. In one area, high Pb and Cr was attributed to use of agro-chemical [29]. Concentrations of Pb were reported "occasionally" exceeding the guideline [20,30] or not exceeding it at all [25]. The French NGO, CRIIRAD, reported water and soil from mine areas in Niger to be contaminated with dangerously high radioactivity levels [31], though values and numbers of samples were not given. In general, limited water quality data are available for these parameters in Mali and Niger.

It is difficult to evaluate rates of cancers directly resulting from exposure from drinking water. Detailed information as to rate, type, and survival of cancer in developing countries is scarce. A global review of cancer rates estimated that cancer rates in developed countries are twice those of developing countries with the discrepancy, in part, attributed to risk factors, diagnosis, and treatment [32]. In order of estimated rate of diagnosis in developing countries are: lung, stomach, and liver for men; and breast, cervix/uterus, and lung for women [32]. The National Cancer Registry in the Gambia shows liver cancer to be the most common cancer among men (62%) and second-most common cancer among women (28% following cervical cancer, which is 29%), with occurrence attributed to prevalence of chronic hepatitis in the population [33].

Finally, Barium at levels above the guidelines may increase the risk of hypertension, but neither mutagenic not carcinogenic impacts are reported in the literature [2]. Barium exceeding the guidelines was found in 6.6% of samples from Ghana, 2.1% from Mali, and 4.2% from Niger. In Ghana, reported concentrations of Ba in groundwater (mean value 125.3 μg/L, n = 60) and localized [20]. No similar studies as to Al concentrations in Mali or Niger were found.

4.3. Caveats

Several caveats are noted for this study. Twelve parameters exceeded WHO drinking water guidelines. While exposure to any of these parameters including Cd, Mo, Ni, Se, Zn, is not trivial, the impact on human health of less-frequently-occurring parameters may be outweighed by reduction of water-related infectious diseases from microbially contaminated surface water [2]. As an example, the prevalence of liver cancer due to water-borne hepatitis may be reduced if alternative water sources are available to the population of The Gambia.

Another consideration is that most drinking water quality studies referencing these guidelines (including this one) focus on the amount of a parameter that is ingested in water rather than total exposure [12], multiparameter effects, and likely underestimate the amount of water consumed in very hot, arid climates. Multi-parameter effects, either adverse or beneficial, are unclear. For example, As toxicity is aggravated by Sb, and As increases toxicity of Ni, Cd, Co, and Cr to where the WHO guideline should be 1/20 of the suggested value; conversely, toxicity of As may be reduced by intake of Se and Zn, though in amounts beyond those typically occurring in drinking water alone [2].

In this study, no samples exceeded Sb guidelines, and restricting criteria to where As exceeded the guideline and Sb exceeded the detection limit yielded seven samples, all from Ghana. Overall, number of samples exceeding guidelines for Ni was less than 0.01 and less than 1% for Cd. No samples exceeded guidelines for Cr and there are no guidelines for Co. Restricting criteria to where As exceeded the guideline and Ni was 1/20 of the guideline value (3.5 μg/L) yielded 78 samples, all from Ghana. This exercise could not be done for Cd, as 1/20 of the 3 μg/L WHO drinking water guideline is below the detection limit. Restricting criteria to where As exceeded the guideline and Cr was 1/20 of the guideline value (2.5 μg/L) yielded 15 samples, all from Ghana.

With respect to mitigating toxicity, the highest value of Zn in a sample with As exceeding the guideline was 697 μg/L. Depending on water intake, other food sources, and weight of the consumer, the WHO recommended daily intake value of Zn of 15 mg may or may not be met. The latter is more likely. Restricting criteria to where both As and Se exceeded the guideline yielded 23 samples, all from Ghana. The WHO recommended daily intake value of Se is 60 μg/kg for women and 70 μg/kg for men. As with Zn, meeting the daily intake of Se will depending on other sources. This finding supports the notion that there is not likely to be enough Zn in drinking waters to reduce toxicity of As.

Countries with very hot climates and high incidence of fluorosis, such as China and India, have established stricter standards than the WHO. The Chinese and Indian standards for fluoride in drinking water are less than 1 and 1.2 mg/L, respectively [34,35]. In determining the optimal fluoride concentration in drinking water for South India, Viswanathan et al. [36] considered average amounts of drinking water, tea or coffee, and food consumed and determined maximum fluoride concentration should be 0.5 mg/L, one third of the WHO guideline. In a similar study in northern Ghana, Apambire [37] used climate data to estimate the optimum concentration for fluoride in drinking water to be 0.4 to 0.6 mg/L.

It is noted that establishing, and amending, drinking water guidelines is difficult for a number of reasons. Data on human health effects are collected from studies carried out laboratory animals and the human population. Results of laboratory studies using high doses of an element on animals yields uncertainty as to the actual effects on human health. Results of population studies are complicated by duration of studies (on the order of decades), scarce information regarding exposure rates, simultaneous presence of other toxic agents, and the fact that some trace elements cause adverse effects only after extended exposure (years rather than months). In general, exhaustive data on health effects are often simply not available. As an example, underlying causes of mental health disorders are difficult to determine under the best of circumstances with respect to clean drinking water and proper nourishment.

5. CONCLUSIONS

The purpose of this study was to evaluate drinking water quality, assess risk as percent of population exposed to parameters, provide overview of human health effects, and evaluate evidence of health effects of drinking water quality in the population, if available. When risk was evaluated by three-country region, F was the most widespread parameter (9.7%) followed by: Mn, Al, Ba, U, and As. In Ghana, F was the most widespread parameter (11.4%) followed by: Mn, As, Ba, Al, and U. In Mali, Pb was the most widespread parameter, followed by Al and Mn. In Niger, U was the most frequent parameter (17.1%) followed by Pb, Al, Ba, and Mn.

An overview and evidence of health effects depended on the parameter, with relatively more information available for some parameters than for others. For F, for instance, there are reports of skeletal fluorosis in Niger and dental fluorosis in Ghana. For Mn and Al, it is difficult to assess mental health due to lack of baseline data and other underlying causes such as childhood malnourishment. For other heavy metals, including Pb, U, and As, it is difficult to quantify rates of cancers water directly related to exposure from drinking water; detailed information as to rate, type, and survival of cancer in developing countries is scarce.

In general, F, Mn, and Al had relatively larger and more common rates of occurrence and mitigation measures should be carefully considered. Exposure to other parameters including As, Pb, and U is not trivial and the risks to community members should also be considered. Evidence suggests that many of the parameters may be localized; detailed geological and geochemical mapping is crucial to understand water-rock interactions and, in turn, identify areas of higher risk. It is anticipated that information from this study may be used to support water resources management, borehole drilling activities, and planning process of new drinking water projects. This is particularly important as the demand for drinking water from groundwater is expected to increase.

ACKNOWLEDGEMENTS

Alexandra Lutz, Samuel Diarra, W. Braimah Apambire, James M. Thomas, Jarvis Ayamsegna The authors wish to thank field staff and the various communities in which sampling took place. Funding for the project was provided, in part, by the Conrad N. Hilton Foundation.

REFERENCES

1. H. Rossiter, P. Owusu, E. Awuah, A. Macdonald and A. Schäfer, "Chemical Drinking Water Quality in Ghana: Water Costs and Scope for Advanced Treatment," Science of the Total Environment, Vol. 408, No. 11, 2010, pp. 2378-2386. doi:10.1016/j.scitotenv.2010.01.053
2. J. Buschmann, M. Berg, C. Stengel, L. Winkel, M. Sampson and P. Trang, "Contamination of Drinking Water Resources in the Mekong Delta Floodplains: Arsenic and Other Trace Metals Pose Serious Health Risks to Population," Environment International, Vol. 34, No. 6, 2008, pp. 756-764.
3. P. Gyau-Boakye, "Sources of Rural Water Supply in Ghana," Water International, Vol. 26, No. 1, 2008, pp. 96-104.
4. W. Edmunds and P. Smedley, "Groundwater Geochemistry and Health: An Overview," In: J. Appleton, R. Fuge and G. McCall, Eds., Environmental Geochemistry and Health, Geological Society of London, London, 1996, pp. 91-105.
5. World Health Organization, "Guidelines for Drinking Water Quality," 2011. http://www.who.int/water_sanitation_health/publications/2011/dwq_guidelines/en
6. United States Environmental Protection Agency, "Standard Methods for the Examination of Water and Wastewater," 18th Edition, United States Environmental Protection Agency, Washington DC, 1992.
7. United States Environmental Protection Agency, "Methods for the Determination of Metals in Environmental Samples Supplement 1," 2012. http://www.nemi.gov
8. P. Smedley and D. Kinniburgh, "A Review of the Source, Behaviour, and Distribution of Arsenic in Natural Waters," Applied Geochemistry, Vol. 17, No. 5, 2002, pp. 517-568. doi:10.1016/S0883-2927(02)00018-5
9. C. Dowling, R. Poreda, A. Basu and S. Peters, "Geochemical Study of Arsenic Release Mechanisms in the Bengal Basin Groundwater," Water Resources Research, Vol. 38, No. 9, 2002, pp. 12-1-12-8.
10. P. Smedley, J. Knudsen and D. Maiga, "Arsenic in Groundwater from Mineralised Proterozoic Basement Rocks of Burkina Faso," Applied Geochemistry, Vol. 22, No. 5, 2007, pp. 1074-1092. doi:10.1016/j.apgeochem.2007.01.001
11. J. Bunnell, R. Finkelman, J. Centeno and A. Selinus, "Medical Geology: A Globally Emerging Discipline," Geologica Acta, Vol. 5, No. 3, 2007, pp. 273-281.
12. J. Fawell, K. Bailey, J. Chilton, E. Dahi, L. Fewtrell and Y. Magara, "Fluoride in Drinking water," World Health Organisation, Geneva, 2006.
13. W. Apambire, D. Boyle and F. Michael, "Geochemistry, Genesis and Health Implications of Fluoriferous Groundwater in the Upper Regions of Ghana," Environmental Geology, Vol. 35, No. 1, 2007, pp. 13-24.
14. S. Ayoob and A. Gupta, "Fluoride in Drinking Water: A Review on the Status and Stress Effects," Critical Reviews in Environmental Science and Technology, Vol. 36, No. 6, 2006, pp. 433-487
15. S. Arji, "Hundreds of Children Poisoned by Tap Water," 2012. http://www.fluoridealert.org/news/hundreds-of-children-poisoned-by-tap-water

16. British Geological Survey, "Groundwater Quality: Mali," 2012. http://www.bgs.ac.uk/downloads [Citation Time(s):1]
17. M. Amini, K. Abbaspour, M. Berg, L. Winkel, S. Hug, E. Hoehn, H. Yang and C. Johnson, "Statistical Modeling of Global Geogenic Arsenic Contamination in Groundwaters," Environmental Science and Technology, Vol. 42, No. 10, 2008, pp. 3669-3675. doi:10.1021/es702859e
18. P. Ferreira, A. Piai-Kde, A. Takayanagui and S. SeguraMuñoz, "Aluminum as a Risk Factor for Alzheimer's Disease," Revista Latino-Americana Enfermagem, Vol. 16, No. 1, 2008, pp. 151-157.
19. G. Wasserman, X. Liu, F. Parvez, H. Ahsan, D. Levy and P. Factor-Litvak, "Water Manganese Exposure and Children's Intellectual Function in Araihazar, Bangladesh," Environmental Health Perspectives, Vol. 114, No. 1, 20006, pp. 124-129.
20. K. Pelig-Ba, "Trace Elements in Groundwater from Some Crystalline Rocks in the Upper Regions of Ghana," Water, Air and Soil Pollution, Vol. 103, No. 1-4, 1998, pp. 71-89. doi:10.1023/A:1004968109028
21. C. Ferri, D. Chisholm, M. Van Ommeren and M. Prince, "Resource Utilization for Neuropsychiatric Disorders in Developing Countries: A Multinational Delphi Consensus Study," Social Psychiatry and Psychiatric Epidemiology, Vol. 39, No. 3, 2004, pp. 218-227.
22. World Health Organization, "Atlas: Mental Health Resources in the World 2001," WHO, Geneva, 2001. http://www.who.int/mental_health
23. M. Blössner and M. de Onis, "Malnutrition: Quantifying the Health Impact at National and Local Levels," Environmental Burden of Diseases Series, Vol. 12, No. 12, 2005, pp. 1-43. http://www.who.int/quantifying_ehimpacts/publications/eb12/en/index.html
24. United Nations International Children's Emergency Fund, "State of the World's Children, 2012," 2012. http://www.unicef.org/sowc2012
25. British Geological Survey, "Groundwater Quality: Ghana," 2000. http://www.bgs.ac.uk/downloads
26. K. Pelig-Ba, A. Parker and M. Price, "Trace Elements from the Birrimian Metasediments of the Northern Region of Ghana," Water, Air and Soil Pollution, Vol. 153, No. 1-4, 2004, pp. 69-93. doi:10.1023/B:WATE.0000019933.83346.1d
27. M. Amini, K. Mueller, K. Abbaspour, T. Rosenberg, M. Afyuni, K. Møller, M. Sarr and A. Johnson, "Statistical Modeling of Global Geogenic Fluoride Contamination in Groundwaters," Environmental Science and Technology, Vol. 42, No. 10, 2008, pp. 3662-3668. doi:10.1021/es071958y
28. P. Milvy and R. Cothern, "Scientific Background for the Development of Regulations for Radionuclides in Drinking Water," In: R. Cothern and P. Rebers, Eds., Radon, Radium, and Uranium in Drinking Water, Lewis Publishers, New York, 1990, pp. 1-16.
29. K. Obiri-Danso, M. Adonadaga and J. Hogarh, "Effect of Agrochemical Use on the Drinking Water Quality of Agogo, a Tomato Growing Community in Ashanti Akim, Ghana," Bulletin of Environment, Contamination, and Toxicology, Vol. 86, No. 1, 2011, pp. 71-77.
30. K. Pelig-Ba, C. Biney and L. Antwi, "Trace Metal Concentration in Borehole Waters from the Upper Regions and the Accra Plains of Ghana," Water, Air and Soil Pollution, Vol. 59, No. 3-4, 1991 pp. 333-345. doi:10.1007/BF00211841
31. Author Unknown, "NIGER: Uranium—Blessing or Curse?" 2012. http://www.irinnews.org/Report/74738/NIGER-Uranium-blessing-or-curse

A. Jemal, F. Bray, M. Center, J. Ferlay, E. Ward and D. Forman, "Global Cancer Statistics," Cancer Journal for Clinicians, Vol. 61, No. 2, 2011, pp. 69-90.

32. D. Sighoko M. Curado, D. Bourgeois, M. Mendy and P. Hainaut, "Increase in Female Liver Cancer in the Gambia, West Africa: Evidence from 19 Years of Population- Based Cancer Registration (1988-2006)," PLoS One, Vol. 6, No. 4, 2010, Article ID: e18415.
33. Q. Ziang, L. Liang, L. Chen, C. Wang, B. Chen, X. Chen and M. Zhou, "Effect of Fluoride in Drinking Water on Children'S Intelligence," Fluoride, Vol. 36, No. 2, 2003, pp. 84-94.
34. Bureau of India Standards, "India Standards for Drinking Water," Bureau of India Standards, New Delhi, 1992.

35. G. Viswanathan, A. Jaswanth, S. Gopalakrishnan and S. Siva Ilango, "Mapping of Fluoride Endemic Areas and Assessment of Fluoride Exposure," Science of The Total Environment, Vol. 407, No. 5, 2009, pp. 1579-1587. doi:10.1016/j.scitotenv.2008.10.020
36. W. Apambire, "Geochemical Modeling and Geomedical Implication of Fluoriferous Groundwater in the Upper Regions of Ghana," Unpublished Dissertation, University of Nevada, Reno, 2000.

CHAPTER 12

Research on the Risk Associated with Rural Drinking Water Safety Based on Catastrophe Theory

Chao Wang[1], Fuquan Ni*[1,2], Yu Deng[2], Linlin Jiang[3]

[1]*College of Information and Engineering, Sichuan Agricultural University, Ya'an, China*

[2]*Provincial Key Laboratory of Agricultural Environmental Engineering, Sichuan Agricultural University, Ya'an, China*

[3]*College of Civil Engineering, Tianjin University, Tianjin, China*

ABSTRACT

In this study, based on the analysis of the catastrophic risk of rural drinking water, with the representative villages in Ya'an as study objects and setting the four evaluation standards: water quality, water quantity, the Guarantee Probability for water Supply and the Convenience Level of the Access to Water Supply as the basic framework, a Rural Drinking Water Safety Assessment was developed (RDWSA). Research showed that Catastrophe Theory was applicable in the RDWSA. Adding RDWSA based on Catastrophe Theory to the supportive system for decision-making in Ecological Hydrographic Management Decision Support System of Ya'an helped obtain the rank and results of RDWSE by the automatic calculation of programs, which could assist the risk assessment and risk management associated with rural drinking water in Ya'an.

Keywords: Catastrophe Theory, Risk, Rural Drinking Water Safety, Ya'an

1. INTRODUCTION

Water is one of the most important natural resources for human development. But with the development of the social economy, the discharge of industrial wastes and domestic pollutants has worsened the water environment. Less water remains suitable for human consumption. More and more countries and relevant organizations have paid growing attention to drinking water issues, especially drinking water safety (DWS). Ensuring the safety of drinking water has become a focus drawing attention of the whole world [1,2]. Over the past decades, the Chinese government has made significant efforts in expanding supply of safe drinking water in rural areas. However, over 300 million rural residents still have no access to safe drinking water, facing problems of shortge as well as the increasing contamination and subsequent health threats [3]. Acceptable levels for DWS are set so that drinking water risk that is low enough to be ignored. Drinking such water will not lead to severe disease or an increased risk of death [4].

In 2004, China Ministry of Water Resources and Ministry of Health published the rural drinking water safety and health assessment index system which composed by four indicators including water quality (WQL), water quantity (WQT), Convenience, and Guaranteed Rate, thereby judging the safety of drinking water [5]. Rural DWS is not only closely bound up with the residents' production activity and the domestic life but also plays a key role in the state economic development, the stability and sustainable development of the society. In the past, methods of subjectively determine the weight of the indexes including Entropy Weight [6] Analysis of Hierarchy Process (AHP) [7] were mainly applied. Compared with the previous assessment methods, Catastrophe Theory does not determine the index weight and reduce the subjective factor, and the calculation is simple and convenient. Catastrophe Theory is a well-established approach with applications in different research areas such as physics [8], building [9], environment [10], as well as be applied in business [11].

In this study, based on Catastrophe Theory, a hierarchical structure evaluation indexes system was derived. And with the aid of VB program language, Rural Drinking Water Safety Assessment was developed (RDWSA) based on Catastrophe Theory was inserted into the supportive system for decision-making in ecological hydrographic risk management and the evaluation of the DWS of resources in rural Ya'an was performed. In this way, the method of determining the values by weight, which is obviously subjective, is avoided, and the results are more scientific and reasonable.

2. MATERIALS AND METHOD

2.1. Overview of the Studied Area

Located on the western edge of the Sichuan basin, Ya'an lies between longitudes 101°56' and 103°20'E, and latitudes 28°51' and 30°56'N, joining the Sichuan basin and the Tibetan plateau. Besides Yucheng city zone, there are another seven counties in Ya'an including Mingshan county, Yingjing county, Hanyuan county, Shimian county, Tianquan county, Lushan county and Baoxing county.

The ground features of Ya'an are quite complicated for the criss-crossing mountains and the sharply cut landform. Ya'an has a very rich water reserve, with a total amount of 3.6 × 1010 m^3. The per capita water possession there is 4.6 times that national average and 3.9 times that provincial average [12]. Ya'an enjoys exceptional advantage in exploiting water resources. However the present state of the rural DWS is unfavorable. According to statistics, 580,200 residents distributed in 383 villages have a difficult access to drinking water or drink insanitary water, amounting to 46.8% of the agricultural population.

The main problems of the rural DWS are embodied in the following aspects: first, some water resources are with high fluoride content causing high morbidity rate of water-borne Skeletal Fluorosis, Kaschin-Beck disease and dental fluorosis, etc.; second, some towns in Lushan county and Tianquan county are known as heavy bilharzias-affected areas whose untreated water is severely schistosome-polluted; third, the ground water in Mingshan county and Yucheng city zone has high mirabilite content, which is undrinkable; fourth, it is extremely inconvenient to access drinking water for those who live in the terrain of medium and high mountains due to the widely spread population and the lack of transportation infrastructure and water resource facilities; fifth, some residents in the Yucheng city zone drink directly those untreated water from streams, reservoirs, pools and rivers; sixth, a heavy rain area though Ya'an is, seasonally water shortage and low GPWS in places where well water and stream water are main water resources remain a serious problem due to the uneven precipitation and less rainfall in winter and spring [5].

2.2. WQL Index

In 2005, the method of random sampling was adopted in the procedure of collecting data, based on the survey of the fundamental state and the recent general investigation on the rural drinking water (RDW) by the Health Ministry, considering comprehensively the studied area's topography, hydrography, water system, water resources, the distribution of diseases linked with drinking water and the types of water supply projects. Source water of central water supply, peripheral water, well water, reservoir water, water from waterworks and water resources often used by residents were set as sampling sites totaling 221 (**Figure 1**), and they were tested twice a month.

The physical, chemical, toxicological, bacteriological and microbiological parameters were detected, amounting to twenty-one factors including color, turbidity, odor, macroscopic items, pH, COD, total hardness, total dissolved solid, chloride, sulphate, Fe, Mn, fluoride, As, Hg, Cd, Cr, Pb, nitrate, total coliforms and total bacteria count. This study selected water resources in eight representative villages as key studying objectives in this evaluation of the rural DWS based on TC, as shown in **Table 1**.

2.3. Catastrophe Theory

Catastrophe Theory, which originated with the work Structural Stability and Morphogenesis of the French mathematician HHHRené ThomHHH in 1972 [13], is a mathematical theory studying non-continuous phenomena based on Topology and Structural stability theory.

It explains the sudden changes of the behavior of a system in nature and society by a series of mathematical models and describes the phenomena where the stability is lost and the system will make a sudden transition to a new, very different behavior [9,14]. It is characterized by classifying the critical points of the potential function to study the features of the instability in the space round them. Cusp catastrophe, Swallowtail catastrophe and Butterfly catastrophe are most commonly used for comprehensive evaluation [10,15]. As shown in **Table 2**, $f(x)$ is the state variable's potential function of the system and the coefficients of X, a, b, c and d are controlled variables of the state variable.

The potential function, $f(x)$, assembles all the critical points into a profile of equilibrium. By derivation, the bifurcation equation can be obtained. The bifurcation equation indicates that when controlled variables satisfy it, the system will suddenly transit into a new, very different behavior.

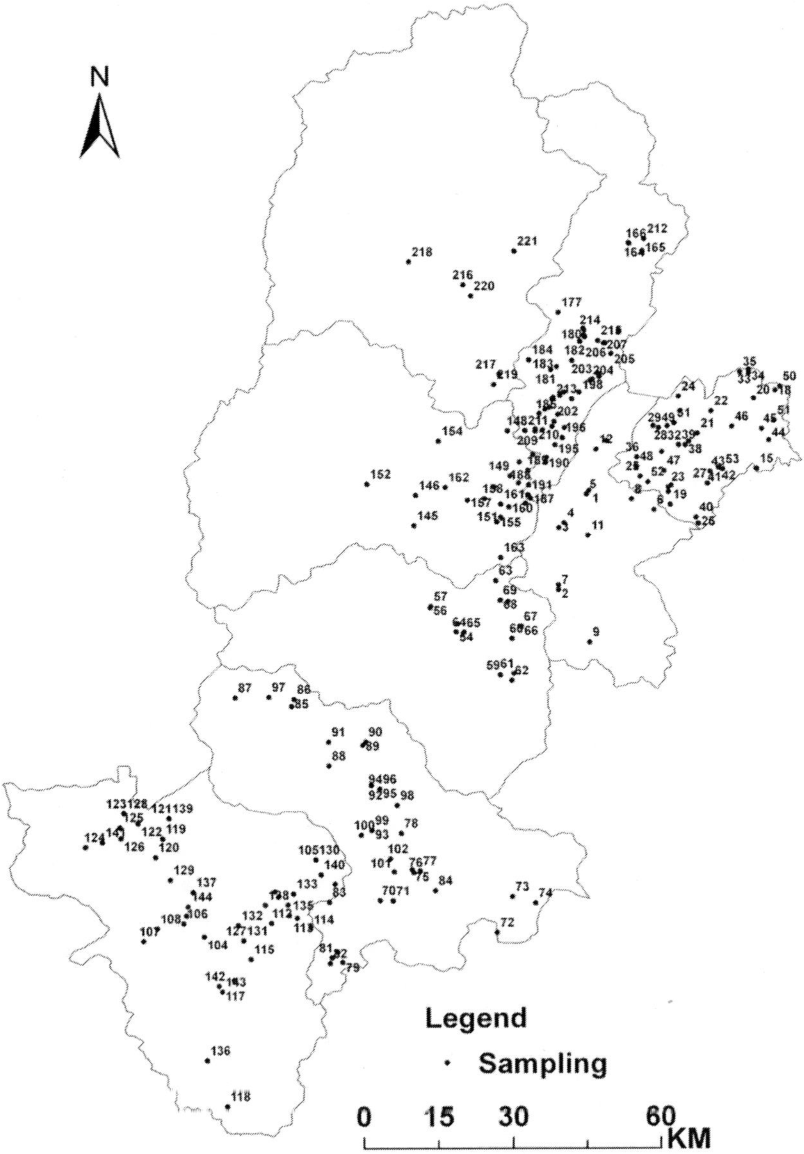

Figure 1. The distribution of sampling sites.

Table 1. The initial data of special samples.

1PPstPP layer	2PPndPP layer	3PPrdPP layer	4PPthPP layer	Liba Village. Yucheng City	Baizhang reservoir. Mingshan County	Wannian Village. Yingjing County	Dashu water factory. Hanyuan County	Zaiyang water supply station. Shimian County	Xiaohe Qiufeng Village. Tianquan County	Taiping River. Lushan County	Dayu Village. Baoxing County
			D(mg/L	174	0	100	0	97	435	200	0
		*C*1	*D*2	176.2	102	70	166	80	230	168	114
			*D*3(mg/L)	2	1.4	1.95	1	1.4	0	0.8	0
			D4	8.57	8.1	7.37	8.2	7.33	7.2	7.55	7.9
	*B*1	*C*2	*D*5 (mg/L)	0.2	0.08	0.04	0	0.2	0.2	0.05	0.04
			*D*6 (mg/L)	0.1	0.05	0.04	0	0.05	0.05	0.04	0.04
		*C*3	*D*7 (mg/L)	0	1.92	5	11.2	4.7	10.5	3	4.1
			*D*8 (mg/L)	22.8	18	6.23	40	10	120	22	37.5
		*C*4	*D*9	5	5	8	0	4	4	4	4
*A*1			*D*10	3	2	3	1	1	0.5	3	2
			*D*11 (mg/L)	0.12	0.2	0.015	0.1	0.2	0.8	0.2	0.08
		CRR_5	*D*12 (mg/L)	0.1	0.1	0.72	0	5.2	1.9	4	0
			*D*13 (mg/L)	0.004	0	0.008	0	0.04	0.04	0.015	0.015
	*B*2		*D*14 (mg/L)	0.01	0	0.008	0	0.04	0.008	0.008	0.008
		*C*6	*D*15 (mg/L)	0.006	0	0.045	0	0.045	0.045	0.008	0
			*D*16 (mg/L)	0.005	0	0	0	0.008	0.004	0.008	0
			*D*17 (mg/L)	0.001	0	0	0	0.0008	0.0008	0.0008	0
	*B*3		*D*18(CFU/mL)	260	100	203	2	34	100	40	210
			*D*19(CFU/mL)	230	3	360	0	2	235	10	110
		*A*2		0.728	1	0.962	0.972	0.498	1	1	0.985
		*A*3		0.868	0.878	1	1	1	1	1	0.940
		*A*4		0.965	0.950	1	0.979	1	1	1	0.923

Table 2. Catastrophe Theory Model.

Contents	Cusp catastrophe	Swallowtail catastrophe	Butterfly catastrophe
Potential function f(x)	$1/4xPF^4PP + 1/2axPP^2PP + bx$	$1/5xPP^5PP + 1/3axPP^3PP + 1/2bxPP^2PP + cx$	$1/6xPP^6PP + 1/4axPP^4PP + 1/2bxPP^3PP + 1/2cxPP^2PP + dx$
Bifurcation equation	$A = -6xPP^2PP$, $b = 8xPP^3$	$A = -6xPP^2PP$, $b = 8xPP^3PP$, $c = -3xPP^4$	$A = -6xPP^2PP$, $b = 8xPP^3PP$, $c = -3xPP^4PP$, $d = 4xPP^5$
Normalizing formula	$Xa = aPP^{1/2}PP$, $Xb = bPP^{1/3}$	$Xa = aPP^{1/2}PP$, $Xb = bPP^{1/3}PP$, $Xc = cPP^{1/4}$	$Xa = aPP^{1/2}PP$, $Xb = bPP^{1/3}PP$, $Xc = cPP^{1/4}PP$, $Xd = dPP^{1/5}$
Controlled variables	2	3	4

2.4. Method for Rural DWS Assessment

1) The establishment of the indexes of the comprehendsive evaluation of rural DWS

In this study, based on the intensive study of the features of one city zone and seven counties of Ya'an involved in DWS, referring to multiple theses that employed TC in studying water resources and combined with the data obtained in the experiment, the four standards: WQL, WQT, Convenience, and Guaranteed Rate, were set as the basic framework. Evaluation of drinking WQL is an important item in the evaluation of rural DWS. As an exploratory experiment, this study referred to the Sanitary Standard for Drinking Water (GB 5749 - 2006), established an index system with WQL as its mainstay and performed evaluation of rural DWS.

2) The standardization of indexes

Membership degree function value was adopted for the calculation of the four indexes: forward formula was employed to calculate the WQL index and the formula is $x_{ij}= y_{ij}/\max(y_{ij})$: inverse formula was used to calculate WQT index, Convenience, and Guaranteed Rate, and the formula is $x_{ij} = \min(y_{ij})/y_{ij}$. $I = A, B, C, D; j = 1, 2, \cdots, m$ (m refers to the numbers of indexes), therefore ensures the final evaluation value tending to be on the small side. Convert the original data in the bottom level into catastrophic fuzzy membership function value within [0,1].

3) The normalization calculation

Based on the numbers of the indexes, the catastrophe model to which the controlled variables correspond was employed. Applying the normalizing formula to do comprehensive quantitative recursion calculation and obtain the total catastrophic membership function value. When applying the normalizing formula to calculate the corresponding value, X, of each controlled variable of the same object, the minimum value or the average value was taken: given that those controlled variables aren't substitutable, i.e. they can't complement each other and are in non-complementary relation, the minimum value will be taken to obtain the total catastrophic membership function value; on the contrary, if they are in complementary relation, the average value will be taken to obtain the total catastrophic membership function value. Based on the analysis of the result, the comprehensive evaluation value varies within [0, 1], the larger the value is, the less safe the rural drinking water is.

3. RESULTS AND ANALYSIS

3.1. Computation

According to the steps of the TC-based evaluation in **Table 1**, the normalization formula was applied to the comprehensive calculation of the indexes system established in **Figure 2**, from the lowest layer upward, until the catastrophe progression of the highest layer is got. Take the Liba Village in Yucheng City zone as example and the calculation procedure as follows:

Step 1, Normalization

$$X_{D1} = y_{ij} / \max(y_{ij}) = 174/435 = 0.4$$

$$X_{D2} = 176.2/230 = 0.766$$

$$X_{D3} = 2/2 = 1$$

Step 2, Catastrophe calculation As shown in **Table 2**, *D1, D2* and *D3* in the bottom layer, which construct a Swallowtail catastrophe, using the corresponding equation: X_{D1} = (0.4)1/2 = 0.6324, X_{D2} = (0.766)1/3 = 0.9149, X_{D3} = (1)1/4 = 1. In this equation, *D1, D2* and *D3* complement each other, acting together on the upper assessment indicators. Therefore, according to the principles of mean value, *C1* = (X_{D1} + X_{D2} + X_{D3}) /3 = 0.8492. Where, *D4, D5* and *D6* construct a Swallowtail catastrophe and *C2* = 1. In this equation, *C3, C4* constitute the cusp catastrophe and Butterfly catastrophe respectively, *C3* = 0.2874 and *C4* = 0.4476. In addition, according to the principles of mean value, *B1* is equal to 0.8763. Using the same method, *B2* is equal to 0.6393 and *B3* is equal to 0.9307, and so *A1, A2, A3*, and *A4* are equal to 0.9266, 0.6841, 1and 0.9567 respectively. The total catastrophe fuzzy membership function, A, is equal to 0.9587.

3.2. Analysis

With the support of the supportive system for the decision-making in ecological hydrographic risk management of Ya'an and applying VB program language, RDWSA based on Catastrophe Theory to was added to the system module.

This submodule was employed in the evaluation of the water safety of 221 water resources. This study will present in detail the evaluation results of drinking water safety in eight representative villages. Results were ranked and their catastrophic risk values are shown in **Figure 3**.

The catastrophic risk value 0.8762 of Dashu water factory in Hanyuan County ranks top indicating the rural DWS here is at the lowest risk among the eight representative villages. Hanyuan County is a mountainous area having complex and varied terrains and landforms. It is topographically high in the northwest and low in the southeast with deep mountainous river downcuts and huge longitudinal slopes. Influenced by Qionglai Mountain Ridge and river, the precipitation is uneven and the river valley is relatively dry and hot with an annual precipitation of 730.8 mm and an annual flow through of 731.3 mm, lower than the average level of Ya'an. The drinking water is with low heavy metal content while the fluoride content is quite high causing high morbidity rate of water-borne Skeletal Fluorosis, Kaschin-Beck disease and dental fluorosis.

The catastrophic risk value 0.9107 of Baizhang reservoir water in Mingshan County ranks second showing the rural DWS is at a relatively low risk level. Located on the western edge of the Sichuan basin, Mingshan landform is a syncline trough between the anticline of Mengshan Mountain and Zonggang Mountain, high in the northwest and low in the southeast. It consists mainly of hills and hummocks and Mingjiang River and the upstream of Qingyi River run through the terrain. Having a humid subtropical monsoon climate, the county is abundant in rainfall, with an annual precipitation of 1501.5 mm.

Waters that permeate through the quaternary loose sand gravel and the fractures in bedrock are the main type of groundwater in the studied area. All the WQL indexes are within the safety range required. Some areas within Mingshan County are with high iron and manganese content because the background value of iron and manganese in groundwater underneath different soil material and soil type varies significantly [16]. Rural residents are widely spread, the economic development lags relatively behind and the infrastructure facility is weak, therefore, groundwater remains the main drinking water resource for domestic use. That as much as 86% of the residents rely mainly on groundwater reduces the safety guarantee of drinking water in Mingshan County.

The catastrophic risk value 0.9163 of Taiping Great River in Lushan County ranks third, indicating the rural DWS is at medium or low risk. Lushan County has an annual precipitation as much as of 1313.1 mm, thus the rural residents here have a high Guaranteed Rate. And the PH value of the water is 7.55, very close to that of neutral water. However the County is a heavy bilharzias-affected area and its untreated water is severely schistosome-polluted.

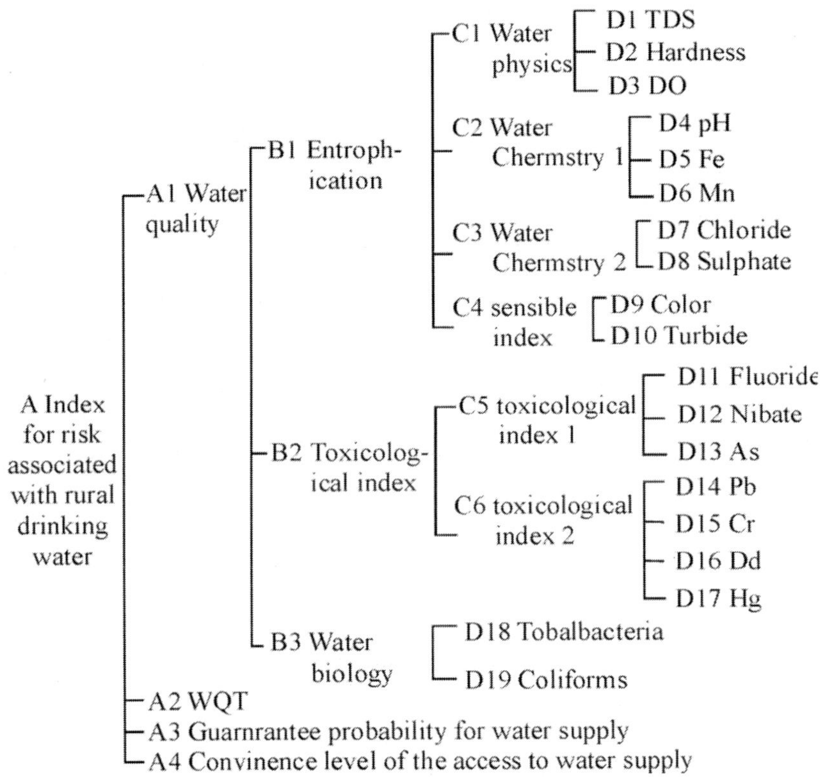

Figure 2. The index for risk associated with rural drinking water.

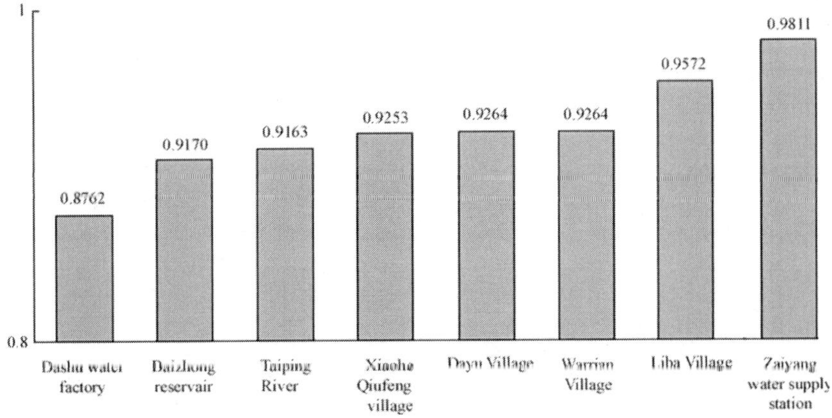

Figure 3. The catastrophe graphe of the case calculation.

The catastrophic risk value 0.9253 of Qiufeng group three, Xiaohe village in Tianquan County ranks fourth indicating the rural DWS is at medium risk. The landform is medium-deeply cut or deeply cut, high in the northwest and low in the southeast. Tianquan County is abundant in rainfall with an annual precipitation of 1660 mm. It rains at an extremely high frequency of 73%, over 200 d/a. The average annual surface runoff volume within the terrain in years is 3.771 billion m^3. The county has a forest coverage amounting to 50.23%, a total water resource of 6.714 billion m^3 and a per capita water possession of 2,724 m^3. The soil type here mainly comprises purple soil, yellow soil and yellow-brown soil, close to acid soil. The water is with high Calcium carbonate concentration resulting in high hardness which is almost 3.5 times that in Yingjing County, and it is with the highest content of dissolved solid and sulphate, causing high morbidity of gallstone.

The catastrophic risk value 0.9264 of Dayu Village, Lingguan Town in Baoxing County ranks fifth indicating the rural DWS is at medium or high risk. Mountains cover 99.7% of the area and with an annual precipitation of 953 mm, water shortage is likely to happen. Folds are highly concentrated within the terrain, the fracture development and construction impacts significantly on the topography and the soil material varies. Dissolved solids were absent and toxicological indexes were low in detection while the total bacteria are 210 CFU/mL, 2.1 times of the limit set by the Sanitary Standard for Drinking Water.

The catastrophic risk value 0.9264 of Wannian Village, Shizi Town in Yingjing County ranks sixth indicating the rural DWS is at high risk with low safety level. It is abundant in rainfall. The precipitation is uneven and flooding remains serious in those mountainous areas. The county has rich water resources with an annual average volume of 3.483 billion m^3, and the runoff is 1702 mm deep and the flow coefficient is 0.86. *E.* coli is 360 MPN/100mL·L, significantly exceeding the standard.

The catastrophic risk value 0.9572 of Liba Village, Yucheng city zone ranks seventh indicating the rural DWS is at a relatively high risk. Located in the central region of the Qingyi river Valley, Yucheng city zone is a broad area high in the west and low in the east. It is topographically an intermountain basin and a valley terrace of low and medium mountains. The city zone is abundant in orographic rainfall with an annual average precipitation of 1732 mm and annual rainy days of 218. As a relatively densely inhabited district and impacted by human activities, industrial wastewater and domestic wastewater, the amount of direct drinking water is insufficient causing drinking water shortage.

In this case, the catastrophic risk value of Liba Village is 0.9587 by hand calculation while it is 0.9572 when calculated by program. The 0.0015 difference results from rounding errors: the four digits after the decimal point were kept as valid figure in the hand calculation while in the program calculation the figures of far more than four digits were kept as valid figures, therefore, 0.0015 is a permissible error.

The catastrophic risk value 0.9811of the water supply station in Zaiyang town, Shimian County ranks eighth indicating the rural DWS here is at the highest risk among all the sampling sites. The exploitable water resources in Shimian County are quite rich while the precipitation and water suitable for direct drinking are in insufficient. During the low flow period of winter and spring, the water reserve in those villages relying on well water and stream water are in severe shortage, causing seasonal water shortage and low GPWS. The residents drank those substandard untreated water from the streams, reservoirs, pools and rivers, thus, the drinking water safety is at quite low level.

Although evaluation methods based on Catastrophe Theory are unavoidably subjective to certain extent, similar to that of other mathematical methods, it is still widely applied in the field of natural science. The calculation is simpler and more precise, therefore, this method is highly recommended by studies.

Based on the regional condition of the rural drinking water, apply practicable structural measures to increase resources of favorable quality, design a rational layout of the rural drinking water projects distribution and make a scientific planning of the engineering construction, taking into consideration of different geographic features of the mountainous areas, bugori, plains and suburbs. And particular emphasis should be given to the distribution of the population, the need for water supply and the economic development.

4. CONCLUSIONS

The attempt of employing it in the risk evaluation of the rural drinking water safety has provided a new method in the risk evaluation of drinking water safety. In this study, the magnitude of the evaluation indexes was determined and quantified by the normalization formula referring to the relationship among the impacting factors, which is relatively scientific and practicable. In this way, the method of determining the values by weight, which is obviously subjective, is avoided. And the normalization formula is more applicable, easier realized by programs.

Substandard water quality is most tropical in the rural drinking water safety problems in Ya'an, among which water with high iron content, high manganese content and high fluoride content are the most common problems. Therefore, the key issue to ensure the rural DWS in Ya'an is to solve its specific water quality problems.

ACKNOWLEDGEMENTS

Education Reform Project Fundation of the Sichuan Education Department (Grant No. P09150), the National Innovative Project for College Students (Grant No. 00509204), the Youth Science and Technology Innovation Fund of Sichuan Agricultural University (Grant No. 00530300), the Talent Introduction Fund of Sichuan Agricultural University (Grant No. 00530301), and the Double-Support Plan of Sichuan Agricultural University (Grant No. 00570103).

REFERENCES

1. J. Bartram, J. Cotruvo, M. Exner, C. Fricker and A. Glasmacher, "Heterotrophic Plate Count Measurement in Drinking Water Safety Management," International Journal of Food Microbiology, Vol. 92, No. 3, 2004, pp. 241-247. doi:10.1016/j.ijfoodmicro.2003.08.005
2. B. Ngwai, A. Sounyo, M. Fiabema, A. Agadah and O. Ibeakuzie, "Bacteriological Safety of Plastic-Bagged Sachet Drinking Water Sold in Amassoma, Nigeria," Asian Pacific Journal of Tropical Medicine, Vol. 3, No. 7, 2010, pp. 555-559. doi:10.1016/S1995-7645(10)60134-4
3. "Improved Water Resources Management and Drinking Water Safety in Rural China." http://www.undp.org.cn/projects/56957.pdf.
4. E. Hrudey, J. Hrudey and J. T. Pollard, "Risk Management for Assuring Safe Drinking Water," Environment International, Vol. 32, No. 8, 2006, pp. 948-957.doi:10.1016/j.envint.2006.06.004
5. F. Q. Ni, Y. L. Yang and H. Z. Ren, "The Study on the Assembled Gis-Based Supportive for the Rural Drinking Water Safety Decision-Making System," The 5th China Water Forum: Environment Variation and the Water Safety, Zhenzhou, 2007, pp. 221-224.
6. C. F. Yuan , H. J. Lei , H. W. Wang , H. W. Pan and J. X. Xu, "Evaluation on Rural Drinking Water Safety of Henan Province," Journal of Irrigation and Drainage, Vol. 28, No. 6, 2009, pp. 59-63.
7. B. H. Li, B. Y. Zhang and H. F. Qiao, "Study on Safety Index Evaluation for Rural Drinking Water in Jianghan Plain," Journal of Yangtze University, Vol. 5, No. 2, 2008, pp. 67-70.
8. D. J. Wales, "A Microscopic Basis for the Global Appearance of Energy Landscapes," Science, Vol. 293, No. 5537, 2001, pp. 2067-2670.
9. K. T. W. Yiu and S. O. Cheung, "A Dynamic Nonlinear Model of Online Retail Competition Using Cusp Catastrophe Theory," Building and Environment," Vol. 41, No. 4, 2006, pp. 438-447. doi:10.1016/j.buildenv.2005.01.007
10. J. T. Liu, J. F. Gao, J. H. jiang, J. C. Huang and J. H. Zhao, "Castrophe Theory-Based Risk Evaluation of Blue Green Algae Bloom in Different Regions of Lake Taihu," Journal of Lake Science, Vol. 22, No. 4, 2010, pp. 488-494.
11. W. Dou and S. Ghose, "A Dynamic Nonlinear Model of Online Retail Competition Using Cusp Catastrophe Theory," Journal of Business Research, Vol. 59, No. 7, 2006, pp. 838-848. doi:10.1016/j.jbusres.2006.02.003
12. M. X. Liu, "The Water Resources Management and Its Countermeasures," Chengdu Water Conservancy, 2003, pp.18-19.
13. R. Thom, "Structural Stability and Morphogenesis," Addison Wesley, Boston, 1975.
14. W. J. Wang, S. L. Liu, S. S. Zhang and J. W. Chen, "Assessment of a Model of Pollution Disaster in Near-Shore Coastal Waters Based on Catastrophe Theory," Ecological Modelling, Vol. 59, 2010, pp. 1-6.

15. Kuijper and M. J. Florack, "Using Catastrophe Theory to Derive Trees from Images," Kluwer Academic Publishers, London, 2004, pp. 1-32. doi:10.1007/s10851-005-0481-9
16. Y. F. Hu, L. J. Deng, S. R. Zhang, F. Q. Ni and J. Zhang, "Spatial Variability of Iron and Manganese Contents in Shallow Groundwater in the West of Sichuan Basin," Acta Ecologica Sinica, Vol. 29, No. 2, 2009, pp. 797-803.

CHAPTER 13

Filamentous Fungi in Drinking Water, Particularly in Relation to Biofilm Formation

Virgínia M. Siqueira [1], Helena M. B. Oliveira [2], Cledir Santos [1], R. Russell M. Paterson [1], Norma B. Gusmão [2] and Nelson Lima [1,*]

[1] *IBB-Institute for Biotechnology and Bioengineering, Centre of Biological Engineering, University of Minho, Campus de Gualtar, 4710-057 Braga, Portugal*

[2] *Department of Antibiotics, Federal University of Pernambuco, Av. Prof. Morais Rego, 1235, 50670-901, Recife, Pernambuco, Brazil*

ABSTRACT

The presence of filamentous fungi in drinking water has become an area worthy of investigation with various studies now being published. The problems associated with fungi include blockage of water pipes, organoleptic deterioration, pathogenic fungi and mycotoxins. Fungal biofilm formation is a less developed field of study. This paper updates the topic and introduces novel methods on fungal biofilm analysis, particularly from work based in Brazil. Further recommendations for standard methodology are provided.

Keywords: fungal biofilms; drinking water; contaminants; filamentous fungi; water distribution system

1. INTRODUCTION

The issue of filamentous fungi in drinking water has matured. This is a direct result of the work of key researchers in the area [1-11], with a crucial issue being identifying isolated strains to the species, rather than simply the genus level. When species identification is undertaken more can be deduced about changes in biodiversity within locations and with time. The properties of the isolates, such as mycotoxin, pigment, and odour formation, are more reliably ascertained because these are associated more with the individual species than the genus. Furthermore, water contaminated with fungi is of importance to hospitals, where immunocompromised patients undergo treatment [12]: *Aspergillus* accounts for the majority of infections, with *Aspergillus fumigatus* accounting for 90% of cases.

Kelley *et al.* [6] provided a comprehensive investigation of filamentous fungi in US distribution systems. In this work, *Alternaria alternata*, *Aspergillus niger*, *Cladosporium* sp., *Epiccocum nigrum* and many *Penicillium* sp. were prevalent. *Penicillium brevicompactum* was rather frequently detected and *P. expansum* considerably less so. However, none of these were detected from the two tap water sites, although *Fusarium* sp. were common together with an unnamed zygomycetes. Very few *Aspergillus* sp. were detected, including *A. fumigatus*. Portuguese tap water was studied more recently [8] and Norwegian and Australian drinking water studies have provided important additional knowledge [9,10,13]. To assist the reader, the most relevant reports from water distribution systems are listed on Table 1.

Table 1. Surveys of fungi in drinking water.

Country, Place, Year	Period of time	Type of water	Main isolation method	Most frequent fungal isolates	Refs.
United Kingdom, 1996	Autumn and Spring	Surface water and network	Membrane filtration, Direct plating and Bating	*Aspergillus*, *Cladosporium*, *Epicoccum*, *Penicillium* and *Trichoderma*	[4]
Greece, Thessaloniki, 1998	One collection (126 samples)	Tap water (hospital and community)	Membrane filtration	*Penicillium*, *Aspergillus* and *Acremonium*	[14]
Greece, 85 haemodialysis units, 1998	One collection (255 samples)	Municipal water supplies of haemodialysis centres	Membrane filtration	*Penicillium* and *Aspergillus*	[15]
Germany, North Rhine-Westphalia, 1998/9	12 months	Drinking water	Pour-plating	*Acremonium*, *Exophiala*, *Penicillium* and *Phialophora*	[5]
Norway, 14 networks, 2002/3	December, June and September	Drinking water (surface and groundwater)	Membrane filtration	*Penicillium*, *Trichoderma* and *Aspergillus*	[9,10]

Portugal, Braga, 2003/4	12 months	Tap water	Membrane filtration	*Penicillium* and *Acremonium*	[8]
Pakistan, Karachi, 2007	One collection (30 samples)	Water (and fruit juice)	Direct plating	*Aspergillus niger* and *A. clavatus*	[16]
Australia, Queensland, 2007/8	18 months	Municipal water	Membrane filtration	*Cladosporium, Penicillium, Aspergillus and Fusarium*	[13]
Brazil, Recife, 2009/10	5 months	Water treatment plant; tap water	Membrane filtration	*Penicillium, Aspergillus* and *Phoma*	[17]
Portugal, Lisbon, 2010	4 months	surface water; spring water; groundwater	Membrane filtration	*Aspergillus, Cladosporium, Penicillium*	[18,19]

2. METHODS TO STUDY FUNGI IN TAP WATER

One of the most important problems for studying fungi is the use of isolation media which tends to vary among researchers. Examples of media employed are listed in Table 2. DRBC and DG18 are recommended as general media for the purpose of isolation and enumeration of fungi in foods of high water activity (aw > 0.90), and is therefore suitable for water analyses. DG18 has several advantages; the characteristic colony appearance on agar plates facilitates differentiation of species which makes secondary selection and re-isolation of pure cultures easier to accomplish, compared to DRBC. DG18 also inhibits overgrowth of fast growing fungi such as members of the Mucorales and *Trichoderma*.

Table 2. Principal media used for the analysis of water for fungi.

Medium	Composition	
CMA/2—Corn meal agar half-strength	Corn meal	8.5 g
	Agar	8.5 g
	Distilled water	1000 mL
	pH	5.8–6.2
CZ—Czapek Dox agar	Czapek solution	10 mL
	K_2HPO_4	1 g
	Saccharose	30 g
	Agar	15 g
	Distilled water	1000 mL
	pH	6.0–6.4
	Czapek solution	
	$NaNO_3$	30 g
	KCl	5 g
	$MgSO_4 \cdot 7H_2O$	5 g
	$FeSO_4 \cdot 7H_2O$	0.1 g
	$ZnSO_4 \cdot 7H_2O$	0.1 g
	$CuSO_4 \cdot 5H_2O$	0.05 g
	Distilled water	100 mL

DG18—Dichloran 18 % glycerol agar	Peptone	5 g
	Glucose	10 g
	K_2HPO_4	1 g
	$MgSO_4 \cdot 7H_2O$	0.5 g
	Dichloran (0.2% in ethanol)	1 mL
	Glycerol	220 g
	Chloramphenicol	0.1 g
	Agar	15 g
	Distilled water	1000 mL
	pH	5.4–5.8
DRBC—Dichloran Rose Bengal chloramphenicol agar	Peptone	5 g
	Glucose	10 g
	K_2HPO_4	1 g
	$MgSO_4 \cdot 7H_2O$	0.5 g
	Dichloran	0.002 g
	Rose Bengal	0.025 g
	Chloramphenicol	0.1 g
	Agar	15 g
	Distilled water	1000 mL
	pH	5.4–5.8
MEA—Malt extract agar	Malt extract	20 g
	Peptone	1 g
	Glucose	20 g
	Agar	20 g
	Distilled water	1000 mL
	pH	5.0–5.5
NGRBA—Neopeptone glucose Rose Bengal aureomycin	Neopeptone	5 g
	Glucose	10 g
	0.67 % (w/v) aureomycin solution	5 mL
	1 % (w/v) Rose Bengal solution	3.5 mL
	Agar	15 g
	Distilled water	1000 mL
	pH	6.3–6.7
PDA—Potato dextrose agar	Potato extract	4 g
	Glucose	20 g
	Agar	15 g
	Distilled water	1000 mL
	pH	5.4–5.8
SDA—Sabouraud dextrose agar	Mycological peptone	10 g
	Glucose	40 g
	Agar	15 g
	Distilled water	1000 mL
	pH	5.4–5.8

Finally, DG18 does not readily produce substances toxic to the fungus when exposed to light: Media containing Rose Bengal are light sensitive and produce inhibitory compounds in significant concentrations after two hours exposure to light. The current authors recommend low and high nutritional media to obtain the complete spectrum of fungi present in water.

The phenotypic identification of fungi to species level is based on morphological characteristics such as (a) septation of hypha, and (b) the formation, morphology, patterns and branching frequency of fruiting-bodies and conidiophores. Also, the size, form, colour and ornamentation of the conidia are important characters (Figure 1).

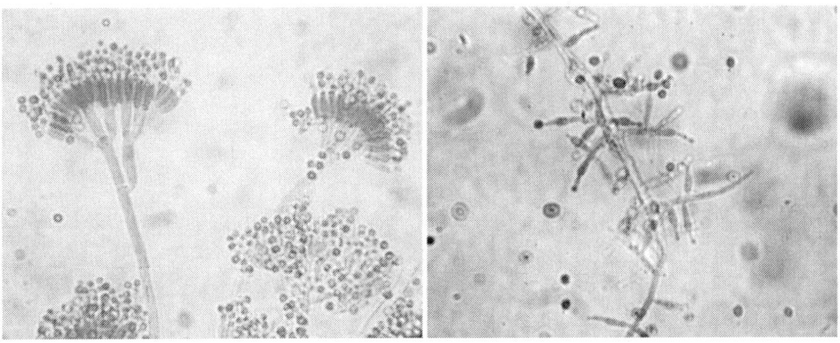

Figure 1. *Penicillium brevicompacum* (left) and *Trichoderma* sp. (right) isolated from tap water grown on MEA and observed under optical microscope after stained with Lactophenol Cotton Blue.

Macroscopic characters such as colony form, structure, size and colour are also important. In addition, appearance and growth on different culture media are used for differentiation (Figure 2). The morphological approach to species identification benefit from supplementation with physiological, biochemical and molecular characters. Ideally, a polyphasic approach to identification is undertaken to cover all aspects of species identity [20,21].

3. FUNGAL WATER BIOFILMS

Biofilms are problematic in water distribution systems as they permit a high concentration of organisms to occur, with the potential to cause some of the problems described above. Their presence can represent a health threat since biofilms play a role in the accumulation, protection and dissemination of pathogens [11,22]. However, the involvement of filamentous fungi in biofilms has not yet been demonstrated satisfactorily due to a dearth of useful techniques.

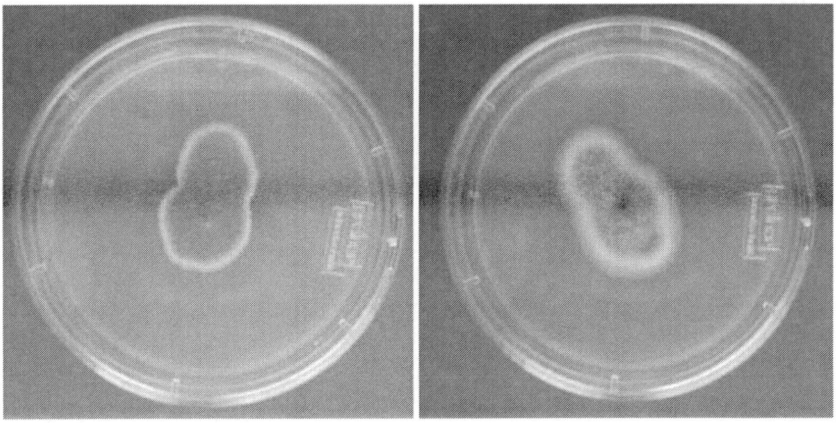

Figure 2. *Penicillium brevicompacum* (left) and *Penicillium aurantiogriseum* (right) isolated from tap water grown on MEA (N.B. it is significant that species names have been obtained rather than simply naming them as *Penicillium* spp.).

3.1. Sampling Biofilms

The methods which have been employed were often indirect, where the involvement of the fungus cannot be proved categorically. An example of this is placing the inner part of pipelines in contact with fungal growth medium and allowing fungi to emerge (Figure 3).

Nevertheless, many studies on drinking water distributed systems (DWDS) have shown that the major part of the biomass is attached to pipe surfaces in a biofilm [23]. The studies of biofilms in DWDS commonly face problems such as limited access to the biofilm and difficulty in repeating experiments, since pipe sections have to be replaced after sampling. A solution is to construct sampling devices that allow for *in situ* investigation of the attached microbial community.

Examinations of DWDS reveal the complexity that will be required of such a system. Several factors are known to influence biofilm development, including temperature, nutrients, residual disinfectant, the hydraulic regime and the characteristics of the substratum [24]. Moreover, various ecological interactions will occur between microorganisms in biofilms.

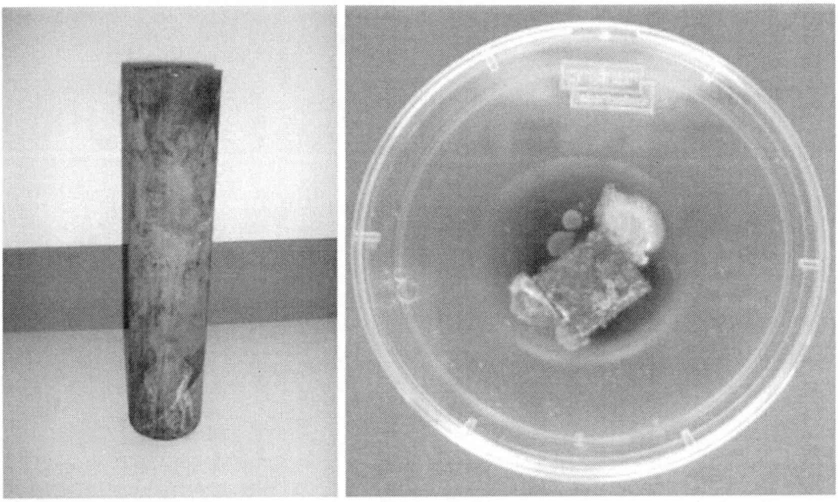

Figure 3. Cast iron tube from a tap water distribution system (left) and a NGRBA plate (see Table 2) with fungi growing from the biofilm present on the internal part of the tube (right).

Since biofilms are complex, laboratorial systems can be very helpful in determining the influence of individual parameters (e.g., medium composition and biocides application). Various attempts have been made to culture biofilms under controlled but realistic conditions, e.g., the Robbins device [25], the RotoTorque reactor [26], and flow cells [27]. However, do they reflect the *in vivo* situation?

We recommend the use of "sampler devices" to investigate natural filamentous fungi biofilms which were produced for a water distribution system called *"Alto do Céu"* in Recife, Brazil. They were developed largely because of the limitations inherent in the alternative methods mentioned previously. The devices can be employed to: (1) mimic the real conditions of the water network and yet be straight forward to insert and handle, (2) be convenient for transportation and storage, (3) maintain the integrity of biofilms and (4) allow *in situ* analyses of the biofilms. The core of the sampler devices consists of hollow polyvinyl chloride (PVC) pipes within polyethylene or acetate coupons held in place to allow biofilm growth (Figure 4).

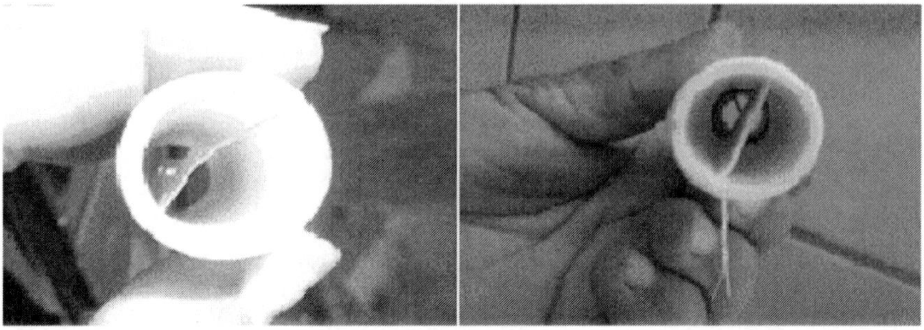

Figure 4. Sampler devices consisted of a hollow PVC pipe within acetate (left) or polyethylene (right) coupons.

The samplers had a diameter of 1.5 cm and a length from 7 to 10 cm. The ends of each sampler form a screw to connect multiple samplers or to close the device with a cap after removal from the water network (Figure 5).

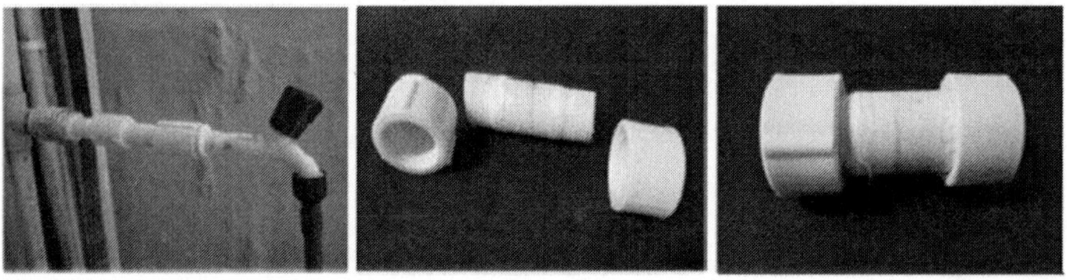

Figure 5. From left to right: (1) sampler device in place, (2) units of samplers to be connected to each other and (3) closed devices for transportation after a suitable period of time.

These features facilitated insertion, handling and removal of each sampler device after collection. The "caps" prevent contact with the external environment during transport. Finally, the pipes can be filled with water, thus maintaining moisture and preserving the integrity of the biofilms formed in the coupons. To study natural biofilms, these sampler devices were installed at different sampling points in the water network: a water treatment plant (reservoir; 20,000 m^3) and three branch connections within the municipal DWDS were selected (Figure 6). The coupons were removed for examination every 15 days for a period of 6 months.

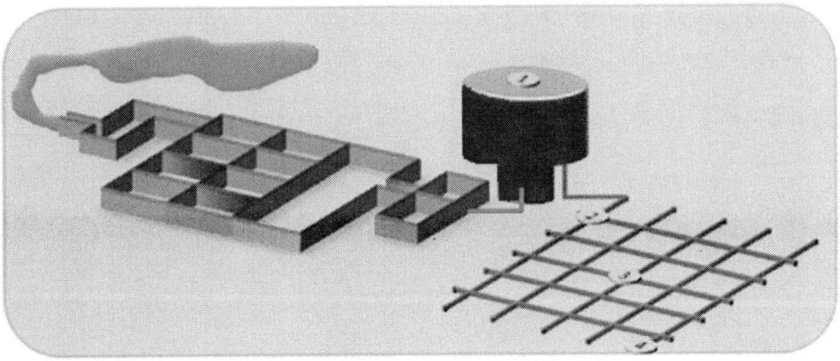

Figure 6. Drinking water distribution system. 1—Water treatment plant and reservoir; 2—beginning; 3—middle and 4—end of the network.

3.2. Biofilm Detection

The development of non-invasive and non-destructive techniques such as fluorescence microscopy, enables *in situ* monitoring of microbial biofilm communities. Fluorescence microscopy provides information on cell morphology, metabolism and biodiversity. Simultaneously, data concerning biofilm matrix structure and architecture are provided when used in conjunction with fluorescent molecular probes [28,29]. These data are crucial and are unattainable by conventional approaches (*i.e.*, culturing methods). They contribute to the understanding of "real time" microbial ecology within biofilms. In this context, samplers and real samples collected from *Alto do Céu* were used. The suitability of Calcofluor White MR2 (CW) staining for morphological characterisation, FUN1 for viability and Fluorescent *in situ* Hybridization (FISH) for diversity studies were investigated following the protocol presented in Figure 7.

A description of the various stains employed now follows: (1) CW is a fluorescent probe capable of making hydrogen bonds with β-(1→4) and β-(1→3) polysaccharides. CW consists of a symmetrical molecule with two triazole rings and two primary alcohol functions on both sides of an ethylene bridge. The fluorophore shows a high affinity for chitin forming hydrogen bonds with free hydroxyl groups which stains fungal cell walls blue. (2) FUN1 stain discriminates fungal dead cells which have a diffuse yellow-green fluorescence, from the metabolically active cells which have red Cylindrical Intra-Vacuolar Structures (CIVS). (3) For taxon diversity the (a) universal rRNA probe specific for Eukarya EUK516 (5′-ACCAGACTTGCCCTCC-3′, MWG Biotech, Ebersberg, Germany) labelled with the red Cy3 at the 5′ terminal and (b) FUN1429 probe specific for Eumycota (5′-GTGATGTACTCGCTGGCC-3′, MWG Biotech, Ebersberg,

Germany) labelled with Oregon-Green at the 5′ terminal for FISH were employed.

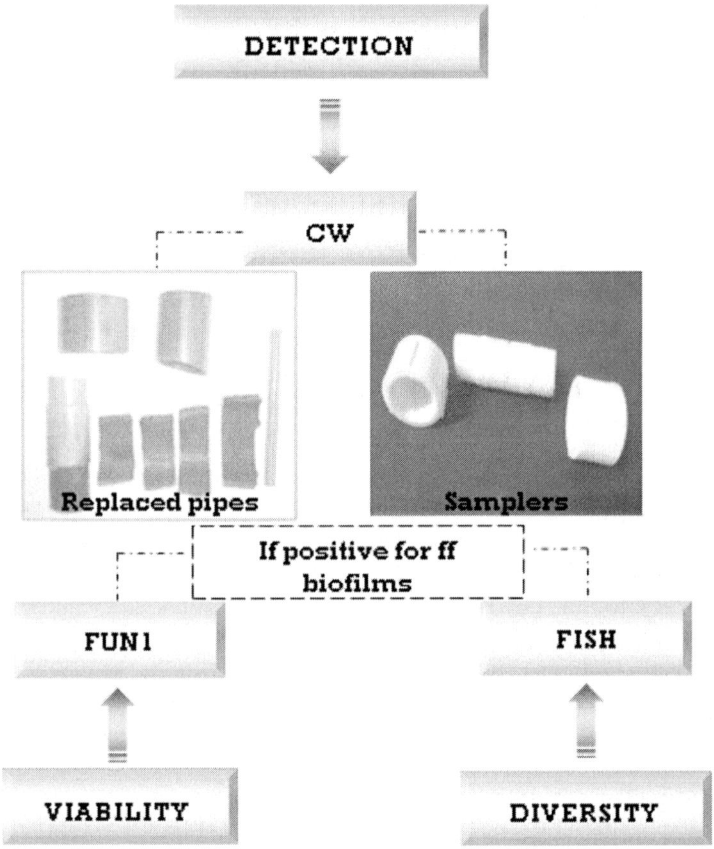

Figure 7. Stepwise approach used to study in situ filamentous fungal (ff) water biofilms using fluorescent staining techniques (for a description of terms see preceding paragraph).

A Olympus BX51 epifluorescent microscope, using UV light equipped with 40x/0.30 and 10x/0.65 objectives and a filter set (EX 450-490 nm, EM 520), was used for detecting *in situ* biofilms after fluorescent staining techniques were applied. The images were acquired with a Zeiss AxioCam HRc colour camera using the software CellB®. Storage and handling of reagents were performed as recommended by the supplier. The sampler study shows presumptive filamentous fungi stained with CW colonising the polyethylene coupons after 90 days in contact with water (Figure 8).

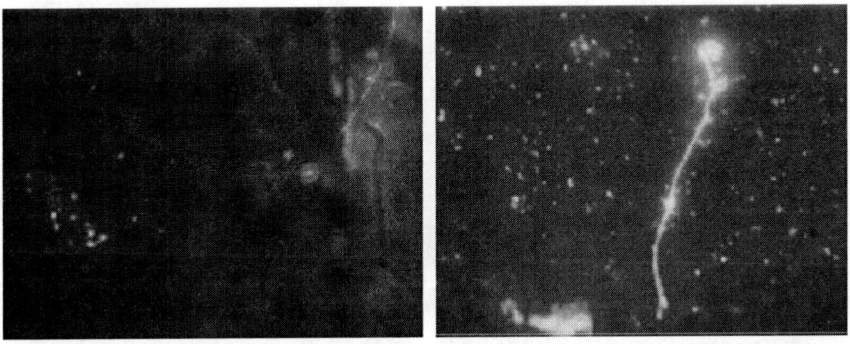

Figure 8. Presumptive filamentous fungal biofilm detection after CW staining on a polyethylene coupon.

Due to the polyethylene PVC autofluorescence, the coupons inside of the samplers were replaced by acetate coupons which have much less autofluorescesce. Notwithstanding this, biofilms were not detected on acetate surfaces of coupons after CW staining.

These unexpected results could occur for several reasons. Acetate coupons are not the optimal surfaces to promote biofilm adhesion, perhaps due to their hydrophobic properties. Alternatively, the length of time of the exposure may be inadequate for the described circumstances. These possibilities require further investigations and other materials also are required to be investigated. In contrast, when CW was applied to the actual replaced pipe samples, filamentous fungi were clearly observed (Figure 9).

Since the real pipes demonstrated filamentous fungi within the biofilm and deposits, the stepwise approach, as defined in Figure 7, was continued. The FUN1 staining for viability showed red CIVS inside of the fungal vacuoles (Figure 10) demonstrating that the fungi were viable in the biofilm. The CIVS are ATP-dependent which is correlated with fungal viability.

To confirm that the filamentous structures in the pipes were fungi, the samples were submitted to analysis by FISH probes. Figure 11 demonstrates a clear relationship between the CW filamentous structures presented on biofilms with the positive signals for FISH probes. The EUK516 probe shows that eukaryotic microorganisms were present and the FUN1429 probe confirm that these were fungi.

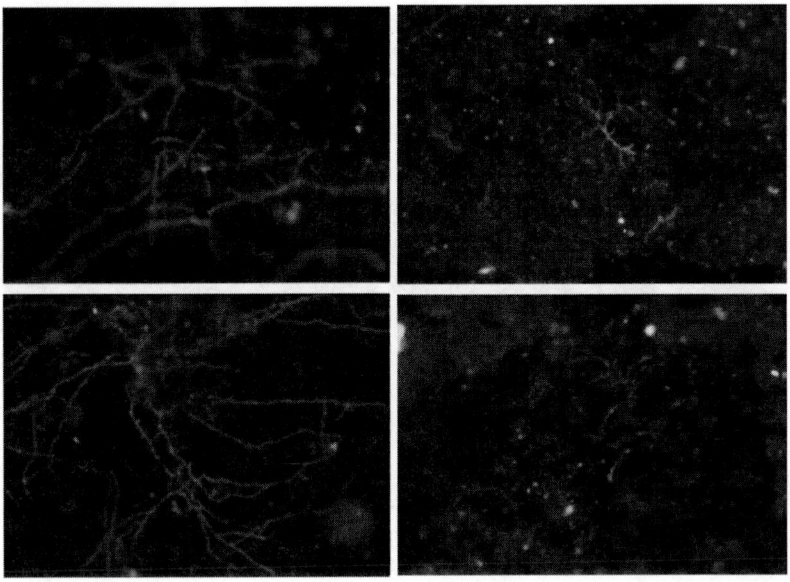

Figure 9. Filamentous fungal biofilms detected after CW staining in replaced pipe samples.

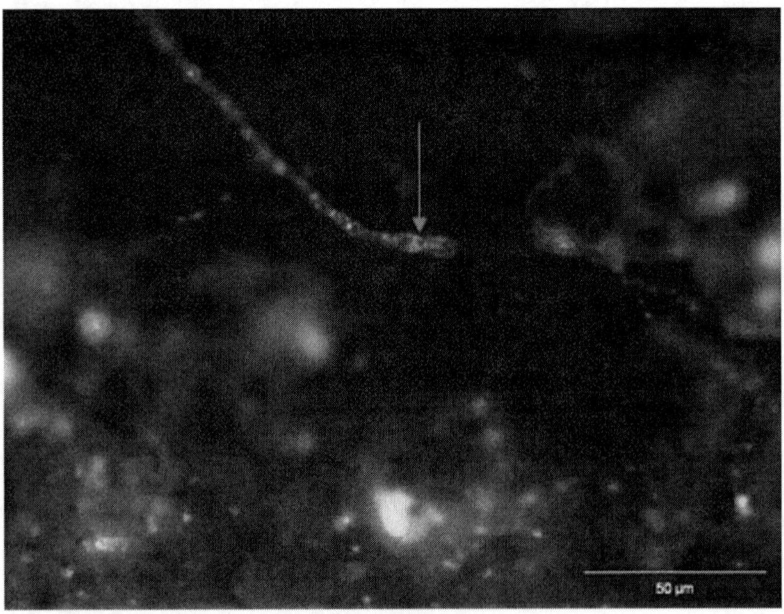

Figure 10. Visualization of red CIVS (arrow) and green diffuse hypha after FUN1 staining biofilm in real replaced pipe samples.

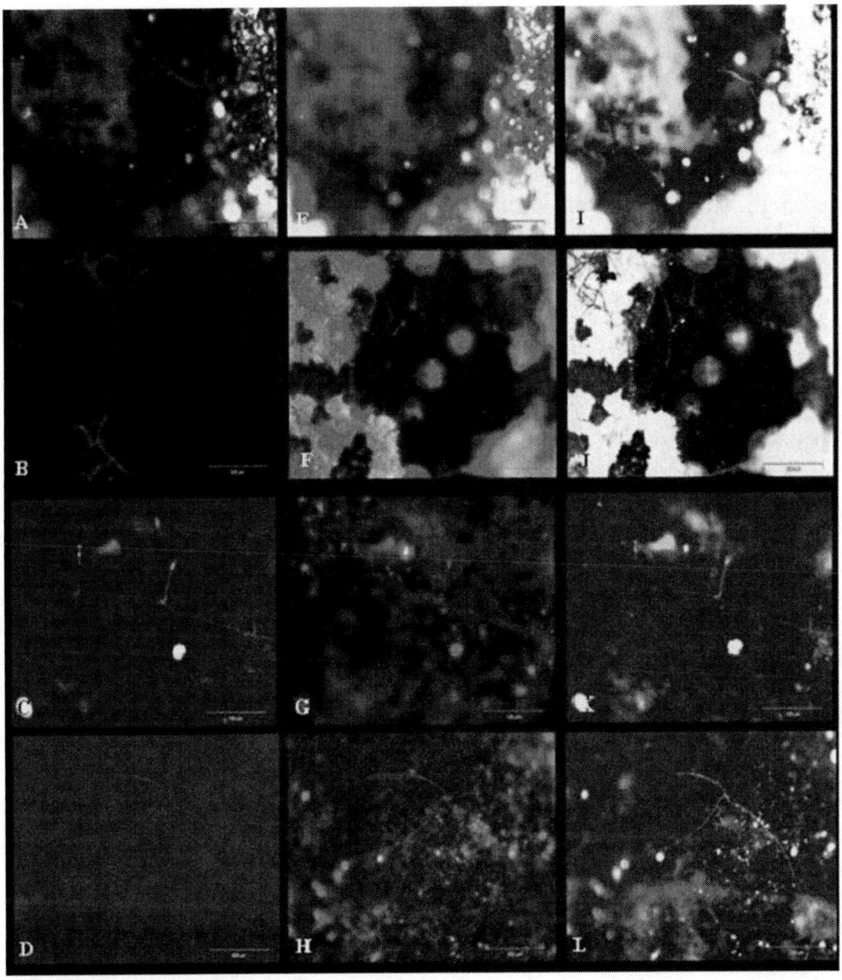

Figure 11. Detection of filamentous fungal water biofilm after CW staining (A-D), and analysis with FISH with EUK516 (E-H) and FUN1429 (I-L) probes in real replaced pipe samples.

4. CONCLUSIONS AND RECOMMENDED PROCEDURES

The present work is an updated overview concerning the main topics related to filamentous fungal contamination of water distribution system and in relation to biofilm formation. Standard methods are required, equivalent to those that exist for bacterial contamination of drinking water. The methods developed for fungal water biofilms demonstrated the presence of these microorganisms in the biofilms. Recommended standard methods for fungi have been developed and tested by the authors, and may be used to analyse water distribution systems for the occurrence of filamentous fungi [11]. The isolation method is based on NS 4716 [30] with some modifications, consist of filtering 100 mL of the water through a 0.45 μm membrane, placing the membrane on dichloran 18% glycerol agar (DG18) plates and incubating the plates at 20 ºC for 1–2 weeks. The plates must be checked visually every week, the colonies counted and the results presented as colony forming units (CFU)/100 mL water. Isolation of pure single colonies is required for identification, and pure cultures must be subcultured on potato dextrose agar (PDA) or malt extract agar (MEA) (Table 2).

To determine total filamentous fungi, the water sample must be filtered through two 0.45 μm membranes. The membranes are to be placed on neopeptone glucose rose Bengal aureomycin agar (NGRBA) with 5 replicates for filamentous fungal isolation and colony counting. The second filter should be placed onto half strength cornmeal agar (CMA/2) with five replicates for the same purposes. The method devised by us for the analysis of fungal biofilms in Brazil is also recommended more generally. Finally, even more progress will be made only if future studies ensure that strains are identified to species and standard techniques are employed universally.

ACKNOWLEDGMENTS

The authors acknowledge Brazilian Companhia Pernambucana de Saneamento (COMPESA) to support and make possible the work performed by H.M.B. Oliveira. V.M. de Siqueira is supported by the grant SFRH/BD/43719/2008 from Fundação para a Ciência e Tecnologia (FCT), Portugal.

REFERENCES

1. Nagy, LA; Olson, BH. The occurrence of filamentous fungi in drinking water distribution systems. *Can. J. Microbiol* **1982**, *28*, 667–671.
2. Nagy, LA; Olson, BH. Occurrence and significance of bacteria, fungi and yeasts associated with distribution pipe surfaces. Proceedings of the American Water Works Association, Water Quality Technology Conference, Houston, TX, December 1985; American Water Works Association: Denver, CO, USA, 1985; pp. 213–238.
3. Franková, E; Horecka, M. Filamentous soil fungi and unidentified bacteria in drinking water from wells and water mains near Bratislava. *Microbiol. Res* **1995**, *150*, 311–313.

4. Kinsey, GC; Paterson, RR; Kelley, J. Methods for the determination of filamentous fungi in treated and untreated waters. *J. Appl. Microbiol. Symp. Suppl* **1999**, *85*, 214S–224S.
5. Göttlich, E; van der Lubbe, W; Lange, B; Fiedler, S; Melchert, I; Reifenrath, M; Flemming, H-C; Hoog, S. Fungal flora in groundwater-derived public drinking water. *Int. J. Hyg. Environ. Health***2002**, *205*, 269–279.
6. Kelley, J; Kinsey, G; Paterson, R; Brayford, D; Pitchers, R; Rossmore, H. *Identification and Control of Fungi in Distribution Systems*; AWWA Research Foundation and American Water Works Association: Denver, CO, USA, 2003.
7. Paterson, RRM; Lima, N. Fungal Contamination of Drinking Water. In *Water Encyclopedia*; Lehr, J, Keeley, J, Lehr, J, Kingery, TB, III, Eds.; John Wiley & Sons: New York, NY, USA, 2005; pp. 1–7.
8. Gonçalves, AB; Paterson, RRM; Lima, N. Survey and significance of filamentous fungi from tap water. *Int. J. Hyg. Environ. Health* **2006**, *209*, 257–264.
9. Hageskal, G; Knutsen, AK; Gaustad, P; de Hoog, GS; Skaar, I. Diversity and significance of mold species in Norwegian drinking water. *Appl. Environ. Microbiol* **2006**, *72*, 7586–7593.
10. Hageskal, G; Gaustad, P; Heier, BT; Skaar, I. Occurrence of moulds in drinking water. *J. Appl. Microbiol* **2007**, *102*, 774–780.
11. Hageskal, G; Lima, N; Skaar, I. The study of fungi in drinking water. *Mycol. Res* **2009**, *133*, 165–172.
12. Paterson, RRM; Hageskal, G; Skaar, I; Lima, N. Occurrence, problems, analysis and removal of filamentous fungi in drinking water. In *Fungicides: Chemistry, Environmental Impact and Health Effects*; De Costa, P, Bezerra, P, Eds.; Nova Science Publishers: New York, NY, USA, 2009; pp. 379–399.
13. Sammon, NB; Harrower, KM; Fabbro, LD; Reed, RH. Incidence and distribution of microfungi in a treated 9 municipal water supply system in sub-tropical Australia. *Int. J. Environ. Res. Public Health* **2010**, *7*, 1597–1611.
14. Arvanitidou, M; Kanellou, K; Constantinides, TC; Katsouyannopoulos, V. The occurrence of fungi in hospital and community potable waters. *Lett. Appl. Microbiol* **1999**, *29*, 81–84.
15. Arvanitidou, M; Spaia, S; Velegraki, A; Pezarloglou, M; Kanetidis, D; Pangidis, P; Askepidis, N; Katsinas, Ch; Vayonas, G; Katsouyannopoulos, V. High level of recovery of fungi from water and dialysate in haemodialysis units. *J. Hosp. Infect* **2000**, *45*, 225–230.
16. Nazim, S; Dawar, S; Tariq, M; Zaki, MJ. Quantitative estimation of mycoflora in drinking water and fruit juices of Karachi. *Pakistan J. Bot* **2008**, *40*, 1263–1268.
17. Oliveira, HMB. Fungos filamentosos na água e em biofilmes na rede de distribuição de água potável do sistema Alto do Céu, Recife-PE, Masters Thesis, Biologia de Fungos do Departamento de Micologia do Centro de Ciências Biológicas da Universidade Federal de Pernambuco (UFPE), Recife, Brazil. 2010.
18. Pereira, VJ; Basílio, MC; Fernandes, D; Domingues, M; Paiva, JM; Benoliel, MJ; Crespo, MT; San Romão, MV. Occurrence of filamentous fungi and yeasts in three different drinking water sources. *Water Res* **2009**, *43*, 3813–3819.
19. Pereira, VJ; Fernandes, D; Carvalho, G; Benoliel, MJ; San Romão, MV; Barreto Crespo, MT. Assessment of the presence and dynamics of fungi in drinking water sources using cultural and molecular methods. *Water Res* **2010**, *44*, 4850–4859.
20. Santos, C; Paterson, RRM; Venâncio, A; Lima, N. Filamentous fungal characterisations by matrix-assisted laser desorption/ionisation time of flight mass spectrometry. *J. Appl. Microbiol* **2010**, *8*, 375–385.
21. Santos, C; Fraga, ME; Kozakiewicz, Z; Lima, N. Fourier transform infrared as a powerful technique for the identification and characterisation of filamentous fungi and yeasts. *Res. Microbiol* **2010**, *161*, 168–175.
22. Huq, A; Whitehouse, CA; Grim, CJ; Alam, M; Colwell, RR. Biofilms in water, its role and impact in human disease transmission. *Curr. Opin. Biotechnol* **2008**, *19*, 244–247.
23. Boe-Hansen, R; Martiny, AC; Arvin, E; Albrechtsen, H-J. Monitoring biofilm formation and activity in drinking water distribution networks under oligotrophic conditions. *Water Sci. Technol***2003**, *47*, 91–97.
24. Niquette, P; Servais, P; Savoir, R. Impacts of pipe materials on densities of fixed bacterial biomass in a drinking water distribution system. *Water Res* **2000**, *34*, 1952–1956.

25. McCoy, WF; Bryers, JD; Robbins, J; Costerton, JW. Observations of fouling biofilm formation. *Can. J. Microbiol* **1981**, *27*, 910–917.
26. Van der Wende, E; Characklis, WG; Smith, DB. Biofilms and bacterial drinking water quality. *Water Res* **1989**, *3*, 1313–1322.
27. Caldwell, DE; Lawrence, JR. Study of attached cells in continues-flow slide culture. In *CRC Handbook of Laboratory Systems for Microbial Ecology Research*; Wimpenny, JWT, Ed.; CRC Press: Boca Raton, FL, USA, 1988.
28. Li, Y; Dick, WA; Tuovinen, OH. Fluorescence microscopy for visualization of soil microorganisms—a review. *Biol. Fert. Soils* **2004**, *39*, 301–311.
29. Gonçalves, AB; Santos, IM; Paterson, RRM; Lima, N. FISH and Calcofluor staining techniques to detect *in situ* filamentous fungal biofilms in water. *Revista Iberoamericana de Micologia* **2006**, *23*, 192–196.
30. Determination with the membrane filters method, NS 4716. In *Norwegian Standard for Water Analysis: Microfungi in Water*; The Standardization Organizations in Norway: Oslo, Norway, 1991; pp. 1–12. (In Norwegian)

CHAPTER 14

Effect of Disinfectants on Bacterial Quorum Sensing in Drinking-Water Distribution System

Qipei Jiang

Institute of Environmental Protection of Baoding, Baoding, China

ABSTRACT

Several kinds of disinfectants were investigated in this paper for their ability to inhibit quorum sensing activities in water in distribution system. Among these disinfectants, free chlorine is the strongest inhibitor for QS activities and all disinfectants have inhibiting effect on QS activities in water. The results offered a novel approach for preventing formation of biofilm in distribution system.

Keywords: Disinfectant, Inhibit, Quorum Sensing, Water of Distribution System, Formation of Biofilm

1. INTRODUCTION

Water is an important part of life and the development of human beings. Microbial safety of drinking water has always been a focus in the field of drinking water research. In recent studies of distribution systems, it was found that biofilm microorganisms predominated over corresponding plank tonic cells. A possible reason was the ability of biofilm microorganisms to metabolize recalcitrant organic compounds and their greater resistance to chlorine and other biocides [1]. Disinfection with chlorine dioxide and chlorite, for example, can reduce the concentration of planktonic bacteria, but has little effect on the concentration of biofilm bacteria. Lapses in chlorination can lead to regrowth of biofilm communities and increased resistance of biofilm bacteria to chlorine. Many "emerging pathogens" in drinking water systems have increased threats to the health of human beings. A large number of these pathogens can form biofilms. Such findings indicated the importance of biofilms to the microbial safety of drinking water.

Many studies suggest that bacterial quorum sensing (QS) plays an important role in the formation of biofilms [2][3]. To understand the formation of biofilms, it is important to study bacterial quorum sensing of microorganisms in the distribution system. Through controlling quorum sensing, biofilm bacteria can be turned into plank tonic cells that are easily killed by dissociative chlorine. Thus, it is possible to inhibit the formation of biofilms by controlling quorum sensing. The microbiological safety of drinking water could be enhanced using disinfectants to control cell-communications among microorganisms in distribution network systems.

In order to biocontrol communications among bacterial cells in the distribution system, some inhibitor was selected to decrease QS activities in water. Several kinds of oxidants and disinfectants were investigated in this paper for their ability to inhibit QS activities in water. Hydrogen peroxide, potassium permanganate, ozone, chloramines hypochlorite, and chlorine dioxide were chosen because they are all commonly used in drinking water treatment. The objective of this work was to find an ideal approach to inhibit the formation of biofilms in the distribution systems.

2. MATERIALS AND METHODS

2.1. Materials

2.1.1. Reporter Strain

Reporter strain KYC55 was obtained from Professor Jun Zhu, Nanjing Agricultural University. P. R. C [4].

2.1.2. Reagents

N-buty-DL-homoserine lactone (C4-HSL), N-OXO-hexanoyl-DL-homoserine lactone, (C6-OXO-HSL), N-OXO- dodecanoyl-DL-homoserine lactone (C8-OXO-HSL), and N-dodeca-noyl-DL-homoserinelactone (C12-HSL) were bought from Sigma.

2.1.3. AT Medium

The reporter strain KYC55 was cultured in AT medium containing KH_2PO_4 10.7 g; $MgSO_4{\cdot}7H_2O$ 148 mg; $CaCl_2{\cdot}2H_2O$ 8.8 mg; $FeSO_4{\cdot}7H_2O$ 7.5 mg; $MnSO_4H_2O$ 1.2 mg; $(NH_4)_2SO_4$ 2 g; glucose 5 g; NaOH 1.76 g and distilled water 1000 ml.

2.2. Methods

2.1.1. Empirical Method

Studying the effects of disinfectants on activities of bacterial quorum sensing in water: AHL powder was diluted to 1 mmol/L with sterile distilled water and was stored at −20°C. The AHL solution was diluted to 1 μmol/L before used. The disinfectants, which are often used in drinking water treatment were diluted and added to a 2-ml cuvette with AHL liquid. Then timed. The reaction liquid was called liquid 1 when the reaction time was arrived. The QS activities of liquid 1 was determined immediately after the reaction was complete [5].

2.1.2. Determination Method

QS activities were estimated using β-galactosidase activity method as described previously (2). Frozen KYC55 was melted in a vial and then 10 μl was used to inoculate 10 ml of AT medium in a cuvette without antibiotics, the 100 μl of AHL C6or liquid 1 was also added. Then The KYC55 cells were incubated for 16 hour at 28°C, until the OD 600 reached 0.2 to 1. The sample at this point was referred to liquid 2. OD600 readings were taken for the total cell suspensions. 0.2 ml of liquid 2 in AT medium were combined with 0.8 ml Z buffer, 10 μl of 0.1% SDS solution, and 15 μl of chloroform in 5 ml test tubes. Tubes were shaked vigorously for at least 10 sec and then 0.4 ml of 4 mg/ml ONPG was added. This point was recorded as time zero (T0). When the solution turned yellow, 0.6 ml of 1 M Na_2CO_3 was added to stop the reaction, and the time was recorded in minutes (Ts). An OD420 reading was taken for the reaction liquid was measured OD420. Results were calculated results in Miller units as follows:

$$\text{miller unit} = \frac{1000 * OD420}{OD600 * T * 0.2}$$

In this paper, β-galactosidase activity was indicated QS activities.

3. RESULTS AND DISCUSSIONS

3.1. Denotation QS Activities

Reporter strain KYC55 reacting different ranges of AHL molecules were indicated in Figure 1; Different types of AHL showed different β-galactosidase activity. 10 nmol/L C6 AHL showed the highest activity and 10 nmol/L C4, C12 demongstrated much lower.

Under normal conditions, AHL molecules are mixture in water in distribution systems. However, β-galacto- sidase activity on certain ratio AHL mixture was not stable. Therefore, 10 nmol/L C6 AHL was selected to measure QS activities in water in distribution systems. As indicated in Figure 1, QS activities of 10 nmol C6 was approximately 1000 miller units. This result was reliably obtained in several parallel experiments. In the experiments below, the effect of different disinfectants on 10 nmol/L C6 AHL liquid will be studied to denote inhibiting QS activities (Figure 2).

3.2. Chlorination

To supply safe, microbe-free drinking water, a chlorination process is usually conducted. Chlorination is the predominant disinfection method applied in water treatment due to its low cost, ease of application, and ability to inactivate a wide variety of pathogenic microorganisms.

However, an important drawback is the formation of harmful chlorination disinfection by-products (DBPs). Some of these DBPs have been linked to cancer and reproductive defects [6][7].

The effect of active chlorine on QS activities in water is shown in Figure 3. Free chlorine from Javel water can evidently inhibit QS activities in water. Even at low concentrations 0.05 mg/L free chlorine, it can significantly inhibit QS activities in water at 12 h. AHL molecules can be hydrolyzed by alkali in Javel water. MX, a kind of chlorination disinfection by-products and an AHL analog, are likely to be the reasons for decreasing QS activities.

3.3. Chloramination

There has been a strong trend toward replacing free chlorine with chloramine in drinking water treatment in order to meet stricter limits on chlorine disinfectant byproducts. Chloramine is generally thought to produce a more stable residual than free chlorine and thus provide lasting protection against bacteria regrowth in distribution systems. In addition, chloramine is believed to penetrate more deeply than chlorine within the biofilmthus providing for greater inactivation [8].

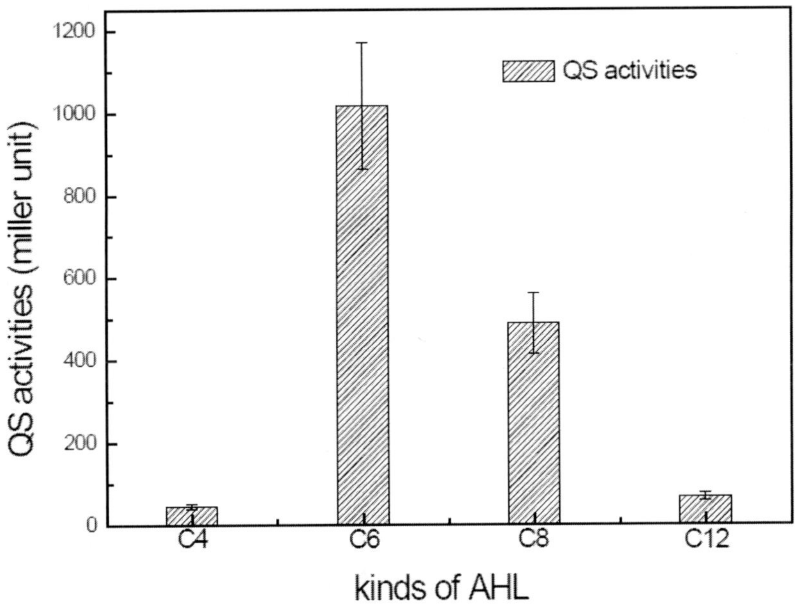

Figure 1. Effect of kinds of AHL on QS activities in water.

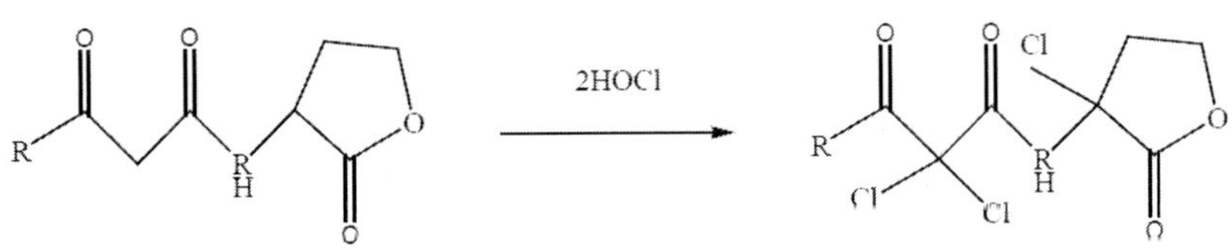

Figure 2. Reaction of C6 AHLs and HOCl.

Chloramine was generated using Javel water (1000 mg/L) and ammonium chloride solution (NH'$_4$?N 1000 mg/L) at Cl_2:N ratio of 4:1 (w/w), with a final pH = 9.0 and low temperature.

The effect of chloramines on QS activities in water is shown in Figure 4. Similar to Chlorine, QS activities in water decreased greatly with chloramination at higher dosages and longer times. QS activities was much lower, when concentration of chloramines reached over 5 mg/L and 12 hour. There are no activities over 20 mg/L in 12 hour. This indicated that chloramines need a long time to react with AHL molecules.

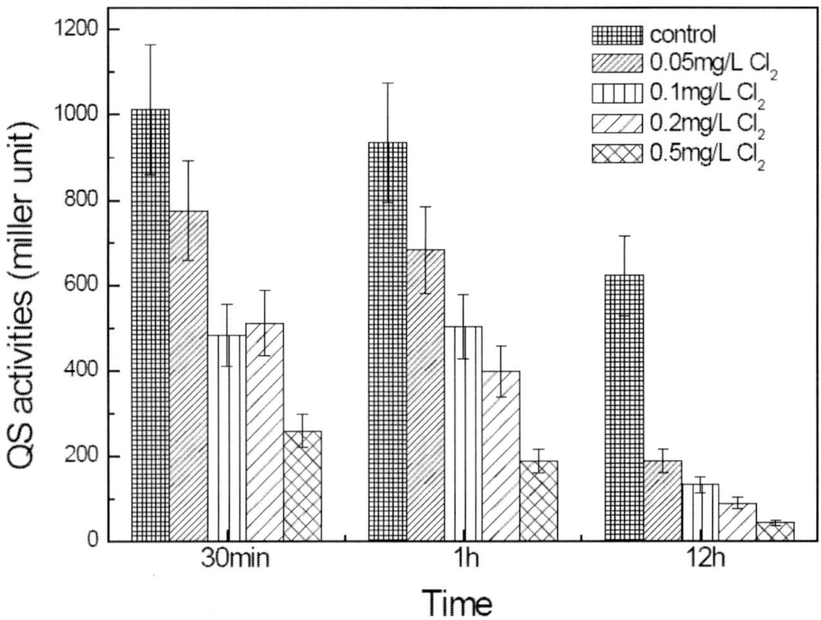

Figure 3. Effect of active chorine on QS activities in water.

3.4. Chlorine Dioxide

Chlorine dioxide is a powerful hypochlorite and an alternative disinfectant agent to chlorine for disinfection and oxidation during the treatment of drinking water. Chlorine dioxide readily dissolves in water to form a solution that is a biocide to a wide range of microorganisms. It is also known that chlorine dioxide undergoes disproportionations in water, producing chlorite and chlorate. The major advantage of chlorine dioxide is that the disinfection process does not generate large quantities of halogenated by-products that are common with chlorination. Another advantage is that chlorine dioxide significantly reduces odor and color [9].

As indicated in Figure 5, 5 mg/L chlorine dioxide has strong inhibitory effects on QS activities in water. 2 mg/L of chlorine dioxide after 12 hours decreased QS activities greatly. Low concentrations of ClO_2 play little part in communication of water in distribution system. The range of chlorine dioxide concentrations that are used in water treatment is 0.02 - 0.8 mg/L according to national regulation. Therefore, chlorine dioxide is not an ideal disinfectant to inhibit QS activities in water.

3.5. Potassium Permanganate ($KMnO_4$)

$KMnO_4$ is a strong and active oxidant, which reacts with a wide variety of organic and inorganic substances. In drinking water treatment technology, it is primarily used for taste and odor control. Its secondary role is to control the formation of trihalomethanes (THMs) and other DBPs oxidizing of their precursors. Previous studies have demonstrated that potassium permanganate may be a feasible oxidants for the removal of MCs present in drinking water resources, although theoxidation might lead to toxic oxidation products [10].

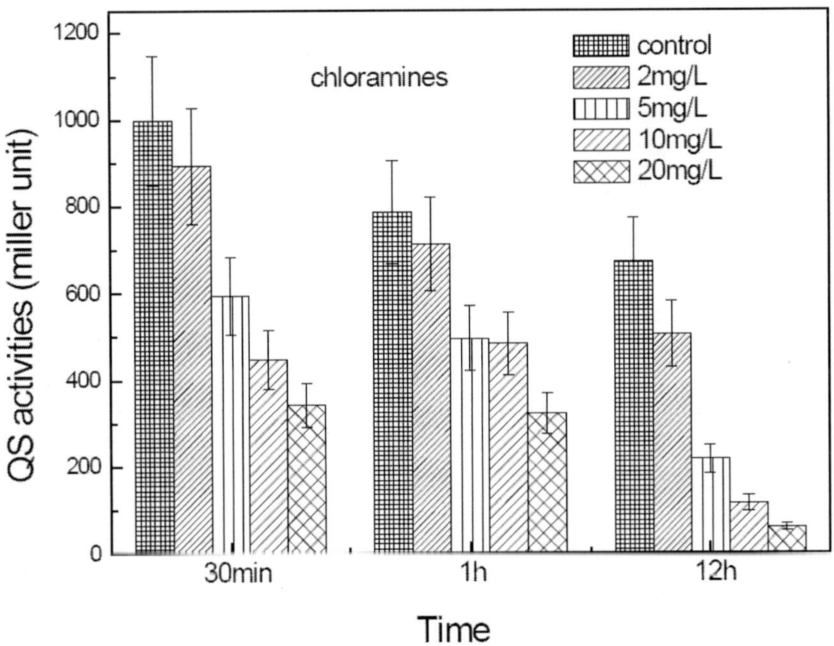

Figure 4. Effect of chloramines on QS activities in water.

The variation of QS activities at different $KMnO_4$ concentration is indicated in Figure 6; at low concentration, $KMnO_4$ significantly decreased QS activities in water. However, it was not evident that QS activities further decreased with increases of $KMnO_4$ contentration. Long treatment times with $KMnO_4$ did not dramatically decrease QS activities. Therefore, short treatment times and at a low dosage of $KMnO_4$ was satisfactory for inhibiting QS activities in water.

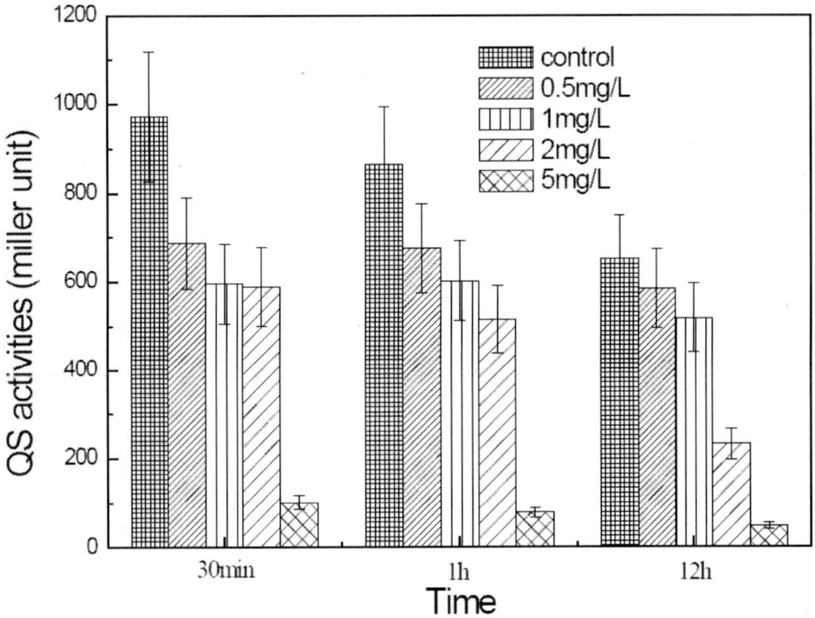

Figure 5. Effect of chlorine dioxide on QS activities in water.

3.6. Hydrogen Peroxide

Hydrogen peroxide, as well as other non-chlorine disinfectants, is usually coupled with ozone or chloramines in drinking water pretreatment. Its advantages over other oxidants are that it is nontoxic, harmless, and generates environmentally acceptable by products [11].

The effect of hydrogen peroxide on QS activities in water is showed in Figure 7. The QS activities decreased with increasing hydrogen peroxide concentrations. High concentrations of hydrogen peroxide treatment (over 5 mmol/L) notably decreased β-galactosidase activity in water. Therefore, decreasing QS activities requires long treatment times and high doses of hydrogen peroxide.

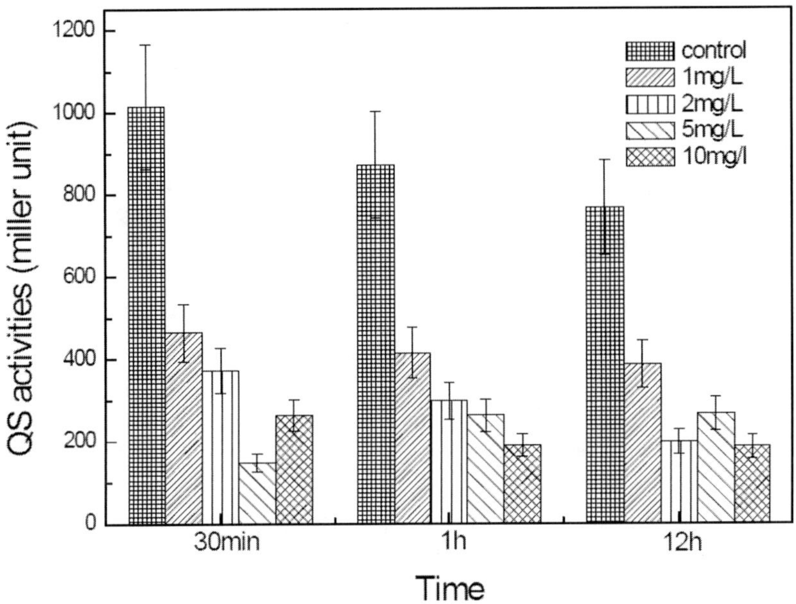

Figure 6. Effect of KMnO4 on QS activities in water.

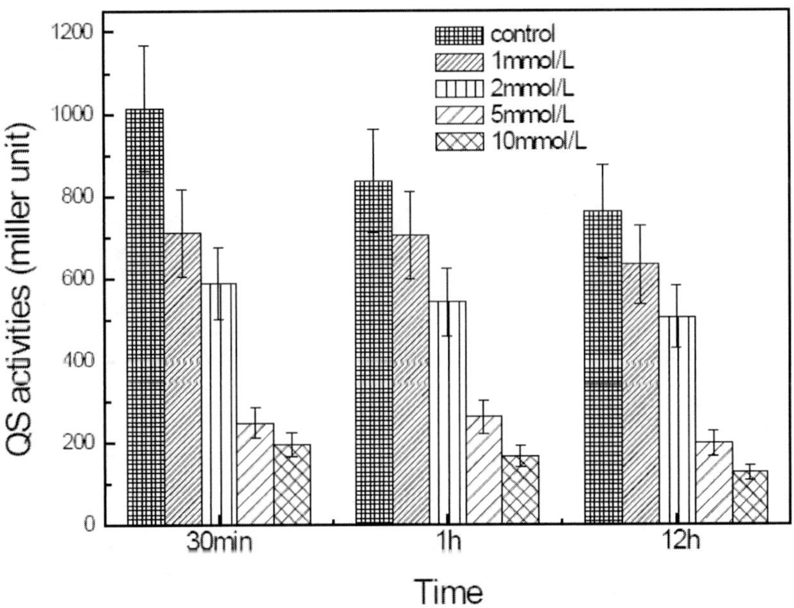

Figure 7. Effect of hydrogen peroxide on QS activities in water.

4. CONCLUSIONS

(1) Reporter strain KYC55 is most sensitive to C6 AHL; therefore, C6 AHL was selected as an indicator of QS in water.

(2) All disinfectants tested in this study were able to inhibit QS activities in water to a certain extent. Javel water was the strongest oxidant and the lowest oxidant was chlorine dioxide.

REFERENCES

1. Szewzyk, U. and Szewzyk, R. (2000) Microbiological Safety of Drinking Water. Annual Review of Microbiology, 54, 81-127. http://dx.doi.org/10.1146/annurev.micro.54.1.81
2. Joelsson, A.C. and Zhu, J. (2006) LacZ-Based Detection of Acyl-Homoserine Lactone Quorum-Sensing Signals. Current Protocols in Microbiology, 1C.2.1-1C.2.9.
3. Waters, C.M. and Bassler, B.L. (2005) Quorum Sensing: Cell-to-Cell Communication in Bacteria. Annual Review of Cell and Developmental Biology, 21, 319-346. http://dx.doi.org/10.1146/annurev.cellbio.21.012704.131001
4. Zhu, J., Chai, Y.R., Zhong, Z.T., Li, S.P. and Winans, S.C. (2003) Agrobacterium Bioassay Strain for Ultrasensitive Detection of N-Acylhomoserine Lactone-Type Quorum-Sensing Molecules: Detection of Autoinducers in Mesorhizobium huakuii. Applied and Environmental Microbiology, 69, 6949-6953.
5. Kumari, A., Pasini, P. and Deo, S.K. (2006) Biosensing Systems for the Detection of Bacterial Quorum Signaling Molecules. Analytical Chemistry, 78, 7603-7609. http://dx.doi.org/10.1021/ac061421n
6. Butterfield, P.W. and Camper, A.K. (2002) Chlorination of Model Drinking Water Biofilm: Implications for Growth and Organic Carbon Removal. Water Research, 36, 4391-4405. http://dx.doi.org/10.1016/S0043-1354(02)00148-3
7. Leea, W. and Westerhoffb, P. (2009) Formation of Organic Chloramines during Water Disinfection—Chlorination versus Chloramination. Water Research, 43, 2233-2239.
8. Huaa, G.H. and Reckhowa, D.A. (2007) Comparison of Disinfection Byproduct Formation from Chlorine and Alternative Disinfectants. Water Research, 41, 1667-1678. http://dx.doi.org/10.1016/j.watres.2007.01.032
9. Navalon, S., Alvaro, M. and Garcia, H. (2009) Chlorine Dioxide Reaction with Selected Amino Acids in Water. Journal of Hazardous Materials, 164, 1089-1097. http://dx.doi.org/10.1016/j.jhazmat.2008.09.010
10. Lee, C., Yoon, J. and Von Gunten, U. (2007) Oxidative Degradation of N-Nitrosodimethylamine by Conventional Ozonation and the Advanced Oxidation Process Ozone/Hydrogen Peroxide. Water Research, 41, 581-590. http://dx.doi.org/ 10.1016/j.watres.2006.10.033
11. Kepa, U. and Stanczyk-Mazanek, E. (2008) Ozone + Hydrogen Peroxide System for the Removal of Cyanide from Water. Desalination, 223, 187-193. http://dx.doi.org/ 10.1016/j.desal.2007.01.215

CHAPTER 15

Removal of arsenic from drinking water using rice husk

Zunaira Asif[1], Zhi Chen[1]

1 Environmental Engineering, BCEE, Faculty of Engineeringand Computer Sciences, Concordia University, Montreal,Canada

ABSTRACT

Rice husk adsorption column method has proved to be a promising solution for arsenic (As) removal over the other conventional methods. The present work investigates the potential of raw rice husk as an adsorbent for the removal of arsenic [As(V)] from drinking water. Effects of various operating parameters such as diameter of column, bed height, flow rate, initial arsenic feed concentration and particle size were investigated using continuous fixed bed column to check the removal efficiency of arsenic. This method shows maximum removal of As, i.e., 90.7 % under the following conditions: rice husk amount 42.5 g; 7 mL/min flow rate in 5 cm diameter column at the bed height of 28 cm for 15 ppb inlet feed concentration. Removal efficiency was increased from 83.4 to 90.7 % by reducing the particle size from 1.18 mm to 710 μm for 15 ppb concentration. Langmuir and Freundlich isotherm models were employed to discuss the adsorption behavior. The effect of different operating parameters on the column adsorption was determined using breakthrough curves. In the present study, three kinetic models Adam-Bohart, Thomas and Yoon–Nelson were applied to find out the saturated concentration, fixed bed adsorption capacity and time required for 50 % adsorbate breakthrough, respectively. At the end, solidification was done for disposal of rice husk.

Keywords: Adsorption, Arsenic, Rice husk, Kinetic models, Fixed bed column

1. INTRODUCTION

Drinking arsenic-contaminated water leads to a series of health problems. Arsenic is a toxic metalloid element that is now recognized to be an important contaminant in drinking water in many countries of south-East Asia including Bangladesh (Sohel et al. 2009), Pakistan (Islam-ul-haq et al. 2007), China (Li et al. 2011), etc. The effects of arsenic are not immediately visible as people can absorb significant quantities of arsenic without any immediate health complications. Treatments are required to remove arsenic and provide treated water for consumption. Some cost effective filtration systems have been developed in past but the main problem is that the socio-economic conditions of the effected people do not allow use of these filters easily in developing countries. There is need to provide low cost technique for removal of arsenic so that safe water is provided at consumer's end (António et al. 2009).

Drinking water is now recognized as a major source of human intake of arsenic in its most toxic (inorganic) form (Dinesh et al. 2007). Arsenic is found in natural environment and behaves as a metalloid. It can exist in organic and inorganic (i.e., most toxic) forms. About one-third of the arsenic in the atmosphere comes from high temperature natural processes, such as coal-fired power plants, burning vegetation and volcanic activity. Rest comes from human activities. Due to natural geological contamination, high levels of inorganic arsenic can be found in drinking water that has come from deep drilled wells (Bundschuh et al. 2011). This is particularly true for Bangladesh, India and Taiwan (Garelick et al. 2005).

In 2010, the World health organization (WHO) estimates that due to long-term arsenic exposure, thousands of people from Pakistan to Vietnam died of cancer each year. Cardiovascular disease and mental development of children are also caused by arsenic (Scott et al. 2010). In the light of potency of arsenic, United States Environmental protection agency (US-EPA) concluded that arsenic is *Group A* carcinogen, known to trigger skin, bladder, and lung cancers and thus has become a metaphor for poison. Hundreds of millions of people daily use drinking water with arsenic concentrations several times higher than the World Health Organization

(WHO) recommended limit of 10 millionths of a gram per liter of water (10 μg/L) (Garelick et al. 2005; Moffat et al. 2014).

The predominant forms of As in water supplies mostly are inorganic species arsenate [As(V)] and arsenite [As(III)]. Arsenic is exclusively sensitive to mobilization (pH 6.5–8.5) and exists under both oxidizing and reducing conditions among heavy metalloids. The As(V) species are mostly stable in oxygen-rich environment and exist as oxyanions ($H_2AsO_4^-$ and $HAsO_4^{2-}$) at neutral pH, whereas As(III) species predominant in moderately reducing anaerobic conditions exist as H_3AsO_3 at neutral pH. As(III) is more toxic and more difficult to remove with the conventionally applied physiochemical treatment methods than As(V) (Dinesh et al. 2007).

A large number of remediation technologies for arsenic removal have been developed, but mostly are available for central water treatment plants at industrial level. However, options for local treatment of water are not fully developed. The available methods include arsenic removal by ion exchange, removal by filtration and precipitation/co-precipitation, coagulation/electro coagulation, removal by adsorption onto artificial and natural sorbents, water softening with lime, membrane processes and bioremediation (António et al. 2009; Dinesh et al. 2007; Dhiraj et al. 2008). These technologies mostly have cost issue and are greatly used for heavy metal removal from wastewater (Dinesh et al. 2007). Many researchers have tried to use naturally occurring materials as low cost adsorbents for removing heavy metals. The metal adsorption capacities of adsorbents depend upon different factors such as concentration of metal ions, pH, contact time and characteristics of individual adsorbent. One of the cost effective and highly efficient sources for adsorption technology is waste of agriculture (Roya et al. 2013). Studies reveal that a variety of agricultural waste materials such as wheat bran, wheat husk (Roya et al. 2013), bark of the trees, coconut shells, ground net shell, walnut shells, tea waste (Dhiraj et al. 2008), coffee beans (Boonamnuayvitaya et al. 2004), maize corn cob (Sekhula et al. 2012), sugar cane bagasse, apple, banana, orange peels (Annadurai et al. 2003), soya bean hulls, sugar beet pulp, sunflower stalks (Hashem et al. 2007), cotton stalks, etc. have been tried to remove metal ions either in their natural form or after some physical or chemical modification (Dinesh et al. 2007). However, a very few studies were done on raw rice husk (Nurul Amin et al. 2006) as a remediation material for removal of arsenic from groundwater.

The purpose of the present work is to explore the potential of raw rice husk as adsorbent in a fixed bed column as well as batch studies for removing As(V) from drinking water. Rice husk is available easily at no or very low cost as agro waste. Moreover, the effect of various parameters such as flow rate, adsorbent volume, and diameter of column, particle size of adsorbent and initial metal concentration on the removal of arsenic was investigated to reveal the performance of column.

2. MATERIALS AND METHODS

2.1 Reagents adsorbent preparation

The reagents that have been used in the present study were of analytical grade and used to make calibration standards. Stock solutions (1000 μg/L) of As(V) were prepared in distilled water from solid KH_2AsO_4. Rice husk was the adsorbent used for the removal of arsenic ions from their aqueous solution. Rice husk was taken from a local market. Then, it was ground and sieved to a size of 1.18 mm (No# 16), 850 μm (No# 20) and 710 μm (No# 25) to have uniform size of adsorbent. Three different sizes were used to study the effect of particle size of rice husk on removal efficiency of arsenic, whereas to study the effect of other parameters such as flow rate, initial arsenic concentration feed and bed height, only one of the sizes (i.e., 1.1 mm No# 16) was used. The adsorbent was then washed with deionized water and dried in an oven at 100 °C for 48 min to remove moisture contents. After that, adsorbent was stored in air tight jars for further use.

2.2 Fixed bed column setup

The continuous flow adsorption experiments were conducted using glass columns of 2.5 and 5 cm inside diameter and height of 1.5 m. In the lower end of the column, a 0.5 mm sieve was used to adjust the required height of the adsorbent media (rice husk) and to support the adsorbent in the column. To adjust the height of rice husk, a wire is fixed with the sieve that allows the sieve to move in the column. A peristaltic pump was used for injecting As(V) solutions to the top of the column from the feed tank. It has a capacity of pumping liquids from 2 to 25 mL/min at different pressures. The flow rate can be controlled or adjusted with the help of flow meter as shown in Fig. 1.

2.3 Analytical procedure

Flow was directed downward as feed (initial arsenic solution) was applied from the top. Continuous study was carried out to determine the effect of different operating parameters like flow rate, height of adsorbent bed, effect of column diameter, initial metal concentration dose and effect of particle sizes of adsorbent. Pre-weighed amount of adsorbent was filled in column. It was air tighten with the help of screws present at inlet of the column so that water must flow under pressure. At the required flow rate, metal solution of known concentration was fed from the feed tank with the help of peristaltic pump. Effluent samples were collected at different time intervals to achieve exhaust time (The time at which adsorbent get saturated and the time at which adsorbent has maximum capacity to absorb the metal). Then, samples were analyzed for their absorbance using atomic adsorption spectrophotometer (AAS).

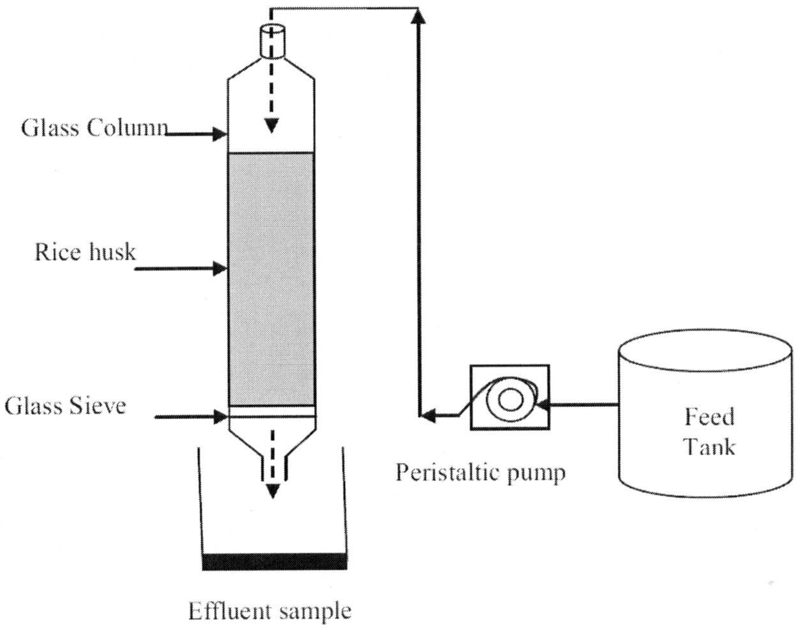

Figure. 1 Schematic diagram of the experimental setup for fixed bed column adsorption

Removal efficiency of metal ions was calculated using the following Eq. (1)

$$\text{Removal Efficiency (R.E)}\,\% = \frac{C_0 - C}{C_0} \times 100 \quad (1)$$

whereas C_0 and C are the concentrations of As in sample water before and after the treatment, respectively.

3. EXPERIMENTAL STUDY RESULTS

3.1 Effect of column diameter and bed height

As shown in Fig. 2, different removal efficiencies were achieved at different diameter columns, i.e., 5 cm and 2.5 cm for the fixed adsorbent quantity of rice husk (47.5 g) and influent metal concentration of 15 ppb by keeping the flow rate constant (7 mL/min). The whole experiment was repeated for the influent metal concentrations of 30 ppb and 50 ppb. It was observed that the maximum average removal of 80.7 % was achieved in 5 cm diameter column at bed height of 28 cm with 47.5 g adsorbent quantity. The reason is that larger diameter with greater column depth increased the contact between metal solution and adsorbents, allowing metal ions to better adsorbed at available binding sites of the adsorbents.

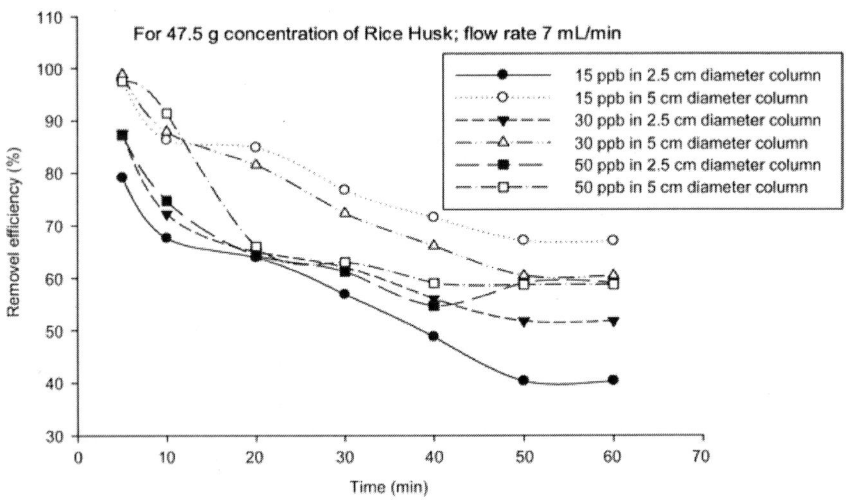

Figure. 2 Effect of diameter on removal efficiency at 7 mL/min in 2.5 and 5 cm diameter columns

The effect of bed height on the continuous column adsorption can also be determined using the time to reach breakthrough point and the shape of the breakthrough curve as shown in Fig. 3. An increase in bed height resulted in more contact time being available for As(V) to interact with the rice husk (Palas et al. 2013). This phenomenon has allowed As(V) to diffuse deeper into the rice husk. Subsequently, the percentage of arsenic removal increased when the bed height was increased.

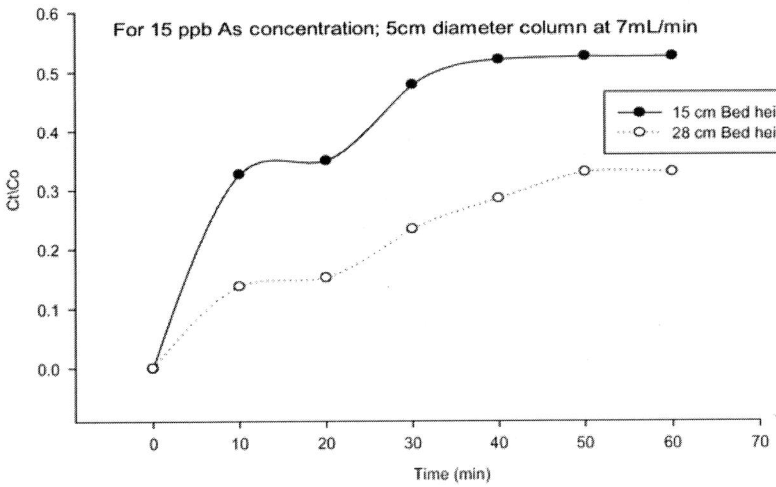

Figure . 3 Breakthrough *curve* showing the effect of bed height in 5 cm column at 7 mL/min for ppb As(V) concentration

3.2 Effect of flow rate

The adsorption behavior of rice husk was investigated by altering the flow rate in the range of 7 to 20 mL/min. The breakthrough curves obtained by keeping constant the metal concentration of 15 ppb at bed height of 28 cm in 5 cm diameter column are shown in Fig. 4. It was observed that quicker breakthrough was achieved for higher flow rates of As(V) solution. The reason is that at low flow rate, rice husk takes more time to achieve saturation. This can be explained by the fact that at higher flow rates, the residence time of the adsorbate in the column would decrease but at lower flow rates metal ions have more time to diffuse into the pores of adsorbent through intra-particle diffusion. Conversely, for the higher flow rate, the arsenic solution will leave the column bed before the equilibrium can be reached. Nurul-Amin et al. (2006) reported the same trend that removal efficiency for As(V) decreased as the treatment flow rate was increased.

3.3 Effect of initial metal concentration

The effect of initial metal concentration was investigated on column by altering influent concentrations of 15, 30, 50 and 70 ppb. The bed height, diameter of the column and flow rate were kept constant at 28 cm (47.5 g rice husk), 5 cm and 7 mL/min, respectively, as shown in Fig. 5. It was revealed that the treated volume was higher at low concentration.

However, the adsorption capacity increased with increasing the initial As(V) concentration (ppb) due to a raise in the driving force of the concentration gradient. At low influent concentration, the driving force of adsorbent was reduced due to low concentration gradient. It is also evident from the breakthrough curve Fig. 5 that maximum removal was 15 ppb and minimum removal was 70 ppb. The slow transport of As(V) onto rice husk column was due to the lower concentration gradient and resulted in a slower breakthrough curve. Conversely, a higher concentration of As(V) has been shown to lead to a higher driving force for the As(V) ions to overcome the mass transfer resistance in the liquid phase. Consequently, quick saturation of the available binding sites for rice husk has caused the breakthrough time to decrease with the increasing inlet As(V) concentration (Palas et al. 2013).

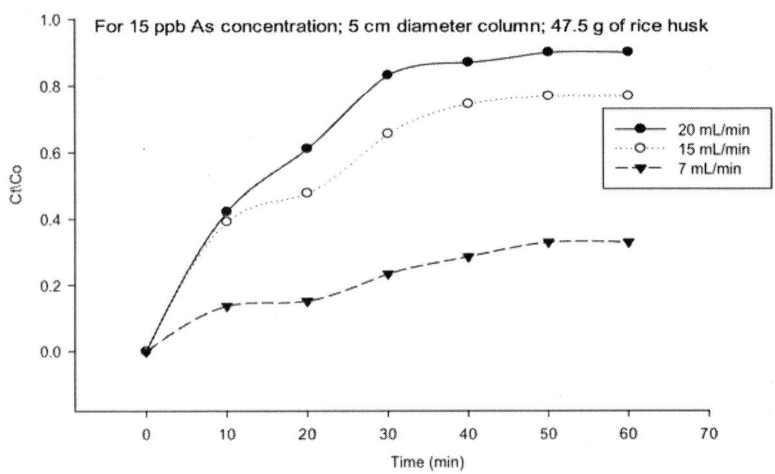

Figure. 4 Breakthrough *curve* showing the effect of flow rate in 5 cm column for 15 ppb arsenic concentration

3.4 Effect of particle size

The effect of particle sizes was observed in column adsorption by varying sizes of 1.18 mm (16 mesh), 850 μm (20 mesh) and 710 μm (25 mesh). For 15 ppb initial influent concentration, experiment was conducted at a flow rate of 7 mL/min in 5 cm column at bed height of 28 cm as shown in Fig. 6. It was concluded that by reducing the particle size of rice husk, removal efficiency increased from 7 to 20 %. This might be due to the fact that by increasing the surface area of particles and compaction between the particles, removal efficiency increased. The more the binding sites available, the more efficient will be the adsorption process (Neeta and Jatinder 2008). Likewise, breakthrough curves showing the same effect of particle sizes for 70 ppb As(V) concentration under the same operating conditions are illustrated in Fig. 7.

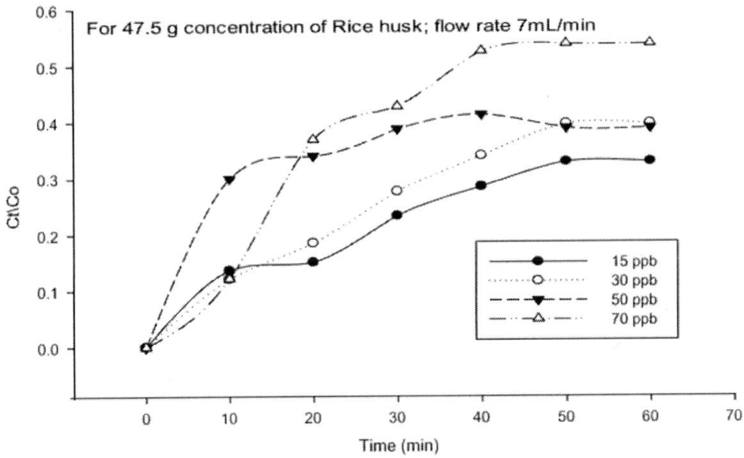

Figure . 5 Breakthrough *curve* showing the effect of feed concentration in 5 cm column at bed height of a 28 cm at 7 mL/min for 15, 30, 50 and 70 ppb As(V) concentration

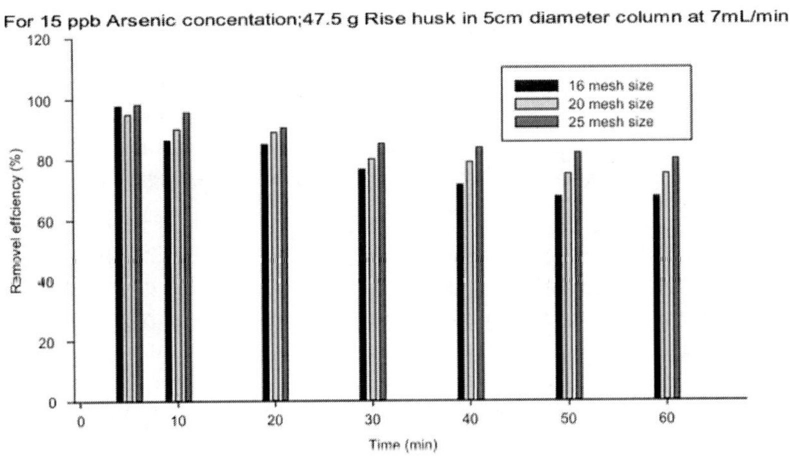

Figure. 6 Removal efficiency showing the effect of particle sizes in 5 cm column at bed height of a 28 cm at 7 mL/min for 15 ppb As(V) concentration

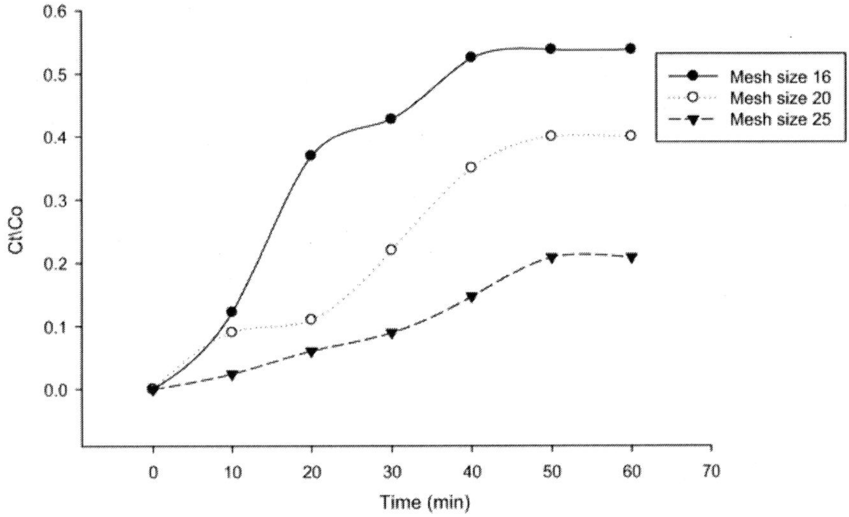

Figure. 7 Breakthrough curve showing the effect of particle size in 5 cm column at bed height of 28 cm for 70 ppb As(V) concentration at 7 mL/min

3.5 Effect of pH

pH is one of the most vital parameters controlling the metal ion sorption process. Figure 8depicts the effect of pH on As(V) removal with removal efficiency.

The removal efficiency of As(V) increased slightly with increasing pH, was nearly constant above pH 8, and decreased tremendously at pH > 12. For the removal of As(V), the efficiency curve was basically a plateau in the pH range of 8–10, and then the efficiency leaned to decrease with increasing pH.

The poor As(V) removal efficiencies at high pH can be attributed to the following reasons: Essentially, chemical species for As(V) in this pH region is oxy-anions. Moreover, hydroxyl groups are more abundant on the surface of rice husk with increasing pH. Hence, the maximum removal was achieved at pH 8. Nurul-Amin et al. (2006) also reported that maximum removal for As(V) was observed in comparative neutral region, i.e., 8.

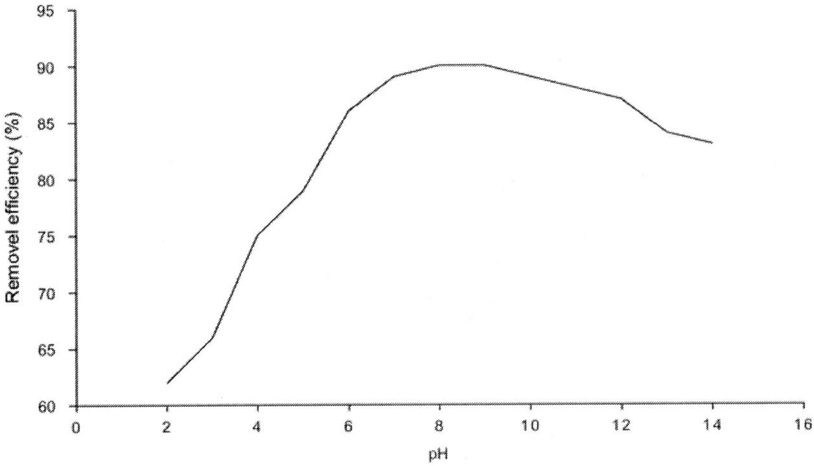

Figure. 8 Removal efficiency showing the effect of pH in 5 cm column at 7 mL/min for 15 ppb As(V) concentration

4. QUANTITATIVE ANALYSIS DISCUSSION

4.1 Adsorption isotherms

4.1.1 Langmuir isotherm

The Langmuir's isotherm model is given by the following Eq. 2 (Hameed et al. 2007):

$$\frac{C_e}{q_e} = \frac{1}{Q_o b} + C_e \frac{1}{Q_o} \tag{2}$$

where C_e is the equilibrium concentration of the adsorbate (μg/L), q_e is the amount of adsorbate adsorbed per unit mass of adsorbate (μg g^{-1}), and Q_o and b arc Langmuir constants rated to adsorption capacity and rate of adsorption, respectively. When C_e/q_e was plotted against C_e, straight line with slope $1/Q_o$ was obtained as shown in Fig. 9, indicating that the adsorption of As(V) on rice husk follows the Langmuir isotherm. The Langmuir constants 'b' and 'Q_o' were calculated from this isotherm and their values are given in Table 1.

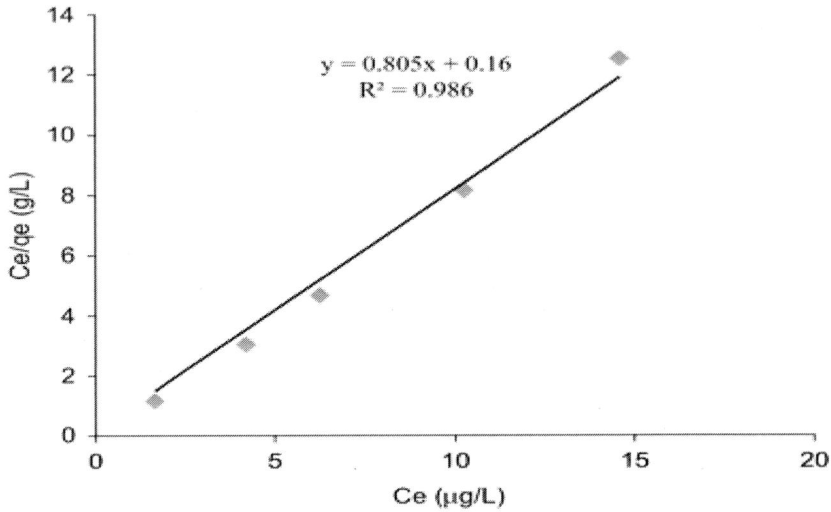

Figure. 9 Langmuir adsorption isotherm of arsenic-rice husk adsorption at room temperature (25 °C)

Table 1 Langmuir and Freundlich isotherm constants for arsenic

Langmuir's isotherm	Values
Q_o µg g^{-1}	225
b µg^{-1}	0.005
R^2	0.986
R_L	0.84
Freundlich's isotherm	
$1/n$	0.344
K_F	2.65
R^2	0.948

The important characteristics of the Langmuir isotherm can be expressed as dimensionless equilibrium parameter (R_L) which is defined by Eq. (3) (Hameed et al. 2007):

$$R_L = \frac{1}{1 + bC_o} \quad (3)$$

where b is the Langmuir constant and C_0 is the maximum arsenic concentration (μg L^{-1}). The value of R_L is indicator showing the type of the isotherm to be favorable ($0 < R_L < 1$), linear ($R_L = 1$), unfavorable ($R_L > 1$) or irreversible ($R_L = 0$). Value of R was found to be 0.84 and confirmed that the rice husk is favorable for adsorption of arsenic under conditions used in this study.

4.1.2 Freundlich isotherm

The logarithmic form of Freundlich model is expressed as the following Eq. (4) (Hameed et al. 2007):

$$\log q_e = \log K_F + (1/n) \log C_e \tag{4}$$

where q_e is the amount adsorbed at equilibrium (μg g^{-1}), C_e is the equilibrium concentration of the adsorbate and K_F and n are Freundlich constants, n value showing how favorable the adsorption process and K_F [mg g^{-1}(l $mg^{-1})^n$] is the adsorption capacity of the adsorbent. $K_F$represents the quantity of As(V) adsorbed onto rice husk adsorbent for a unit equilibrium concentration. The slope $1/n$ ranging between 0 and 1 is a measure of adsorption intensity or surface heterogeneity. A value for $1/n < 1$ indicates a normal Langmuir isotherm while $1/n > 1$ is indicative of cooperative adsorption. The plot of log q_e versus log C_e gives straight lines with slope '$1/n$' (Fig. 10), which shows that the adsorption of arsenic also follows the Freundlich isotherm.

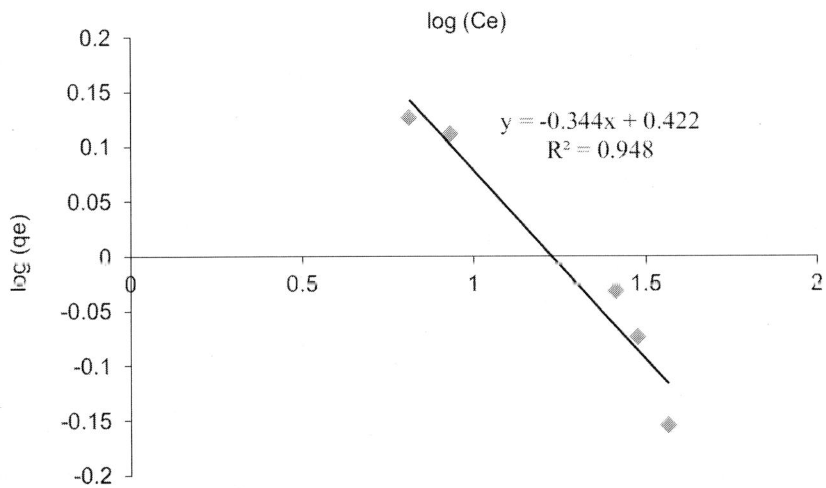

Figure. 10 Freundlich adsorption isotherm of arsenic-rice husk adsorption at room temperature (25 °C)

The value of $1/n$ obtained was 0.344, which indicates favorable adsorption. K_F and n were calculated and recorded in Table 1.

4.2 Kinetic models

4.2.1 Adam-Bohart model

The following expression was used to find out the saturated concentration of adsorbate (Zhao et al. 2011):

$$\frac{Ct}{Co} = \exp\left(K_{AB} Co\, t - K_{AB} N_o \frac{Z}{F}\right) \tag{5}$$

where K_{AB} is the kinetic constant (L/μg.min), F is the linear velocity (cm/min) that is the ratio of flow (mL/min) and cross sectional area (cm^2), N_o is the saturation concentration (μg/L), Z is the bed depth of column (cm) and t is the time (min). K_{AB} and N_o were the parameters used to describe the characteristic operations of the column. Ln(CtCo)Ln(CtCo) was plotted against time (t) using linear regression. K_{AB} and N_o values were determined from the intercept and slope of the graph. Table 2 shows the values of kinetic constants and saturated concentrations calculated using Adam-Bohart model at different bed heights, flow rates and inlet feed concentrations in 5 cm diameter column.

Table 2 Adam-Bohart constants for 5 cm diameter column at different bed heights and flow rate

C_o (ppb)	H (cm)	Q (mL min^{-1})	R^2	K_{AB} (mL μg^{-1} min^{-1})	N_0 ($\mu g.L^{-1}$)
15	28	7	0.959	0.00135	11.74
15	28	20	0.841	0.00179	7.589
15	15	7	0.881	0.00187	11.640
30	28	7	0.956	0.000863	28.323
30	15	7	0.959	0.000993	55.254
50	28	7	0.767	0.000626	42.958
70	28	7	0.733	0.000461	56.172

It is shown from the Table 1 that correlation coefficient R^2 was in the range of 0.733 to 0.959 for 15, 30, 50 and 70 ppb inlet arsenic concentrations. The results indicate that N_o had no definite pattern when the inlet As(V)

concentration, flow rate and bed height were increased. The k_{AB} values decreased from 0.00135 to 0.000461 mL/min.µg when the inlet As(V) concentrations increased from 15 to 70 ppb, whereas it increased from 0.00135 to 0.00179 when the flow rate increased from 7 mL/min to 20 mL/min. Hence, it can be concluded that the overall kinetics in the initial part of the adsorption process were dominated by external mass transfer.

4.2.2 Thomas Model

The following expression was used to find out the maximum adsorption capacity of rice husk (Palas et al. 2013):

$$\frac{C_t}{C_o} = \frac{1}{1 + \exp(\mathrm{KTH}q_o\frac{m}{Q} - \mathrm{KTH}C_o t)} \quad (6)$$

where q_o is the equilibrium As uptake per g of the rice husk adsorbent (µg/g), m is mass of the rice husk in grams (g), Q is the flow rate (mL/min), K_{TH} is the Thomas rate constant (L/µg.min), C_o is the initial metal concentration (ppb) and Ct is the effluent metal concentration (ppb). The values of K_{TH} and q_o were determined from a plot of Ln(C0Ct−1)Ln(C0Ct−1) against time (t) using linear regression analysis. Table 3 presents calculated values of kinetic constants and uptake capacity of arsenic at different flow rates, bed heights and inlet feed concentrations of arsenic in 5 cm diameter column.

Table 3 Thomas model constants for 5 cm diameter column at different bed heights

C_o (ppb)	H (cm)	Q (mL.min^{-1})	R^2	K_{Th}(mL µg^{-1} min^{-1})	q_o (µg g^{-1})
15	28	7	0.9641	0.00235	103.69
15	15	7	0.8893	0.00647	50.428
30	28	7	0.9736	0.0013	368.229
30	15	7	0.9531	0.001317	319.882
50	28	7	0.7685	0.000878	396.355
50	15	7	0.8448	0.000222	387.747
70	28	7	0.7954	0.000699	416.787

It can be seen from the Table 3 that all the factors (the inlet arsenic concentration, flow rate and bed height) affect the Thomas rate constant (k_{TH}) and the equilibrium of arsenic uptake (q_o). The inlet arsenic concentration and bed height influence the k_{TH} and q_0 in a similar pattern. When the inlet As(V) concentration increased from 15 to 70 ppb, the k_{TH} decreased from 0.00235 to 0.000699 mL/min μg while the qe increased from 103.6 to 416.7 μg/g.

Whereas, when the bed height increased from 15 to 28 cm, the k_{TH} decreased from 0.00647 to 0.00235 mL/min μg and the q_o increased from 50 to 103.6 μg/g. These findings could be attributed to the higher driving force of the higher inlet As(V) concentration (Padmesh et al. 2005). The R^2 values were in the range of 0.768 to 0.964.

4.2.3 Yoon and Nelson model

Yoon and Nelson model was based on the assumption that the rate of decrease in the probability of adsorption of adsorbate molecule is proportional to the probability of the adsorbate adsorption and the adsorbate breakthrough on the adsorbent. The following expression was used (Zahra et al. 2013)

$$\text{Ln}\ \frac{Ct}{C_0 - Ct} = K_{YN}t - K_{YN}\text{T} \qquad (7)$$

where K_{YN} (1/min) is the rate velocity constant. T (min) is the time required for 50 % adsorbate breakthrough. K_{YN} and T were determined from intercept and slope from the linear plot of LnCtCo−CtLnCtCo−Ct against sampling time (t). Table 4 represents calculated values of rate velocity constants and time required for 50 % adsorbate breakthrough, using Yoon and Nelson model at different flow rates, bed heights and inlet feed concentrations in 5 cm diameter column.

The results indicate that the k_{YN} value increased from 0.0235 to 0.0489 L/min when inlet As(V) value increased from 15 to 70 ppb. This was due to the fact that by increasing the initial arsenic concentration competition between the adsorbate molecules for the binding sites increased and ultimately resulted in the increased uptake capacity of rice husk. Rate constant (K_{YN}) increased with increase in flow rate, whereas "T" decreased with decreasing bed height. The linear regression coefficient (R^2) was in the range of 0.7954–0.9641.

Table 4 Yoon and Nelson model constants for 5 cm diameter column at different bed heights and flow rate

C_o (ppb)	H (cm)	Q (mL min^{-1})	R^2	K_{YN} (min^{-1})	T (min)
15	15	7	0.9641	0.0235	40.73
15	28	7	0.955	0.0306	71.78
15	28	15	0.9323	0.0444	19.25
15	28	20	0.8893	0.0647	11.97
30	15	7	0.9736	0.039	48.22
30	28	7	0.9531	0.0395	72.35
50	15	7	0.7685	0.0439	46.85
50	28	7	0.8448	0.0111	79.76
70	28	7	0.7954	0.0489	60.40

4.3 Disposal method for rice husk

Once the rice husk become exhausted, the problem of how to dispose of concentrated arsenic product must be addressed. Regeneration of rice husk is not cost effective and recovery of arsenic has limited markets (Verma et al. 2014). One attractive option for treating rice husk is encapsulation through solidification followed by disposal of treated landfills ((Leist et al. 2003). To form satisfactory storable solids, several solidification attempts were made based on literature studies; thus, in this study the effective solidification was done through arsenic concentrates incorporated into Portland cement.

5. CONCLUSIONS

The proposed fixed bed column treatment system is appropriate and suitable domestic approach to arsenic removal in local areas, because of its simplicity, easy operation and handling. The present method is effective for a wide range of arsenic concentrations (i.e., 15–70 ppb), which is quite similar to those observed in contaminated groundwater of south Asian countries. Adsorption isotherm study was carried out on two well-known isotherms models, Langmuir and Freundlich. Langmuir model yielded a somewhat better fit (R^2 = 0.986) than the Freundlich model (R^2 = 0.948). Column operations under different conditions were also employed to study its practical applicability of rice husk in the removal of arsenic. The column experimental data were fitted well to the Adam Bohrat, Thomas and Yoon–Nelson models for arsenic feed concentrations from 15 to 30 ppb at the flow rate of 7 mL/min. No secondary-pollution problem will occur, because solidification of the arsenic is possible through arsenic concentrates incorporated into Portland cement. Moreover, it has recycling potential and can be used as cement/insulation panels and roof sheets.

REFERENCES

1. Annadurai G, Juang RS, Lee DJ (2003) Adsorption of heavy metals from water using banana and orange peels. Water Sci Technol 47(1):185–19
2. António ALS, Sílvia JAC, António JA (2009) Emerging and innovative techniques for arsenic removal applied to a small water supply system. Sustain Sustain Water Manage 1(4):1288–1304
3. Boonamnuayvitaya V, Chaiya C, Tanthapanichakoon W, Jarudilokku S (2004) Removal of heavy metals by adsorbent prepared from pyrolyzed coffee residues and clay. Sep Purif Technol 35(1):11–22
4. Bundschuh J, Bhattacharya P, Sracek O, Mellano MF, Ramírez AE, Storniolo AR, Martín RA, Cortés J, Litter MI, Jean JS (2011) Arsenic removal from groundwater of the Chaco-Pampean plain (Argentina) using natural geological materials as adsorbents. J Environ Sci Health 46(1):1297–1310
5. Dhiraj S, Garima M, Kaur MP (2008) Agricultural waste material as potential adsorbent for sequestering heavy metal ions from aqueous solutions—a review. Bioresour Technol 99:6017–6027
6. Dinesh M, Charles U, Jr Pittman (2007) Arsenic removal from water/wastewater using adsorbents. J Hazard Mater 142(1–2):2–53
7. Garelick H, Dybowska A, Valsami-Jones E, Priest ND (2005) Remediation technologies for arsenic contaminated Drinking waters. J Soil Sediment 5(3):182–185
8. Hameed BH, Din ATM, Ahmad AL (2007) Adsorption of methylene blue onto bamboo-based activated carbon: kinetics and equilibrium studies. J Hazard Mater 141:819–825
9. Hashem A, Abou-Okeil A, El-Shafie A, El-Sakhawy M (2007) Grafting of high α-cellulose pulp extracted from sunflower stalks for removal of Hg(II) from aqueous solution. Polym Plast Technol Eng 45(1):135–141
10. Islam-ul-haq M, Deedar N, Wajid H (2007) Groundwater arsenic contamination—a multi directional emerging threat to water scarce areas of Pakistan. In: 6th International IAHS Groundwater Quality Conference, held in Fremantle, Western Australia
11. Leist M, Casey RJ, Caridi D (2003) The fixation and leaching of cement stabilized arsenic. Waste Manage 23(4):353–359
12. Li G, Sun GX, Williams PN, Nunes L, Zhu YG (2011) Inorganic arsenic in Chinese food and its cancer risk. Environ Int 37(1):1219–1225
13. Moffat CD, Dominik JW, Arun S, Andrew JPW, Pascal S, Ramon V (2014) Molecular recognition and scavenging of arsenic from aqueous solution using dimetallic receptors. Chem Eur J 20(51):17168–17177
14. Neeta S, Jatinder S (2008) Removal of Zn^{+2} ions from aqueous solution using rice (*Oryza Sativa*) husk in a sequential bed adsorption column. Int J Enviro Tec Manage 12(1):944–951
15. Nurul-Amin Md, Kaneco S, Kitagawa T, Begum A, Hideyuki KH, Suzuki T, Ohta K (2006) Removal of arsenic in aqueous solutions by adsorption onto waste rice husk. Ind Eng Chem Res 45:8105–8110

16. Padmesh TVN, Vijayaraghavan K, Sekaran G, Velan M (2005) Batch and column studies on biosorption of acid dyes on fresh water acro alga *Azolla filiculoides*. J Hazard Mater 125(1–3):121–129
17. Palas R, Naba KM, Shreya B, Biswajit D, Kousik D (2013) Removal of arsenic (III) and arsenic (V) on chemically modified low-cost adsorbent: batch and column operations. Appl Water Sci 3:293–309
18. Roya E, Maleki A, Shahmoradi B, Daraei H, Hossein MA, Hoshang BA, Akbar E (2013) Elimination of arsenic contamination from water using chemically modified wheat straw. Desalin Water Treat 51(10–12):2306–2316
19. Scott F, Holly AM, Alexander VG (2010) Spatial and temporal variations of groundwater arsenic in South and Southeast Asia. Science 328(5982):1123–1127
20. Sekhula MM, Okonkwo JO, Zvinowanda CM, Agyei NN, Chaudhary AJ (2012) Fixed bed column adsorption of Cu (II) onto maize tassel-PVA beads. Chem Eng Process Technol 3(2):2157–7048
21. Sohel N, Persson LA, Rahman M, Streatfield PK, Yunus M, Ekstrom EC, Vahter M (2009) Arsenic in drinking water and adult mortality a population-based cohort study in rural Bangladesh. Epidemiology 20(6):824–830
22. Verma P, Agarwal A, Singh VK (2014) Arsenic removal from water through adsorption—a review. Rec Res Sci Technol 6(1):219–226
23. Zahra S, Reyhane S, Reza F (2013) Fixed-bed adsorption dynamics of Pb(II) adsorption from aqueous solution using nanostructured γ-alumina. J Nanostructure Chem 3(48):2193–8865
24. Zhao G, Wu X, Tan Wang X (2011) Sorption of heavy metal ions from aqueous solutions: a review. TOCOLLSJ 4:19–31

CHAPTER 16

Physico-Chemical Quality of Selected Drinking Water Sources in Mbarara Municipality, Uganda

Ben Lukubye*, Morgan Andama

Department of Biology, Mbarara University of Science and Technology, Mbarara, Uganda

ABSTRACT

The study assessed the physico-chemical quality of selected drinking water sources (springs, boreholes, shallow wells and rainfall) in Mbarara municipality with respect to World Health Organization (WHO) drinking water guidelines and other guidelines in light of the increased anthropogenic activities in the municipality. A total of 70 water samples were collected from purposively selected boreholes, springs, wells and rainwater in Nyamitanga, Kamukuzi and Kakoba divisions of Mbarara municipality with various human activities. The samples were analysed for physico-chemical parameters: Temperature, pH, Dissolved Oxygen (DO), Biological Oxygen Demand (BOD), Total Dissolved Solids (TDS), Electrical Conductivity (EC) and Total hardness using American Public Health Association (APHA) standard methods. The mean temperature and pH ranged between 18.07°C - 23.45°C and 5.74 - 7.54, respectively. The mean DO values were found to be between 4.84 and 12.86 mg/l; whereas mean BOD was within the range of 1.83 - 7.71 mg/l. The mean TDS and EC of the water samples ranged, between 33.40 - 569.20 mg/l and 29.30 - 1139.90

μS/cm respectively. Furthermore, the lowest and highest mean total hardness were 70.00 and 264.00 mg/l, respectively. The recorded mean water temperatures for each of the water sources were above the WHO threshold temperature (15°C) which makes drinking water palatable. Boreholes in Nyamitanga and Shuhaddea Secondary Schools, spring in Kiswahili, well in Kisenyi and rainwater in Mbarara University of Science and Technology (MUST) had mean pH below the WHO minimum guideline value (6.5) hence acidic. Borehole in Nyamitanga secondary school, spring in Kisenyi, shallow well in Nyamitanga and the rainwater in MUST had mean DO values below the WHO range (10 - 12 mg/l). Borehole in Shuhaddea Secondary School and the well in Kisenyi had average BOD values above the range of European Union guideline values (3 - 6 mg/l). TDS and EC of all the water sources were below the WHO maximum guideline limits of 1000 mg/l and 1500 μs/cm respectively. Total hardness was also below the WHO harmless limit of 1000 mg/l. However rainwater in MUST was moderately soft while the other drinking water sources exhibited moderate to full total hardness. The physico- chemical parameters of some of the selected water sources in Mbarara municipality have been compromised mainly by the increased human activities especially croplands, latrines, landfills, transportation, animal and municipal wastes at the vicinity of the water sources. Mbarara municipal council should therefore ensure proper sanitation and water safety plans for these drinking water sources to avoid further contamination from the human activities.

Keywords: Drinking Water Sources, Mbarara Municipality, Physico-Chemical Parameters

1. INTRODUCTION

Water is a natural resource which forms an essential component of life [1]. However, the suitability of water for various uses depends on the biological, physico-chemical and radiological properties of water [2]. Over 1 billion people lack access to safe drinking water worldwide [3]. The situation is worse in developing countries like Uganda where many people especially the poor have opted to using the underground drinking water sources like boreholes, springs, shallow wells as a source of drinking water and for other domestic use [4]. Particularly in Mbarara municipality, National water and sewerage cooperation supplies water to only 47.5% of the total population which can afford the piped water with the rest of the 52.5% of the population using the unprotected drinking water sources (boreholes, springs, wells) and rainwater to access drinking water [5]. However these natural drinking water sources are at a high risk of contamination from many sources of contaminants like pit latrines, agricultural pesticides and fertilizers, domestic and industrial wastes, leakages from landfills [6]. Unfortunately the public due to the less sensitization and limited knowledge about the quality of these water sources in Mbarara municipality has continued to use them for drinking and cooking purposes. The study therefore aimed at ascertaining the quality of selected drinking water sources (boreholes, springs, wells, rainwater) in Nyamitanga, Kamukuzi and Kakoba divisions of Mbarara municipality with various human activities.

2. MATERIALS AND METHODS

2.1. Location of Study Area

The study was conducted in Mbarara municipality located in Mbarara District, South Western Uganda, 290 km from Kampala (capital city of Uganda) on Kampala-Kabale road at longitude 30.6582° and latitude 0.6132° [7] [8]. The municipality with a total area of 51.47 km^2 is situated on hilly areas (1432 m above sea level) separated by short, small and shallow valleys and is underlain mainly by precambium parent rock system consisting of gneisses, granitoid rocks, phyllites, shales and mud stones [9].

The fairly densely populated Mbarara municipality of about 194,973 people [10] and growth rate of 2.9% is divided into Nyamitanga, Kamukuzi and Kakoba divisions. With limited access to piped water sources, the dense population greatly exerts pressure on drinking water sources (boreholes, springs, wells) especially those found in densely populated slum areas of Kakoba and Kiswahili [5]. Hence seven drinking water sources were purposively selected from the three divisions in Mbarara municipality viz. 1 spring i.e. Spring (Kise) at geographical coordinates 0.6004°S, 30.6682°E) and 1 well i.e. Well (Kise) (0.6030°S, 30.6655°E) in Kisenyi village (Kakoba), 1 spring i.e. Spring (Kisw) (0.6109°S, 30.6631°E) in Kiswahili village (Kakoba), 2 boreholes in Nyamitanga village at Nyamitanga Secondary School i.e. Borehole (Nyam. S.S) (0.6257°S, 30.6501°E) and Shuhaddea Secondary School i.e. Borehole (Shud. S.S) (0.6262°S, 30.6532°E) (Nyamitanga). One shallow well i.e. S. well (Nyam) (0.6254°S, 30.6607°E) in Kitebero village (Nyamitanga) and rainwater i.e. Rain(MUST) (0.6164°S, 30.6561°E) at Mbarara University of Science and Technology (MUST) located in Kamukuzi division as shown in Figure 1. The study was conducted from January, 2016 to May, 2016.

2.2. Survey Data Collection

A 200 m radius was measured from the water sampling point using a tape measure and selected anthropogenic activities with great influence on water quality within a 200 m radius circumference along established line transects were observed and noted. A 200 m radius from the sampling points of drinking water sources was chosen because it is the maximum distance for protection of a lake [11] which is a static water body just like the drinking water sources. The proximity, distance (m) from the water sampling point to the first identified anthropogenic activity along the line transects were measured as well as the area (m^2) covered by each activity. The total coverage (m^2) of all the identified activities at each of the sampling sites was obtained by summing up the individual areas (m^2) of each activity.

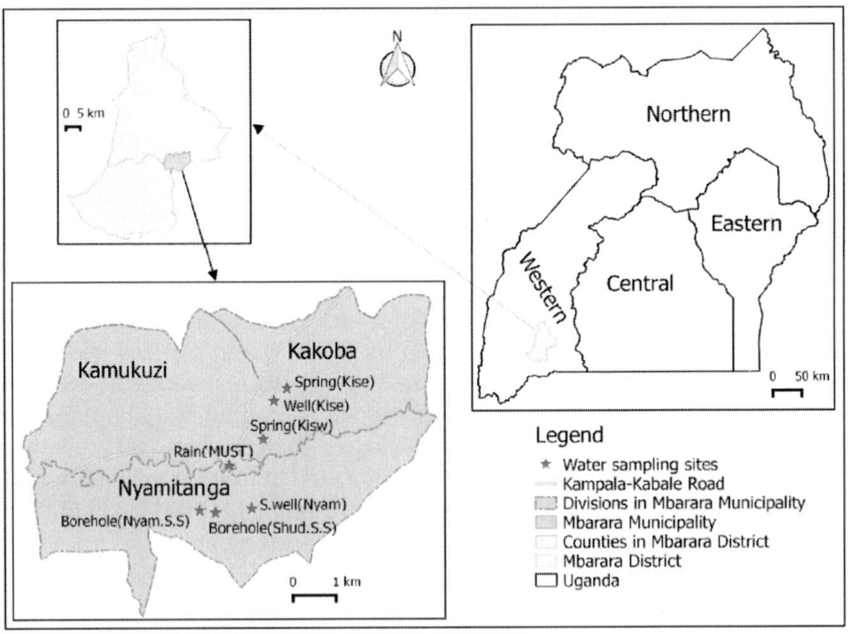

Figure 1. Map of Mbarara municipality showing the location of drinking water sources.

2.3. Collection of Water Samples

A total of 70 water samples were collected from the seven drinking water sources (2 boreholes, 2 springs, 2 wells (shallow well, deep well) and rainwater) with ten samples from each water source. The ten rainwater samples were collected directly from a rainfall after an hour of downpour in an open space to be used as a control. The water samples were collected in 300 ml sterile plastic bottles which were immediately analysed for most of the physico-chemical parameters.

2.4. Analysis of Physico-Chemical Parameters in the Water Samples

The water samples were analysed for physico-chemical parameters; Temperature, pH, Dissolved Oxygen (DO), Biological Oxygen Demand (BOD), Total Dissolved Solids (TDS), Electrical Conductivity (EC) and Total hardness using standard methods [12]. A multi-purpose digital meter was used to determine the TDS, pH, temperature and EC of the water samples. Similarly the DO meter was used to determine the Dissolved Oxygen concentration in the water samples. Total hardness strips were used to determine the total hardness of the water. The water samples were immediately analysed for pH, Temperature, DO, EC, TDS, Total hardness after collection but not BOD.

For BOD, four 300 ml glass stoppered BOD bottles were used to determine the Biological Oxygen Demand in the water samples. To two of the bottles, water sample (10 ml) was added and filled to the mark with dilution water while the other two were filled to the mark only with dilution water to act as blank solutions. All bottles were immediately stoppered and labelled. One bottle with blank solution and another with the water sample were placed in the incubator and kept for 5 days at 21°C. The other two bottles (one with a blank solution and the other with the water sample solution) were immediately analysed for dissolved oxygen (DO) and their values noted down as B_0 and D_0 respectively. After 5 days the blank solution and water sample solution were removed from the incubator and analysed for dissolved oxygen (DO) with their DO values noted as B_5 and D_5 respectively.

The BOD was then calculated using the formular, $BOD = ((D_0 - B_0) - (D_5 - B_5))/P$ where B_0 is dissolved oxygen for blank solution on first day; D_0 is dissolved oxygen for sample solution on first day; D_5 is dissolved oxygen for sample solution on fifth day; B_5 is dissolved oxygen for blank solution on fifth day; P is decimal volumetric fraction sample used..

2.5. Data Analysis

The results of the proximity (m) and coverage (m^2) of the anthropogenic activities as well as the minimum, maximum, standard deviation, mean, standard error of the mean and coefficient of variation values of the measured physico-chemical parameters were tabulated. The physico-chemical parameters were compared with set standards, WHO and European Union guidelines for drinking water. Mean differences of the parameters between water sources were determined using one-way ANOVA (F) test and relationships between the physico-chemical parameters established using Pearson product-moment correlation coefficient (r) at 5% level of significance. The data presentations and analysis were done by using Microsoft Excel 2007 and SPSS 20.0 Statistical packages.

3. RESULTS AND DISCUSSION

This section presents the findings of the study and discussion beginning with the anthropogenic activities in the vicinity of the drinking water sources followed by the physico-chemical quality of the water sources.

3.1. Anthropogenic Activities in the Vicinity of the Drinking Water Sources in Mbarara Municipality

The identified anthropogenic activities within the vicinity of the drinking water sources with great impact on the quality of the water sources are presented in Table 1. The nearest croplands covering 3096 m^2 area were located 1 m from the shallow well in Nyamitanga while the furthest was situated 35 m from the borehole in Shuhaddea Secondary School though it had the highest coverage (3970 m^2) as shown in Table 1. Animal farms were only observed around the well in Kisenyi, a distance of 2 m and a coverage of 27 m^2 and borehole in Shuhaddea Secondary School at a distance of 23 m and covering 8 m^2area. Latrines were found around all the drinking water sources with the closest located 2 m from the shallow well (Nyamitanga) followed by Spring (Kise) and the furthest situated 53 m from borehole (Shuhaddea Secondary School) greatly coinciding with the human settlements (Table 1). The closest landfills were also to the shallow well in Nyamitanga (7 m) with a coverage of 158 m^2 followed by Spring (Kisenyi) recording an overall highest coverage of 442 m^2. The furthest land fills were located 58 m from Spring (Kiswahili) with 387 m^2 area coverage. Brick laying and cloth washing sites were only observed around S. well (Nyam) and Well (Kise) respectively both at a proximity of 10 m from the water sources with each covering 9 m^2 and 21 m^2 in that order. Municipal waste drainage channels were found only around the spring in Kiswahili and well in Kisenyi at distances of 4 m and 26 m respectively.

Table 1. Proximity (m) and coverage (m2) of the selected anthropogenic activities in the vicinity of the drinking water sources.

Activities		Borehole (Nyam. S.S)	Borehole (Shud. S.S.)	Spring (Kisw)	Spring (Kise)	Well (Kise)	S. well (Nyam)
Croplands	Proximity	24	35	12	-	-	1
	Coverage	2589	3970	207	-	-	3096
Animal farms	Proximity	-	23	-	-	2	-
	Coverage	-	8	-	-	27	-
Latrine	Proximity	37	53	17	4	20	2
	Coverage	148	185	370	370	222	256
Settlements	Proximity	16	54	17	8	16	3
	Coverage	-	-	-	-	-	-
Land fills	Proximity	12	33	58	11	12	7
	Coverage	149	67	387	442	211	158
Brick laying	Proximity	-	-	-	-	-	10
	Coverage	-	-	-	-	-	9
Washing sites	Proximity	-	-	-	-	10	-
	Coverage	-	-	-	-	21	-
Municipal waste	Proximity	-	-	4	-	26	-
	Total coverage	2886	4230	964	812	481	3519

Nyam. S.S: Nyamitanga Secondary School; Shud. S.S: Shuhaddea Secondary School; Kisw: Kiswahili; Kise: Kisenyi; Nyam: Nyamitanga; S. well: Shallow well; -: represents no anthropogenic activity.

The increased anthropogenic activities (croplands, animal farms, latrine, settlements, landfills, brick laying, washing sites and municipal waste) within the vicinity (200 m) of the different drinking water sources could be attributed to the increased population as a result of urbanisation. This is in line with [10] which reported that Mbarara municipality is fairly densely populated. Urbanisation is associated with activities such as settlements, landfills, education, infrastructure development, slum development, urban traffic of cars, and transport among others [13].

3.2. Physico-Chemical Quality of the Drinking Water Sources in Mbarara Municipality

The descriptive statistics of the physico-chemical parameters of the selected drinking water sources and the comparisons among the water sources are summarized in Table 2.

3.2.1. Temperature

There were significant temperature variations ($p < 0.05$) in the water sources with low variability within the samples (CV = 1.38% - 4.84%). The highest mean temperature was recorded in spring (Kisenyi) and lowest in rainwater (MUST). Temperature is among the physico-chemical parameters useful in evaluating the quality of drinking water as it influences the overall quality of water (physico- chemical and biological characteristics) including the rate of chemical reactions in the water body, decrease in the solubility of gases and improving the tastes and colours of water [14] [15]. The temperature range of boreholes in Nyamitanga and Shuhaddea secondary schools were within the temperature range (23°C - 25.90°C) of boreholes in Lira district, Uganda obtained by [16] while that of springs in Kiswahili (21.3°C - 23.0°C) and Kisenyi (22.6°C - 24.9°C) were close to the temperate range (23.6°C - 26.4°C) of Katwe and Kisenyi springs in Kampala City, Uganda [17]. However all the recorded water temperatures of the drinking water sources were above the WHO recommended value (<15°C), the threshold temperature which makes drinking water palatable [18] [19]. The high temperatures could be associated with global warming. According to [20] , the global land mean surface temperature increase is 0.85°C over the period from 1850- 2012 and Uganda's National Adaptation Program of Action, reported an average temperature increase of 0.28°C per decade between 1960 and 2010 [21].

Table 2. Physico-chemical parameters in the drinking water sources in Mbarara Municipality.

Parameter (n = 10)	Water source	Min.	Max.	SD	Mean ± SE	CV (%)	p-value for F test	Guideline value
Temperature (°C)	Borehole (Nyam. S.S)	22.30	23.50	0.57	**22.98 ± 0.29**	2.50	0.00	15°C*
	Borehole (Shud. S.S)	21.60	22.90	0.41	**22.34 ± 0.13**	1.84		
	Spring (Kisw)	21.30	23.00	0.52	**22.38 ± 0.16**	2.32		
	Spring (Kise)	22.60	24.90	0.72	**23.45 ± 0.23**	3.07		
	Well (Kise)	18.50	21.70	1.01	**20.79 ± 0.32**	4.84		
	Shallow well (Nyam)	21.70	22.60	0.31	**22.17 ± 0.10**	1.38		
	Rainwater	17.20	18.80	0.54	**18.07 ± 0.17**	2.98		
pH	Borehole (Nyam. S.S)	6.19	6.77	0.22	**6.37 ± 0.07**	3.42	0.00	6.5 - 9.5*
	Borehole (Shud. S.S)	5.50	6.53	0.29	**5.74 ± 0.09**	5.06		
	Spring (Kisw)	6.17	6.29	0.03	**6.21 ± 0.01**	0.56		
	Spring (Kise)	6.65	7.02	0.14	6.77 ± 0.04	2.08		
	Well (Kise)	4.49	6.70	0.64	**6.31 ± 0.20**	10.18		
	Shallow well (Nyam)	7.23	7.70	0.13	7.54 ± 0.04	1.71		
	Rainwater	5.80	6.70	0.32	**6.21 ± 0.10**	5.16		
Dissolved Oxygen (mg/L)	Borehole (Nyam. S.S)	5.30	8.60	1.08	**7.42 ± 0.34**	14.57	0.00	10 - 12 mg/l*
	Borehole (Shud. S.S)	5.10	14.20	2.83	10.41 ± 0.90	27.22		
	Spring (Kisw)	8.00	15.60	3.15	10.28 ± 10	30.61		
	Spring (Kise)	1.90	6.30	1.20	**4.84 ± 0.38**	24.76		
	Well (Kise)	9.00	18.10	3.01	12.86 ± 0.95	23.37		
	Shallow well (Nyam)	3.20	6.30	1.01	**5.07 ± 0.32**	19.86		
	Rainwater	9.30	10.00	0.22	**9.54 ± 0.07**	2.33		
BOD (mg/L)	Borehole (Nyam. S.S)	4.00	7.00	1.10	5.59 ± 0.35	19.60	0.00	3 - 6 mg/l***
	Borehole (Shud. S.S)	3.10	12.00	3.33	**7.71 ± 1.05**	43.19		
	Spring (Kisw)	4.00	9.10	1.69	5.98 ± 0.54	28.34		
	Spring (Kise)	0.50	3.10	0.86	2.16 ± 0.27	39.72		
	Well (Kise)	3.20	8.90	2.09	**6.72 ± 0.66**	31.10		
	Shallow well (Nyam)	1.80	5.40	1.10	4.14 ± 0.35	26.59		
	Rainwater	1.30	2.50	0.36	1.83 ± 0.11	19.79		
Total Dissolved Solids (mg/L)	Borehole (Nyam. S.S)	256.00	380.00	32.74	297.70 ± 10.35	11.00	0.00	1000 mg/l*
	Borehole (Shud. S.S)	272.00	348.00	22.51	298.40 ± 7.12	7.54		
	Spring (Kisw)	345.00	357.00	3.63	350.10 ± 1.15	1.04		
	Spring (Kise)	557.00	579.00	6.61	569.20 ± 2.09	1.16		
	Well (Kise)	347.00	370.00	6.03	357.90 ± 1.91	1.68		
	Shallow well (Nyam)	250.00	268.00	5.53	257.80 ± 1.75	2.15		
	Rainwater	12.00	55.00	13.28	33.40 ± 4.20	39.78		

Electrical Conductivity (µs/cm)	Borehole (Nyam. S.S)	565.00	615.00	14.81	585.60 ± 4.68	2.53	0.00	1500 µs/cm*
	Borehole (Shud. S.S)	580.00	630.00	15.49	601.40 ± 4.90	2.58		
	Spring (Kisw)	685.00	715.00	9.92	698.90 ± 3.14	1.42		
	Spring (Kise)	1113.00	1152.00	10.89	1139.90 ± 3.44	0.96		
	Well (Kise)	708.00	730.00	7.98	717.80 ± 2.52	1.11		
	Shallow well (Nyam)	409.00	553.00	37.17	508.70 ± 11.75	7.31		
	Rainwater	16.00	47.00	10.09	29.30 ± 3.19	34.43		
Total Hardness (mg/L)	Borehole (Nyam. S.S)	215.00	285.00	33.81	264.00 ± 10.69	12.81	0.00	1000 mg/l**
	Borehole (Shud. S.S)	140.00	215.00	38.73	185.00 ± 12.25	20.94		
	Spring (Kisw)	215.00	285.00	33.81	236.00 ± 10.69	14.33		
	Spring (Kise)	140.00	215.00	39.53	177.50 ± 12.50	22.27		
	Well (Kise)	140.00	215.00	31.62	200.00 ± 10.00	15.81		
	Shallow well (Nyam)	140.00	140.00	0.00	140.00 ± 0.00	0.00		
	Rainwater	70.00	70.00	0.00	70.00 ± 0.00	0.00		

Min.: Minimum; Max.: Maximum; SD: Standard deviation; SE: Standard error of the mean; CV: coefficient of variation; *: [18]; **: [22]; ***: European Union guideline as cited in [23].

3.2.2. pH

Statistically significant mean pH variations were observed among the seven water sources ($p < 0.05$) although there were low differences within samples for six water sources (CV < 10%) apart from the well in Kisenyi (CV = 10.18%). The highest mean pH was recorded in shallow well (Nyam) and the least in Borehole (Shud. S.S). The pH of water is a useful parameter as most biological activities take place only within a narrow range. As a result, any pH variations beyond an acceptable limit could be fatal to a particular organism [24]. The mean pH values of spring (kise) and shallow well (nyam) drinking water sources were within the pH range for drinking water sources (6.5 - 9.5) by [18] while the rest of the water sources had mean pH values below the minimum guide line value (6.5). Apart from the shallow well in Nyamitanga with a pH value above 7 (slightly alkaline), the other water sources were a little acidic (pH below 7). The pH range of rainwater (5.80 - 6.70) was close to that given by [25] of freshly fallen rainwater (5.5 - 6.0). The slightly acidic nature of rainwater is mainly attributable to sulfuric and nitric acids formed from increased SO_2 and NO_x emissions into the atmosphere as a result of combustion of fossil fuels from transportation vehicles as mentioned by [26]. According to them, the SO_2 and NO_x (which are transferable from one place to another) react with water and oxygen in the atmosphere to produce sulfuric and nitric acids which form acid rain. Additionally, the rainwater sampling site, Mbarara University of Science and Technology borders the main Kampala-Kabale road (Figure 1) that experiences dense traffic flow of vehicles, a common global problem in the recent decades [26] which produce a lot of SO_2 and NO_x emissions.

Acid rains in Mbarara municipality resulting from SO_2 and NO_x emissions from urban traffic of cars [13] could have contributed to the slightly acidic pH of most of the drinking water sources in addition to other causes. One of the impacts of acid precipitation is acidification of surface and ground waters [26]. Furthermore, the ground water sources (boreholes, springs) including the well in Kisenyi in the present study also had slightly acidic pH values consistent with the findings of [27] that show that ground water is acidic. [28] and [17] associated the low pH values in most wells and springs to carbon dioxide saturation in the groundwater. The physico-chemical nature of the soil of the sampling sites partly dictates the final pH of the water samples as it was the case for the water samples collected and analyzed from Katanga, North of Kampala city [29] [30]. The mean pH of springs in Kiswahili (6.21) and Kisenyi (6.77) were close to the pH range (4.4 - 6.7) of Katwe and Kisenyi springs in Kampala City [17].

However, the slightly higher mean pH value (7.54) in the shallow well in Nyamitanga (a surface water source) than other drinking water sources can be attributed to organic material and sediment runoff from the closely located surrounding croplands (1 m from the water source) as opposed to croplands around other water sources. The deposition of sediments and organic materials into the water causes a change in the carbonate-bicarbonate equilibrium in surface water causing a relatively higher pH beyond neutral [31].

3.2.3. Dissolved Oxygen (DO)

DO concentrations showed significant variations ($p < 0.05$) among the water sources with highly variable values within samples for most of the water sources (CV > 10%) except for rainwater at MUST (CV = 2.33%). The highest and lowest mean DO concentrations were recorded in the well and spring at Kisenyi respectively. Dissolved oxygen (DO) is a very important parameter of water quality and an index of physical and biological process going on in water which favors solubility of oxygen [32]. The DO of some of the water sources i.e. Borehole (Shud. S.S) and spring (Kisw) lie within the acceptable standards for dissolved oxygen in fresh water (10 - 12 mg/l) by [18]. Only one drinking water source i.e. well (Kise) had mean DO value above WHO range. On the other hand, over 50% of the drinking water sources i.e. Borehole (Nyam. S.S), spring (Kise), Shallow well (Nyam) and rainwater had DO values below the WHO range. Similar results were obtained by [33] for drinking water in Bahir Dar, Ethiopia, where the mean DO concentration of the water samples were between 0.45 and 5.27 mg/l.

The low DO in the shallow well in Nyamitanga can be attributed to increased decomposition of organic material from the closely situated surrounding croplands (1 m from the water source) though on the contrary, BOD does not reflect it despite the significant overall positive correlation with DO ($r = 0.683$, $p < 0.05$, $n = 70$). The low DO values in spring (Kise.), Borehole (Nyam. S.S) and Shallow well (Nyam) can be associated with the slightly elevated temperature values in these water sources. This is further supported by the overall significant negative correlation of temperature with DO ($r = -0.454$, $p < 0.05$, $n = 70$). Low DO values in drinking water samples have been attributed to the high temperature of the water [29]. Furthermore, [34] and [33] asserted that the solubility of oxygen in water is a function of its temperature viz. the lower the temperature, the greater the solubility of oxygen in the water and vice versa. Thus the slightly high dissolved oxygen (DO) in Well (Kise) is attributable to the slightly low temperature of the well but not due to increased productivity (photosynthesis) in the well. High photosynthetic rates in the water which reduce the available carbon dioxide (increasing the pH) would liberate oxygen leading to positive correlation between DO and pH. According to [35] photosynthesis of aquatic plants releases oxygen into the water and in low-velocity or still waters with plants, an increase in pH can be expected during the growing season. However there was an overall significant negative correlation between pH and DO ($r = -0.444$, $p < 0.05$, $n = 70$). The slightly low DO in rainwater (MUST) despite the low mean temperature recorded could be due to use of some of the atmospheric oxygen in chemical reactions forming acid rain. This is in agreement with [26] who reported that SO_2 and NO_x emissions from vehicles react with water and oxygen in the atmosphere to produce sulfuric and nitric acids which form acid rain.

3.2.4. Biological Oxygen Demand (BOD)

BOD showed significant variations ($p < 0.05$) among the water sources and were very variable within samples for all the water sources (CV > 10%). The borehole at Shuhaddea Secondary School had the highest mean BOD and the least was in rainwater at Mbarara University of Science and Technology. BOD is the measure of the amount of dissolved oxygen required for the decomposition of organic compounds [36]. The mean BOD after 5 days (BOD5) for most of the water sources i.e. Borehole (Nyam. S.S), Spring (Kisw), Spring (Kise), Shallow well (Nyam) and Rainwater (MUST) fall within the European Union guideline range for fisheries and aquatic life of 3 - 6 mg/l [23]. This guideline was used to assess the water sources due to lack of set guideline for maximum tolerable limit of BOD in drinking water [29]. On the other hand, the water sources i.e. Borehole (Shud. S.S) and Well (Kise) had BOD values above the European Union guideline value attributable to increased organic matter in these water sources as previously reported by [29] in their drinking water sources. The borehole in Shuhaddea secondary school is located in a swampy area with probably organic matter rich underlying soil. Swamps are characterized by high amounts of dissolved organic matter [37]. Due to the swampy nature of the area where Shuhaddea borehole is located, it probably has a lot of micro-organisms which could have resulted into increased microbial activity hence a high BOD. [38] earlier on found out that swamp soils are usually rich in organic matter content and have high numbers of aerobic and anaerobic microorganism that decompose the organic matter. For the case of the well in Kisenyi, the close proximity (2 m) of the animal farms which expose the well to animal wastes (organic matter) can be blamed for the high BOD. According to [39] , the proximity of boreholes to animal droppings being littered around them contaminates the quality of groundwater. Surface waters such as the well in Kisenyi are quite easily contaminated by these animal droppings which directly end up into them. According to [40] , animal wastes greatly affect the quality of ground water in terms of BOD due to the direct disposal of the wastes on the land surface. These wastes could much easily affect surface waters like the well in Kisenyi than ground water.

3.2.5. Total Dissolved Solids (TDS)

Total dissolved solids showed significant variations ($p < 0.05$) among the water sources and were less variable within samples for most of the water sources (CV < 10%) except for rainwater at MUST and borehole in Nyamitanga (CV > 10%). The spring water in Kisenyi also recorded notably the highest total dissolved solids while rainwater (MUST) had the least. Total dissolved solids (TDS) including carbonate, bicarbonate, chloride, sulphate, phosphate, nitrate, calcium, magnesium, sodium, organic ions and other ions determine the general nature of water quality [14]. They affect the taste of drinking water if found at concentrations above the WHO recommended value. Consequently the TDS in all the water sources were below the WHO maximum allowable limit of 1000 mg/l [18] , hence making these water sources suitable for drinking.

3.2.6. Electrical Conductivity (EC)

There was a statistically significant difference ($p < 0.05$) among mean electrical conductivities of the different water sources with spring (Kise) having the highest extreme value and the least value was in rainwater (MUST). The within sample variation of the conductivities for six water sources were small (CV < 10%) except for rainwater at MUST (CV = 34.43%). Electrical conductivity is the ability of any medium, water in this case, to carry an electric current. The presence of dissolved solids such as calcium, chloride, and magnesium in water samples carries the electric current through water [41]. All the mean EC values of the water sources were below the WHO maximum guideline limit of 1500 µs/cm. Similar results were obtained for water samples in Ethiopia by [29]. The mean conductivity of spring in Kiswahili (698.90 µS/cm) was within the range of conductivity (95 - 705 µS/cm) of Katwe and Kisenyi springs in Kampala City

[17]. However the conductivity of the spring in Kisenyi (1139.90 μS/cm) was outside the ranges of Katwe and Kisenyi springs in Kampala. The extremely high conductivity in spring (Kise) compared to the other water sources could be associated with the corrosion of the metallic pipe of the spring that led to enrichment of metals in the water. High conductivity was recorded from tap water sources in Ethiopia attributable to the corrosion of metals leading to accumulation of heavy metals [29]. High concentration of charged ions resulting from leaching and run off into the drinking water sources particularly from latrines and landfills including the underlying rocks could have also contributed.

The notably higher mean conductivity in spring (Kise) than the other water sources corresponds to the high Total Dissolved Solids (TDS) in the spring water. This is further augmented by the overall significant positive correlation between conductivity and TDS ($r = 0.996$, $p < 0.05$, $n = 70$). This confirms that it is mainly the same ions that constitute TDS i.e. carbonate, bicarbonate, chloride, sulphate, phosphate, nitrate, calcium, magnesium, sodium, organic ions and other ions which lead to the conductivity of the water sources. The high temperature (23.45°C) in spring (Kise) could have increased the dissolution of the ions leading to high TDS and conductivity as supported by the overall significant positive correlation ($p < 0.05$) of temperature with TDS ($r = 0.762$) and conductivity ($r = 0.764$).

3.2.7. Total Hardness

There were significant variations ($p < 0.05$) in total hardness of the 7 water sources with borehole (Nyam. S.S) registering the highest value while the least was for rainwater (MUST). Total hardness values were highly variable within samples for most of the water sources (CV > 10%) except for shallow well in Nyamitanga and rainwater at MUST (CV = 0.00%). Calcium and magnesium dissolved in water are the two most common minerals that make water "hard" [42] [43]. These were the same ions that also possibly contributed to the TDS and conductivity of the water sources as total hardness was significantly positively correlated ($p < 0.05$) with TDS ($r = 0.504$) and conductivity ($r = 0.513$). The total hardness of all the drinking water sources were below the harmless limit of 1000 mg/L [22]. Basing on the classification of total hardness of water by [44] , rainwater was moderately soft; S. well (Nyam) and borehole (Shud. S.S) were moderately hard; Spring (Kisw.), Borehole (Nyam. S.S), spring (Kise.) and well (Kise.) waters were hard. Generally, all the other drinking water sources except rain were moderately hard to hard probably due to the presence of high concentration of calcium and magnesium ions in the water as a result of leaching and run off into the water sources especially from croplands, latrines, landfills, animal and municipal wastes within their vicinity as well as the underlying rocks. Calcium ions are usually common in water sources due to leaching of the rocks [45]. However rainwater was not soft as would be expected probably due to atmospheric particulates that had calcium ions. The dissolution of calcium and magnesium minerals which cause hardness of water was enhanced by the relatively high temperatures of the water sources. This is supported by the significant positive correlation of temperature with total hardness ($r = 0.645$, $p < 0.05$, $n = 70$) and the slightly elevated water temperatures of the drinking water sources.

4. CONCLUSIONS

A range of physico-chemical parameters in most of the selected drinking water sources (boreholes, springs, wells, rainwater) in Mbarara Municipality have been compromised by anthropogenic activities (croplands, latrines, landfills, transportation, animal and municipal wastes) situated within the vicinity of these water sources.

The temperature of the water sources were above the WHO threshold temperature for palatable water.

The pH of boreholes in Nyamitanga and Shuhaddea Secondary Schools, spring in Kiswahili, well in Kisenyi and rainwater in Mbarara University of Science and Technology were below the WHO minimum guideline value hence acidic.

The dissolved oxygen levels of borehole in Nyamitanga secondary school, spring in Kisenyi, shallow well in Nyamitanga and the rainwater in Mbarara University of Science and Technology were below the WHO range thus less suitable for other aerobic aquatic life.

BOD values of borehole in Shuhaddea secondary school and the well in Kisenyi were above the European Union guideline value signifying a lot of decomposition of organic matter in these water sources.

The selected water sources (boreholes, springs, and wells) were found to have moderately hard and hard water from the total hardness measure hence undesirable for domestic uses like washing and boiling etc.

Mbarara municipal council should therefore ensure proper sanitation and water safety plans for these drinking water sources to avoid further contamination from the human activities.

ACKNOWLEDGEMENTS

The Authors would wish to acknowledge the support offered to them by the entire Department of Biology of Mbarara University of Science and Technology during the research and preparation of this paper.

REFERENCES

1. Ondor, S.T. and Addo, K.K. (2013) Bacteriological Profile and Physico-Chemical Quality of Ground Water: A Case Study of Borehole Water Sources in a Rural Ghanaian Community. International Journal of Current Microbiology and Applied Sciences, 2, 21-40.
2. Mirribasi, R., Mazloumadeh, S.M. and Rahnama, M.B. (2008) Evaluation of Irrigation Water Quality Using Fuzzy Logic. Research Journal of Environment Sciences, 2, 340-352. https://doi.org/10.3923/rjes.2008.340.352
3. World Health Organisation (2000) The Global Water Supply and Sanitation Assessment Report. World Health Organisation Press, Geneva.
4. Howard, G., Pedley, S., Barett, M., Nalubega, M. and Johal, K. (2003) Risk Factors Contributing to Microbiological Contamination of Shallow Ground Water in Kampala, Uganda. Water Research, 37, 3421-3429. https://doi.org/10.1016/S0043-1354(03) 00235-5
5. Mbarara Municipality (2009) Mbarara Municipality Council Development Plan 2009-2012. Mbarara Municipality Press, Mbarara.
6. Narendra, P. and Sharma, H. (1993) African Water Resources. World Bank Publications, Washington DC.
7. United Nations Human Settlements Programme (2012) Mbarara Municipality Urban Profile. UNON, Publishing Services Section, Nairobi.
8. Egor, M., Mbabazi, J. and Ntale, M. (2014) Heavy Metal and Nutrient Loading of River Rwizi by Effluents from Mbarara Municipality. Journal of Chemistry and Materials Research, 2, 36-47.
9. Uganda National Roads Authority (2010) Feasibility Study, Detailed and Supervision of the Construction of Northern Corridor Route, Mbarara by Pass Road. Republic of Uganda, Kampala.
10. Uganda Bureau of Statistics (2014) National Population and Housing Census. Kampala. Republic of Uganda, Kampala, Uganda.
11. Ministry of Water, Lands and Environment (2000) National Environment Regulations. Kampala, Uganda. Republic of Uganda, Kampala, Uganda.
12. American Public Health Association (2012) Standard Methods for Examination of Water and Waste Water. 22nd Edition, American Public Health Association Press, Washington DC.
13. United Nations (2014) World Urbanisation Prospects.
14. Olajire, A.A. and Imeokparia, F.E. (2001) Water Quality Assessment of Osun River: Studies on Inorganic Nutrients. Environmental Monitoring and Assessment, 69, 17-22. https: //doi.org/10.1023/A:1010796410829
15. Palamuleni, L. and Akoth, M. (2015) Physico-Chemical and Microbial Analysis of Selected Borehole Water in Mahikeng, South Africa. Journal of Environmental Research and Public Health, 12, 8619-8630. https://doi.org/10.3390/ijerph120808619
16. Okot, J. and Otim, J. (2015) The Quality of Drinking Water Used by the Communities in Some Regions of Uganda. International Journal of Biological and Chemical Sciences, 9, 552-562. https://doi.org/10.4314/ijbcs.v9i1.47

17. Haruna, R., Ejobi, F. and Kabagambe, E.K. (2005) The Quality of Water from Protected Springs in Katwe and Kisenyi Parishes, Kampala City, Uganda. African Health Sciences, 5, 14-20.
18. World Health Organisation (1996) World Health Organisation Guidelines for Drinking Water. World Health Organisation Press, Geneva.
19. World Health Organisation (2003) Technical Support Document for Ontario Drinking Water Standards, Objectives and Guidelines. World Health Organisation Press, Geneva.
20. IPCC (2013) Climate Change 2013: The Physical Science Basis. Contribution of Working Group I to the Fifth Assessment Report of the Intergovernmental Panel on Climate Change.
21. NAPA (2007) Climate Change Uganda National Adaptation Programmes of Action.
22. WHO (1993) Guidelines for Drinking Water Quality-Recommendations. 2rd Edition, Vol. 1, World Health Organisation, Geneva, Switzerland.
23. Israel, D. (2007) Assessment of Drinking Water Quality and Pollution Profiles along Awetu Stream Jimma. M.Sc. Thesis, Addis Ababa University, Ethiopia.
24. Trivede, P., Bajpai, A. and Thareja, S. (2010) Comparative Study of Seasonal Variations in Physico-Chemical Characteristics in Drinking Water Quality of Kanpur, India With Reference to 200 MLD Filteration Plant and Groundwater. Natural Sciences, 8, 11-17.
25. Bridge, J. and Demicco, R. (2008) Earth Surface Processes, Landforms and Sediment Deposits. Cambridge University Press, Cambridge. https://doi.org/10.1017/ CBO97 80511805516
26. Dincer, I. and Rosen, M.A. (2007) Exergy: Energy, Environment and Sustainable Development. Elsevier Science Publishers, Amsterdam.
27. Deas, M.L. and Orlob, G.T. (1999) Klamath River Modelling Project. University of California, California.
28. Byamukama, D., Kansiime, F., March, R.L. and Farnleitner, H. (2000) Determination of Escherichia coli Contamination with Chromocult Coliform Agar Showed a high Level of Discrimination Efficiency for Differing Fecal Pollution Levels in Tropical Waters of Kampala, Uganda. Applied and Environment Microbiology, 66, 864-868. https://doi.org/ 10.1128/AEM.66.2.864-868.2000
29. Yasin, M., Ketema, T. and Bacha, K. (2015) Physico-Chemical and Bacteriological Quality of Drinking Water of Different Sources, Jimma Zone, Southwest Ethiopia. BMC Research Notes, 8, 541. https://doi.org/10.1186/s13104-015-1376-5
30. Nshekanabo, N. and Wozei, E. (1997) Spring Water Quality Improvement in Slums. Sanitation and Water for all. Proceedings of 24th WEDC Conference, Islamabad, Pakistan, 1998, 265.
31. Karanth, K.R. (1987) Ground Water Assessment Development and Management. McGraw Hill Publishing Company Ltd., New Delhi.
32. Vyas, V.G., Hassan, M.M., Vindhani, S.I., Parmar, H.J. and Bhalani, V.M. (2015) Physicochemical and Microbiological Assessment of Drinking Water from Different Sources in Junagadh City, India. American Journal of Microbiological Research, 3, 148-154.
33. Tenagne, A.W. (2009) The Impact of Urban Storm Water Runoff and Domestic Waste Effluent on Water Quality of Lake Tana and Local Groundwater near the City of Bahir Dar, Ethiopia. M.Sc. Thesis, Cornell University, New York.
34. Wetzel, R.G. (2001) Limnology: Lake and River Ecosystem. 3rd Edition, Academic Press, San Diego.
35. Alliance for the Chesapeake Bay (2010) Alliance for the Chesapeake Bay's River Trends Manual. http://www.paesta.psu.edu/sites/default/files/paestaidea/water_ quality_ parameter_info_acb.pdf
36. Jena, S.K., Patra, L., Padhy, R., Mahapatra, M., Mohanty, B.K. and Sabat, G. (2016) Physico-Chemical Analysis of Pond Water in Berhampur Town Odisha. International Journal of Recent Research and Applied Studies, 7, 33-37.
37. Fraser, L.H. and Keddy, P.A. (2005) The World's Largest Wetlands: Ecology and Conservation. University Press, Cambridge. https://doi.org/10.1017/ CBO9 780511542091

38. Priester, D.S. and Harms, W.R. (1971) Microbial Populations in Flooded Swamp Soils of South Carolina. Asheville. Department of Agriculture, Forest Service, Southeastern Forest Experiment Station.
39. Bello, O.O., Osho, A., Bankole, S.A. and Bello, T.K. (2013) Bacteriological and Physicochemical Analysis of Borehole and Well Water Sources in Ijebu-Ode, Southwestern Nigeria. International Journal of Pharma and Bio Sciences, 8, 18-25.
40. United States Environment Protection Agency (1990) Citizens Guide to Groundwater Protection. Office of Drinking Water, Washington DC.
41. Rahmanian, N., Ali, S.H.B., Homayoonfard, M., Ali, N.J., Rehan, M., Sadef, Y. and Nizami, A.S. (2015) Analysis of Physiochemical Parameters to Evaluate the Drinking Water Quality in the State of Perak, Malaysia. Journal of Chemistry, 2015, Article ID: 716125, 10 p.
42. Shyamala, R., Shanthi, M. and Lalitha, P. (2009) Physicochemical Analysis of Bore well Water Samples of Telungupalayam Area in Coimbatore District, Tamilnadu, India. Electronic Journal of Chemistry, 5, 924-929.
43. Patil, S.G., Chonde, S.G., Jadhav, A.S. and Raut, P.D. (2012) Impact of Physico-Chemical Characteristics of Shivaji University Lakes on Phytoplankton Communities, Kolhapur, India. Research Journal of recent Sciences, 1, 56-60.
44. Tyson, A. and Harrison, K. (1990) Water Quality for Private Water Systems. University of Georgia, Athens, Georgia.
45. Agunwamba, J.C. (2000) Water Engineering Systems. 2nd Edition, Immaculate Publications Limited, Enugu.

CHAPTER 17

Binational Arsenic Exposure Survey: Methodology and Estimated Arsenic Intake from Drinking Water and Urinary Arsenic Concentrations

Jason Roberge [1,*], Mary Kay O'Rourke [2], Maria Mercedes Meza-Montenegro [3], Luis Enrique Gutiérrez-Millán[4], Jefferey L. Burgess [2] and Robin B. Harris [1]

[1]*Department of Epidemiology and Biostatistics, Mel and Enid Zuckerman College of Public Health, The University of Arizona, 1515 N. Campbell Ave., Tucson, AZ 85724, USA*

[2]*Division of Community, Environment and Policy, Mel and Enid Zuckerman College of Public Health, The University of Arizona, Tucson, AZ 85724, USA*

[3]*Department of Biotechnology and Food Sciences, Instituto Tecnológico de Sonora, 5 de Febrero 818 Sur, Zona Centro, Cd. Obregon, Sonora 85000, Mexico*

[4]*Department of Scientific and Technological Research, University of Sonora, Hermosillo, Sonora 83000, Mexico*

ABSTRACT

The Binational Arsenic Exposure Survey (BAsES) was designed to evaluate probable arsenic exposures in selected areas of southern Arizona and northern Mexico, two regions with known elevated levels of arsenic in groundwater reserves. This paper describes the methodology of BAsES and the relationship between estimated arsenic intake from beverages and arsenic output in urine. Households from eight communities were selected for their varying groundwater arsenic concentrations in Arizona, USA and Sonora, Mexico. Adults responded to questionnaires and provided dietary information. A first morning urine void and water from all household drinking sources were collected. Associations between urinary arsenic concentration (total, organic, inorganic) and estimated level of arsenic consumed from water and other beverages were evaluated through crude associations and by random effects models. Median estimated total arsenic intake from beverages among participants from Arizona communities ranged from 1.7 to 14.1 μg/day compared to 0.6 to 3.4 μg/day among those from Mexico communities. In contrast, median urinary inorganic arsenic concentrations were greatest among participants from Hermosillo, Mexico (6.2 μg/L) whereas a high of 2.0 μg/L was found among participants from Ajo, Arizona. Estimated arsenic intake from drinking water was associated with urinary total arsenic concentration ($p < 0.001$), urinary inorganic arsenic concentration ($p < 0.001$), and urinary sum of species ($p < 0.001$). Urinary arsenic concentrations increased between 7% and 12% for each one percent increase in arsenic consumed from drinking water. Variability in arsenic intake from beverages and urinary arsenic output yielded counter intuitive results. Estimated intake of arsenic from all beverages was greatest among Arizonans yet participants in Mexico had higher urinary total and inorganic arsenic concentrations. Other contributors to urinary arsenic concentrations should be evaluated.

Keywords: arsenic; urine; water; beverages; metabolite; intake; BAsES

1. INTRODUCTION

Arsenic (As) is a widely distributed element found naturally in the Earth's crust [1]. Arsenic is classified as a known human carcinogen and is among the top 20 hazardous materials compiled by the Environmental Protection Agency (EPA) [2,3]. Consumption of As^3 and As^5 at levels above 100 μg/L over long periods of time has been associated with lung, bladder, and skin cancer [4,5,6].

The health effects of arsenic consumption are attributed primarily to exposures from higher arsenic concentrations (>100 μg/L) found in local water supplies [4,6,7,8,9,10]. Yet, the contribution of arsenic from foods and beverages may play a larger role than arsenic from regulated drinking water in defining the total arsenic concentration in urine [11,12]. This may be the case particularly where populations are consuming relatively lower concentrations of arsenic from drinking water. Few studies have examined the effects of the consumption of arsenic from drinking water at the maximum contaminant level, 10 μg/L, established by the United States EPA at in 2006 [13].

In the United States (U.S.), arsenic concentration in groundwater is greatest in the West, although isolated drainage basins in the Midwest and Northeast exceed 10 μg/L is some areas. Public water utilities are mandated to have no more than 10 μg/L of arsenic in water supplied for domestic use. It is a public health concern that owners of private wells are not required to assess and control arsenic or other contaminants from their drinking and cooking water. Limited information is available about water consumption patterns among people in the U.S. and less is known for those in Mexico. National surveys inquire about water consumption, but few distinguish the various sources of water and other beverages consumed in the home.

The Binational Arsenic Exposure Survey (BAsES) was designed to assess arsenic concentrations in drinking water among selected communities in Arizona and Mexico and describe the relationships between estimated As concentrations in water and other beverages and As exposure reflected by urinary biomarkers. This pilot study also sought to build a binational research collaboration across the U.S.–Mexico border. Utilizing terminology as found in the International Society for Exposure Analysis glossary [14], the current study describes BAsES and examines the relationship between arsenic intake from water, other beverages, and urinary arsenic output.

2. METHODS

2.1. Study Design

The Binational Arsenic Exposure Survey was cross-sectional in design and conducted in Arizona and northern Mexico. Communities in these regions were selected based on their arsenic concentrations contained in groundwater and likely represented a high versus low arsenic exposure among Mexicans, and Hispanic and Non-Hispanic participants within the U.S.

Randomly selected households were identified through phone interviews (e.g., random digit dialing) or door-to-door contact. All individuals who had lived in the home for at least one continuous year and were at least

18 years of age were eligible to participate. One person was selected randomly among household participants and identified as the primary respondent.

Interviewers from all sites received the same training regarding questionnaire administration. Interviews were administered in either English or Spanish as determined by the participant. The primary respondent completed a household questionnaire reporting characteristics of the home and individual respondents completed a personal questionnaire. In addition, biological samples, and environmental samples were collected during household visits. A single data entry system was created in Microsoft Access. De-identified data were electronically shared among the sites. The Arizona team served as the data coordinating center. Data were visually verified and outliers were evaluated.

Interviewing and sample collection protocols were reviewed by the Department of Health of Sonora in Mexico, respective university committees, and the Institutional Review Board (IRB) at the University of Arizona. All participants signed consent forms approved by the IRB in either Spanish or English.

2.2. Recruitment—Arizona

In Arizona, 225 people from 152 households were recruited from the communities of Ajo, New River, San Manuel, and Tucson from January through October 2006. Data from Arizona were collected by personnel at the Mel and Enid Zuckerman College of Public Health and the Arizona Cancer Center at The University of Arizona.

The goal was to recruit at least 25 households from each of the four communities. Recruitment primarily occurred through list-assisted random digit dialing. Up to 10 attempts were made to reach an individual at the designated phone number. The 2005–2006 QWEST phonebook was used for each Arizona community to gather phone pre-fixes and random phone numbers were generated.

An additional strategy was used in the rural community of New River. All residential addresses identified on rural delivery mailboxes were recorded and a recruitment letter was mailed to randomly selected addresses from the list. The letter contained a description of the study and a phone number to contact the study coordinator to schedule a date and time for consent and initial interview.

2.3. Recruitment—Mexico

In Mexico, 262 people were recruited from 202 households within two neighborhoods in Hermosillo and from the communities of Tobarito and Guadalupe Victoria in the Yaqui Valley near Ciudad Obregon. Recruitment occurred from January 2006 through February 2007 in Mexico. Data from Hermosillo were collected by personnel at the Universidad de Sonora (UNISON) and data from the Yaqui Valley were collected by personnel at the Instituto Tecnológico de Sonora (ITSON) in Ciudad Obregon.

The aim was to recruit at least 50 households from each community or neighborhood. All wells supplying these neighborhoods/communities were tested for total arsenic to construct arsenic concentration maps. Each well supplied water to a defined neighborhood similar to a U.S. census tract. Seventy-three wells from the Yaqui Valley and 41 wells from Hermosillo were tested. In each region (Hermosillo and Yaqui Valley), one neighborhood with an elevated As concentration in the water supplied by the neighborhood well was selected and matched to another neighborhood with a lower As concentration in the water supply. Neighborhoods were matched on demographic characteristics (e.g., similar incomes, education level, *etc.*). The selected neighborhoods were subdivided into zones based on street boundaries or other map features and recruitment was conducted within a randomly selected zone. Every address within the boundary of the selected zone was recorded. Addresses excluded were clearly and exclusively non-residential. Addresses that were both a business and a residence were also included. Teams of interviewers went to randomly selected households within the recruitment zones to ask for participation. The recruitment sites are illustrated in Figure 1.

2.4. Household Questionnaire

The primary participant was asked to complete a household questionnaire containing questions about characteristics of the dwelling, availability of water, and access to a refrigerator, phone, computer and television.

2.5. Personal Questionnaires and Dietary History

All participants in the household were asked about demographics, smoking history, alcohol consumption, current occupation, job and hobby history, health history (e.g., history of diabetes, respiratory, cancers, *etc.*), medication and supplement use, reproductive history, physical activity, and residential history.

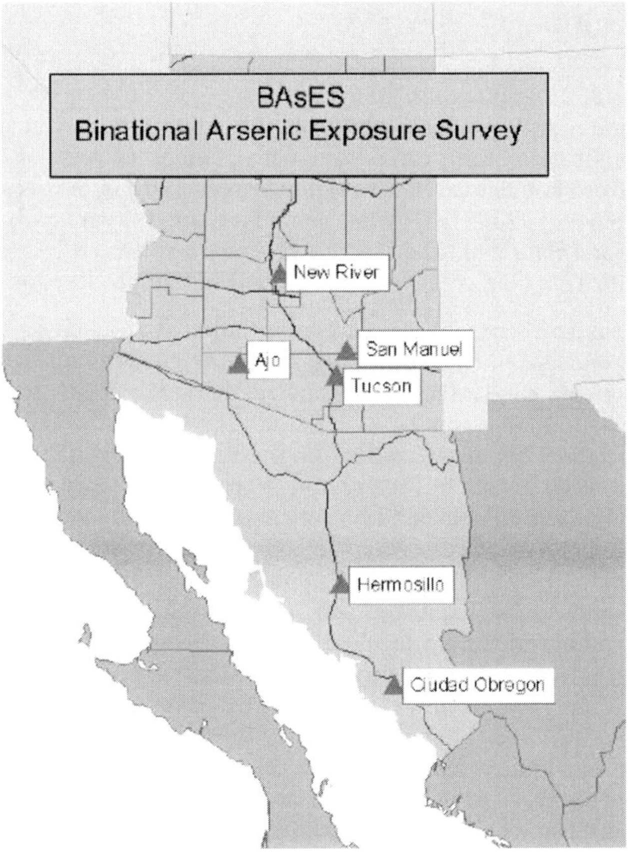

Figure 1. BAsES Recruitment Sites.

Participants reported all water sources used for cooking and drinking in the home (kitchen tap water, bottle water, refrigerator spigot water, *etc.*) and frequency of use of each water source (frequently, moderately, rarely, never). The primary source of running water (well water, municipal water, *etc.*) and the presence of a water filtration device (reverse osmosis, charcoal filter, *etc.*) were recorded.

Each participant completed a single 24 h administered dietary recall with portion size prompt using food models. Participants were asked to describe everything they ate or drank the previous day (including water) from the time they woke up in the morning until the time they went to bed (including anything eaten in the middle of the night). Interviewers recorded responses including amount consumed, specifics of preparation, and time of consumption.

The 24 h dietary recalls were sent to the Arizona Diet, Behavior and Quality of Life Assessment Lab (ADBA) at the Arizona Cancer Center for data entry and nutrient analysis using the Nutrition Data System for Research software version 2005 (NDSR), developed by the Nutrition Coordinating Center (NCC), University of Minnesota, Minneapolis, MN. Beverages (milk, apple juice, grape juice, *etc.*) were assigned total arsenic values as reported by Schoof *et al.* [15] USDA recipes were primarily used to estimate components of beverages that were not available in NDSR. Major components were then assigned arsenic values based on items reported by Schoof *et al.* [15].

The 24 h dietary recall quality control/quality assurance (QC/QA) procedures included a 15% duplicate entry of recalls by a different coder, followed by a comparison of entered recalls and reconciliation. Discrepancies attributable to coder error were addressed by closer surveillance when necessary. All data entry coders went through a formal training and probation period until their work met the ADBA standards. Whenever differing interpretations of a subject's data set were identified, the recall was reviewed to provide consistency of coding.

2.6. Environmental and Biological Samples

Water, urine, and anthropometric measurements were collected from all participants. Water samples were collected from all sources of water reported as consumed in the home. An additional water sample was collected from outside the home (well, spigot, *etc.*) if a filtration system was present filtering the water coming into the home.

Participants were supplied a urine cup and instructions for urine collection prior to the day of the interview. Participants were asked to collect a first morning urine sample. In the U.S., the urine cup was mailed to the participants. In Mexico, the urine cup was delivered by an interviewer the day before the interview. Specific gravity from the urine samples was determined using a refractometer.

Height, weight, hip and waist circumference were collected during the household interview. A calibrated digital scale was used to assess the weight of each participant in pounds. A stadiometer was used to assess the height of the participant. A standard measuring tape was used to assess hip and waist circumference.

All samples were stored in a cooler on icepacks during transport from the interview site to the laboratory. Sample aliquots were stored at −80 °C. Samples of urine and water from Mexico were transported to The University of Arizona for the analysis of arsenic. Water samples were analyzed for total arsenic. Urine samples were analyzed for total arsenic,

As^3, As^5, methylarsonic acid (MMA^5), and dimethylarsinic acid (DMA^5). All samples were analyzed for arsenic by the Southwest Hazardous Waste Program, Hazard Identification Core at The University of Arizona. Water and Urine samples were also analyzed for antimony, barium, beryllium, cadmium, cesium, cobalt, lead, molybdenum, platinum, selenium, thallium, tungsten, and uranium by the Bureau of State Laboratory Services at the Arizona Department of Health Services.

2.7. Laboratory Preparation and Analysis of Samples for Arsenic

Arsenic analyses were performed by the Analytical Section of the Hazard Identification Core from the University of Arizona.

Preparation of urine samples for total arsenic analyses was accomplished by microwave digestion. A 1 mL aliquot of each beverage sample was placed in an acid-washed 7 mL teflon bomb, and 0.5 mL of concentrated nitric acid was added. Samples were sealed and heated in a microwave for 5 min on the medium-high setting. Samples were cooled, vented, and heated for an additional 5 min on the high setting. Additional steps of heating on the high setting were performed as necessary to complete sample digestion. After digestion, samples were brought up to volume with Milli-Q water for inductively coupled plasma mass spectroscopy (ICP-MS) analysis. Preparation of urine samples for arsenic speciation was accomplished by filtering the urine through 0.45 μm nylon centrifuge filters.

Total arsenic concentrations in water and urine were determined using an Agilent 7500ce ICP-MS (Agilent Technologies, Inc., Santa Clara, CA) with a MicroMist nebulizer (Glass Expansion). An ASX500 auto sampler (CETAC Technologies) was used to introduce the samples into the Agilent 7500ce. The operating parameters were as follows: Rf power, 1500 watts; plasma gas flow, 15 L/min; carrier flow, 0.85 L/min; makeup gas, 0.15 L/min. Acquisition parameters were as follows: arsenic measured at *m/z* 75, terbium (internal standard) measured at *m/z* 159, points per peak 3, dwell time for arsenic was 1.5 s, and dwell time for terbium was 1.5 s. Each sample was evaluated seven times and the reported concentration represents the mean of these seven measures.

Concentrations of arsenic species in urine were analyzed using an Agilent 1100 HPLC (Agilent Technologies, Inc., Santa Clara, CA) with an anion exchange column (PRP-X100, 10 μm, 250 × 4.1 mm, Hamilton, Reno, NV) and guard cartridge. The mobile phase was 50 mM ammonium carbonate (pH 9) with 4% (*v/v*) methanol at a flow rate of 1.0 mL/min. Column temperature was maintained at 30 °C and samples were kept at 4 °C in a thermally controlled auto sampler. An Agilent 7500ce ICP-MS with a Conikal nebulizer (Glass Expansion) was used as the detector. The operating and acquisition parameters were the same as for total arsenic analysis.

NIST SRM2670a Toxic Elements in Urine (National Institute of Standards and Technology, Gaithersburg, MD, USA) was used for QA/QC purposes. The standard reference material was run in duplicate with every set of 20 samples. For analysis of total arsenic, the percent recovery was between 96.7% and 107.5% with a coefficient of variance (COV) of 2.7. For arsenic speciation, the percent recovery was between 93.5% and 97.4% with a COV of 1.7.

Results included concentrations of total arsenic, the inorganic species: As^3 and As^5, and the organic species: MMA^5, and DMA^5. Analysis of the arsenic species MMA^3 and DMA^3 was not possible due to their unstable nature. The detection limits for the arsenic species are as follows: total arsenic 0.1 µg/L, As^3 0.12 µg/L, As^5 0.21 µg/L, MMA^5 0.12 µg/L, and DMA^5 0.12 µg/L.

2.8. Estimation of Arsenic Exposure in Water and Beverage

Consumption of water and other beverages was reported in liters per day (L/day) from the 24 h dietary recall. The interviewer asked participants to report all water consumed, regardless of location or source, on the dietary recall. To determine exposure from drinking water, we assumed water consumed outside of the home would have similar arsenic water concentrations as those in the home. This is a reasonable assumption since homes are generally on the same water system as other local venues (*i.e.*, work, school, restaurants, *etc.*) or water is drawn from the same aquifer in most cases. Of the 465 people in the analysis, there were 362 people with water intake volumes recorded on the 24 h dietary recall and 103 people for whom total water volume consumed for the day was not recorded. A single imputation approach was used to estimate the total water volume consumed in a 24 h period for those without a total water volume. For participants lacking a consumption volume, the average total water consumption was substituted based on age, gender, and recruitment site categories. The age categories included: 18–29, 30–39, 40–49, 50–59, 60–69, 70+. The recruitment site categories included Arizona, Hermosillo, and the Yaqui Valley. The imputed water consumption volumes among the categories are not shown but as an example, a female participant from Arizona in her 30 s who had a missing total water consumption volume would be estimated to have consumed 1.02 L of water for the day reflecting the mean consumption of others in this category.

Every water source identified as consumed in the home, including bottled water, was tested for total arsenic. Participants were asked to report usual consumption for each water source. Consumption categories included frequently, moderately, rarely, and never. A weighted average based on these frequencies was used to determine total volume of water consumed from the individual sources. A frequency value was assigned for consumption (e.g., 3 for frequent use, 2 for moderate use, 1 for rare use, and 0 for no use). For example, if a participant drank from three water sources, the first of which was consumed "frequently" (3), the second "moderately" (2), and the third "frequently" (3), the water sources could be assigned proportions of 3/8, 2/8, and 3/8 respectively with the denominator being the sum of the values. The fractions were multiplied by the total daily water volume resulting in a volume for each water source. The arsenic concentration per source was then multiplied by the amount of water consumed yielding the amount of arsenic consumed in a 24 h period (μg/day) per source. The arsenic intake from each water source was then summed to estimate the total arsenic intake from all water sources consumed in the home for the day.

Total arsenic intake from other beverages was calculated by multiplying arsenic concentration of specific beverages as reported in the literature by the reported volume from the 24 h recall. The arsenic intake from each beverage was then summed to estimate the total arsenic intake from all beverages consumed for the day.

2.9. Analysis Strategy

Of the 487 participants, 19 people were excluded in this analysis due to inadequate information regarding water consumption and three people were excluded for consumption reports exceeding 10 L of water a day.

Urinary total arsenic, inorganic arsenic (expressed as the sum of As^3 and As^5), and sum of species (expressed as the sum of As^3, As^5, MMA, and DMA) were reported as continuous variables in micrograms per liter. The relative distributions of arsenic species were reported as the percent MMA (MMA over sum of species) and the DMA/MMA ratio. Half of the detection limit was substituted in instances where arsenic concentration was below the detection level. Urinary inorganic arsenic, sum of species, and total arsenic concentrations were adjusted for specific gravity using the formula reported by Nermell *et al.* using the specific gravity population mean [16]. The distributions of urinary total arsenic, sum of species, inorganic As, DMA/MMA ratio, arsenic concentration consumed from water, and arsenic concentration consumed from all beverages were right skewed and were natural log transformed.

Summary statistics were generated for urinary total arsenic, inorganic arsenic, DMA/ MMA ratio, and percent MMA. A Kruskal-Wallis test was used to detect any differences in the various urinary arsenic measurements between recruitment sites. Random effect models were used to evaluate the association between the concentration of arsenic in urine and the estimated amount of arsenic consumed from water and other beverages. The random effect in all models was defined as household nested within study site. A random effect model was selected to account for the variance within households and across study sites. Other variables were considered in the statistical analyses with their potential to be associated with both the concentration of total arsenic in urine and the estimated amount of arsenic consumed from drinking water.

These variables included: age in years (continuous), gender (male, female), ethnicity (Arizona Hispanic, Arizona non-Hispanic, Mexicano), and current smoking status (current smoker, not current smoker). Ten models were generated. The predictor in models 1–5 was arsenic concentration consumed from water (log arsenic values). The predictor in models 6–10 was log arsenic concentration consumed from all beverages. Each set of five models, adjusted for confounders, contained one of the following as the dependent variable for urinary arsenic concentration: log total arsenic, log inorganic arsenic, log sum of species, log DMA/MMA ratio, and percent MMA. All statistical analyses were completed using SAS (version 9.2; SAS institute, Cary, NC).

3. RESULTS

3.1. Household Recruitment

The goal was to recruit at least 50 households from each neighborhood/community in Mexico and 100 households in Arizona (25 per community). Household participation by community ranged from 18% to 30% in Arizona, as shown in Table 1. Household participation by mail in New River was 20%. In contrast, household participation rates were greater in all Mexican sites ranging from 43% to 93% (Table 2) with participation lower in the more urban neighborhoods of Hermosillo. At least one person per household was asked to participate at all study sites. The mean number of interviewed people per household was 1.7 in Arizona, 1.1 in Hermosillo and 1.5 in Yaqui Valley.

Table 1. Household Participation Rates among Arizona Residents by Location.

Characteristic	Ajo	New River	San Manuel	Tucson
List-assisted dialing				
Phone Numbers Dialed	692	699	974	413
Households Recruited	25	29	31	47
Participants Recruited	31	77	49	68
Households that Refused to Participate	57	75	142	143
Household Unavailable on Day of Sampling	7	4	6	0
Only an Answering Machine	60	215	136	0
Household Never Answered the Phone	41	89	118	116
Business/Fax Machine	64	104	69	19
Disconnected Phone Numbers	438	183	472	88
Household Participation Rate *	30%	28%	18%	25%
Mailing				
Households Mailed a Brochure	-	101	-	-
Households Recruited from Mailing	-	20	-	-
Household Mailing Participation Rate †	-	20%	-	-

* Household Participation Rate = Households recruited/(Households recruited + Households that refused to participate); † Household mailing participation rate = Households recruited from mailing/Households mailed a brochure

Table 2. Household Participation Rates from Door-to-Door Recruitment in Sonora, Mexico by Location.

Characteristic	Hermosillo Community 1	Community 2	Yaqui Valley Guadalupe Victoria	Tobarito
Households Selected for Recruitment	107	122	62	68
Households Recruited	50	50	52	50
Participants Recruited	55	56	61	90
Households that Refused to Participate	41	59	4	8
No Eligible Participant in the Household	3	1	2	2
No Response from Household	13	12	3	8
Household Participation Rate *	55%	46%	93%	86%

* Household Participation Rate = Households recruited/(Households recruited + Households that refused to participate)

3.2. Participant Characteristics and Water Consumption

Summary statistics for participant characteristics are presented in Table 3. More females than males participated across the three recruitment sites: Arizona 56.9%, Hermosillo 73.2%, Yaqui Valley 69.8%. The median age of the participants from Arizona was 56.5 years compared to 39.5

years from Hermosillo and 44.0 years from Yaqui Valley. Hispanics accounted for 24% of the participants from Arizona. There were significant differences ($p < 0.001$) in the current smoking rates across the populations (16.7% from Arizona, 30.3% from Hermosillo, and 11.1% from Yaqui Valley).

Table 3. Demographic and Lifestyle Characteristics of BAsES Participants by Location.

Characteristic	Arizona	Hermosillo	Yaqui Valley	*p*-value
Participants (*n*)	218	108	139	
Age in years				
Mean (s.d.)	55.3 (15.3)	41.5 (13.9)	47.0 (16.4)	< 0.001 *
Median	56.5	39.5	44.0	
Gender, *n* (%)				
Male	94 (43.1)	29 (26.7)	42 (30.3)	< 0.001†
Female	124 (56.9)	79 (73.2)	97 (69.8)	
Ethnicity, *n* (%)				
Arizona Hispanic	52 (23.9)	0	0	-
Arizona Non-Hispanic	166 (76.1)	0	0	
Mexicano	0	108	139	
Current smoker, *n* (%)				
Yes	33 (16.7)	30 (30.3)	15 (11.1)	< 0.001†
No	165 (83.3)	69 (69.7)	120 (88.9)	
Unknown	20	9	4	
Self-reported estimated fluid intake from the 24-hour dietary recall (L/day)				
Drinking water				
Mean (s.d.)	1.67 (1.20)	0.44 (0.36)	0.24 (0.17)	< 0.001*
Median	1.43	0.38	0.24	
n	218	108	139	
Non-water beverages				
Mean (s.d.)	1.77 (1.25)	0.92 (0.85)	0.58 (0.32)	< 0.001*
Median	1.39	0.79	0.52	
n	214	102	130	
Total beverage consumption				
Mean (s.d.)	3.51 (1.89)	1.31 (0.93)	0.79 (0.36)	< 0.001*
Median	2.88	1.18	0.75	
n	218	108	139	

L = Liters; s.d. = standard deviation; * Kruskall-Wallis test of association across location; † Chi-square test of significance across location.

Median water consumption reported in the 24 h dietary recall was 1.43 L/day among Arizona participants, 0.38 L/day among Hermosillo participants, and 0.24 L/day among Yaqui Valley participants ($p < 0.001$). The median non-water beverage consumption was greatest among

Arizona residents at 1.39 L/day and was significantly different among the three sites ($p < 0.001$). Median total fluid intake was significantly different ($p < 0.001$) between participants from Arizona (2.88 L/day), from Hermosillo (1.18 L/day), and from the Yaqui Valley (0.75 L/day).

3.3. Distribution of Arsenic in Water and Beverages

Table 4 contains the concentration of arsenic in unfiltered tap water coming into the home by community. The highest average arsenic concentration was found in the community of New River (98.9 μg/L). The lowest median arsenic concentration was found in Tucson (3.9 μg/L). Also shown in the table are the estimated total arsenic concentrations consumed from water and other beverages by community. Median total arsenic consumed from water by participants in Arizona ranged from 1.7 to 14.1 μg/day, compared to 0.6 to 3.4 μg/day for participants from Mexico. Estimated total arsenic intake via water was greatest among participants from New River and least among participants from Tobarito, Mexico. When examining estimated arsenic intake from all beverages, participants in Mexico had lower daily consumption rates than most participants from Arizona. This corresponds to participants in Mexico reporting less volume per day from beverages leading to a lower estimated intake of As through beverages.

Table 4. Arsenic Levels in Water and Estimated Total Arsenic Intake from Drinking Water and Other Beverages during 24 h by Recruitment Site.

	Arizona				Hermosillo		Yaqui Valley	
	Ajo	New River	San Manuel	Tucson	Community 1	Community 2	Guadalupe Victoria	Tobarito
As levels from an unfiltered water source by household (μg/L)								
Households (n)	25	48	31	47	48	49	48	37
As concentration								
Mean (s.d.)	9.7 (26.2)	98.9 (198.5)	6.9 (2.2)	4.3 (6.6)	24.5 (7.6)	8.6 (2.3)	4.8 (2.5)	7.6 (7.7)
Median	4.2	22.1	6.9	3.9	26.3	9.0	4.1	4.0
Estimated total arsenic intake (μg/day)								
Participants (n)	31	71	48	68	54	54	52	87
From drinking water								
Mean (s.d.)	5.2 (4.3)	166.0 (468.4)	9.0 (2.5)	4.1 (5.8)	5.5 (6.8)	2.4 (2.5)	1.9 (1.5)	1.1 (1.5)
Median	4.2	14.1	1.7	2.8	3.4	1.7	1.6	0.6
From total beverage consumption								
Mean (s.d.)	9.3 (7.3)	176.3 (467.1)	13.0 (9.9)	8.9 (9.3)	9.4 (16.2)	5.4 (4.6)	3.9 (2.7)	3.9 (6.6)
Median	8.3	27.9	11.0	6.1	5.4	4.7	3.0	2.1

s.d. = standard deviation.

3.4. Distribution of Arsenic in Urine

Table 5 presents the distributions of urinary arsenic concentrations adjusted for specific gravity by community. The greatest concentrations of total and inorganic urinary arsenic were among participants in New River. However, except for this one Arizona community, median urinary total As concentration among Arizona participants (ranged from 1.2 to 2.0 μg/L) was less than the median urinary inorganic

As concentration for Mexican participants (range 2.5 to 6.2 μg/L). Among Arizonans, the average sum of species was highest among those from New River (39.9 μg/L). The participants from Arizona communities, except those in New River, had a lower median sum of species level than participants from Mexico communities. The sum of species was not significantly different between communities in the Yaqui Valley, but it was significantly different between the communities in Hermosillo ($p < 0.001$). There were no differences in levels of DMA/MMA or % MMA by community.

Table 5. Urinary Arsenic Concentration Adjusted for Specific Gravity among BAsES Participants by Recruitment Site.

	Arizona				Hermosillo		Yaqui Valley	
	Ajo	New River	San Manuel	Tucson	Community 1	Community 2	Guadalupe Victoria	Tobarito
Participants (*n*)	31	71	48	68	54	54	52	87
Total As (μg/L)								
Mean (s.d.)	35.3 (27.2)	121.7 (221.4)	42.2 (19.6)	47.7 (94.0)	61.6 (43.3)	110.0 (89.9)	119.3 (131.8)	93.0 (108.7)
Median	25.9	33.0	36.3	25.4	50.7	79.9	94.3	64.6
Inorganic As (μg/L)								
Mean (s.d.)	2.9 (3.6)	10.2 (28.5)	2.3 (2.9)	1.9 (3.0)	3.5 (3.1)	7.4 (5.6)	6.7 (4.4)	5.7 (5.8)
Median	2.0	1.7	1.4	1.2	2.5	6.2	6.0	3.7
Sum of Species (μg/L)								
Mean (s.d.)	8.6 (4.2)	39.9 (104.2)	8.1 (5.4)	7.7 (7.0)	13.5 (6.9)	26.1 (25.0)	21.9 (11.4)	21.3 (15.8)
Median	8.5	7.4	6.0	6.3	12.2	20.5	18.8	17.2
DMA/MMA ratio								
Mean (s.d.)	7.1 (4.3)	7.2 (4.2)	7.8 (4.2)	7.6 (3.8)	7.6 (4.5)	8.2 (5.3)	7.8 (4.1)	7.8 (4.5)
Median	5.9	6.2	7.1	6.8	6.3	6.8	6.7	6.7
Percent MMA								
Mean (s.d.)	12.1 (4.8)	12.8 (4.8)	11.6 (4.6)	11.8 (3.9)	11.9 (3.9)	11.1 (4.1)	11.2 (3.8)	11.4 (3.5)
Median	12.3	12.5	10.7	11.7	12.1	11.1	10.5	11.0

s.d. = standard deviation; inorganic As = $As^3 + As^5$; Sum of Species = $As^3 + As^5$ + MMA + DMA; Percent MMA = (MMA/Sum of Species)*100.

3.5. Associations Between Arsenic Exposure and Urinary Arsenic Concentrations

Table 6 presents results of ten random effect models evaluating the association between concentrations of urinary as species and estimated As exposure from water and other beverages. The estimated arsenic intake from water consumption had a statistically significant association with each of the five urinary As variables. Due to the independent and dependent variables being log transformed, inferences of the model estimates come from the interpretations found in *Regression Methods in Biostatistics* [17]. We can infer from model 1 that there is a 7% increase in total urinary arsenic for a 1% increase in arsenic consumption from water ($p < 0.001$), from model 2 that there is a 12% increase in urinary inorganic arsenic for a 1% increase in arsenic consumption from water ($p < 0.001$), and from model 3 that there is a 9.8% increase in urinary sum of species for a 1% increase in arsenic consumption from water. In contrast, no statistically significant association was found between estimated arsenic intake from all beverages and either urinary DMA/MMA ratio or urinary percent MMA. A significant association was found between estimated arsenic intake from all beverages with urinary total As, inorganic As, and sum of species.

Table 6. Random Effect Models Depicting the Association between Urinary Metabolite Concentrations and Arsenic Consumed from Drinking Water and Beverages *.

Dependent variable	Estimate of Log As consumed from water ± s.e.	*p*-value
Model 1: Urinary Log (Total As)	0.074 ± 0.020	<0.001
Model 2: Urinary Log (Inorganic As)	0.123 ± 0.022	<0.001
Model 3: Urinary Log (Sum of Species)	0.098 ± 0.021	<0.001
Model 4: Urinary Log (DMA/MMA ratio)	−0.032 ± 0.012	0.007
Model 5: Urinary (Percent MMA)	0.252 ± 0.103	0.016
	Estimate of Log As consumed from all beverages ± s.e.	***p*-value**
Model 6: Urinary Log (Total As)	0.143 ± 0.032	<0.001
Model 7: Urinary Log (Inorganic As)	0.234 ± 0.036	<0.001
Model 8: Urinary Log (Sum of Species)	0.203 ± 0.034	<0.001
Model 9: Urinary Log (DMA/MMA ratio)	−0.029 ± 0.019	0.130
Model 10: Urinary (Percent MMA)	0.214 ± 0.170	0.213

* The random effect was the variable household nested within study site. Adjusted for age, gender, current smoking status, and ethnicity; s.e. = standard error; inorganic As = $As^3 + As^5$; Sum of Species = $As^3 + As^5 + MMA + DMA$; Percent MMA = (MMA/Sum of Species)*100.

4. DISCUSSION

This paper describes the relationship between urinary arsenic concentrations and estimated arsenic intake from beverages from a cross-sectional survey of community-based residents living in southern Arizona and northern Mexico. Residents from the rural communities in Yaqui Valley and rural areas in Arizona were more likely to participate than residents living in urban areas, and household response rates were greater in Mexico than in Arizona. Anecdotally, participants from Mexico liked the attention received from a study conducted by the local university.

Strikingly, the estimated volume of all beverages consumed among participants from Arizona was nearly four times greater than the amount consumed by participants in Mexico. This difference reflected either reduced consumption of beverages by Mexican participants compared to Arizona participants (0.75 L *vs.* 2.9 L), a misrepresentation of the amount of beverages consumed by Mexicans, or over reporting of beverages consumed by Arizona participants. A reported water volume that was incorrect would alter the estimated arsenic level consumed from water for the day.

The median volume of water consumed by participants from Arizona (1.4 L/day) is similar to the reported consumption noted in the literature [18]. The largest database containing water consumption comes from the National Health and Nutrition Examination Survey (NHANES) from 1999 to 2001 [18]. Nearly 88% of participants reported drinking an average amount of 1.53 L/day, while 12% drank no water [19]. In a pilot study in Canada, researchers found that 56% of the population drank bottled or filtered water the previous day with an overall water consumption rate of 1.6 L/day [20].

Median estimated arsenic intake from water in Arizona communities ranged from 1.7 to 14.1 μg/day compared to 0.6 to 3.4 μg/day in Mexico communities. In contrast to the estimated exposures, the median urinary inorganic arsenic concentration was lowest among participants from Tucson 1.2 (μg/L) and highest among participants from Hermosillo (6.2 μg/L). Median urinary total arsenic concentration was lowest in Tucson participants (25.4 μg/L) and highest among Yaqui Valley participants (94.3 μg/L). The discrepancy between urinary inorganic arsenic concentration and total arsenic concentration among participants could potentially be due to the ingestion of organic forms of arsenic.

Table 4 shows the median arsenic concentration of the drinking water coming into the home and is similar among several of the study sites. If the median volume of water consumed for the day was similar among participants from each study site, then we would expect the intake of arsenic from water between the study sites to be similar. However, we find that the estimated arsenic intake from water is lower among several of the communities in Mexico. It is possible that the total water intake from Mexico participants is underestimated thus resulting in an underestimate of total arsenic intake. Further analysis into reporting of intake volumes of water and beverages is under review.

Strengths of this study include the use of standardized methodologies for sample collection, administration of questionnaires, and data entry. A centralized laboratory was used to analyze urine and water samples from participants as well as centralized analysis of the 24 h dietary recalls.

Limitations in this study include potential selection bias. Participants were mostly female. This could be addressed in future studies conducted in Mexico by recruiting during the evening and on weekends. Other demographic characteristics in the U.S. such as age, ethnicity and race were comparable to percentages found in the Arizona census.

An additional limitation is that studies reporting arsenic concentrations in foods and beverages have primarily been conducted among products sold in the United States. In this analysis, the arsenic concentration in beverages are coming from the paper by Schoof *et al.* [15] These data sources are limited by the types of foods and beverages being analyzed as well as the quantity being analyzed. A single arsenic concentration for one product is being applied to a wide array of beverages. Furthermore, these databases may not represent the food products in subregions of the U.S. or outside the U.S. Few data are available on the concentration of arsenic in beverages sold in Mexico which may resulted in an over or under estimation of the arsenic concentration among beverages consumed by Mexican participants.

Furthermore, in this analysis it was assumed that the arsenic concentrations in urine were attributed to the water and beverages consumed the previous day. Urine was collected on the day of the interview and given the study resources and known recall bias we could only collect a dietary recall for the day prior the interview. It has been reported that people may be able to recall the preceding 24 h of food and beverage consumption with interviewer prompts [21]. It is difficult for people to recall amounts ingested for 2–3 days. Using a prospective design, investigators could monitor food and beverage consumption for 2–3 days prior to collection of the urine sample.

Reported water consumption for the day was missing from 28% of the 24 h dietary recalls. This was due to the interviewers not prompting for water consumption during the interview. We also used qualitative terms in this study to identify amount of water consumed per source based on the total intake of water from the 24 h dietary recall. The Arizona Diet, Behavior and Quality of Life Assessment Lab suggested a conservative scoring system to assign volumes to each water source using a 1 point difference to separate frequent, moderate, and rare consumption. By using a conservative scoring system we avoided over estimating arsenic consumption from water intake. Future studies exploring contaminants in drinking and cooking water via a 24 h dietary recall should specifically ask participants about the volume of water consumed per source avoiding the need to estimate the volume.

Communities in BAsES were in part selected for their groundwater arsenic concentration so that there would be communities with a higher or lower arsenic concentration in the household tap water. The two communities selected in Hermosillo display this relationship. One community had a median tap water arsenic concentration of 26.3 μg/L while the other was at 9.0 μg/L. This contrast was not seen in the two communities of the Yaqui Valley. In Tobarito, the community water source was unavailable and residents retrieved their water in bottles from a well in a nearby community. This change in water source was unknown until after the study had been completed. As a result, fewer people were exposed to elevated concentrations of arsenic via drinking water in the Yaqui Valley.

The statistical model used in this analysis was based on a strategy to describe the association between estimated arsenic intake from water and arsenic output in urine. Among participants in the U.S., the greatest urinary arsenic concentrations were seen in geographic locations with the greatest concentration of arsenic in the drinking water. However; among the participants from Mexico, elevated urinary arsenic levels were present among individuals consuming lower concentrations of arsenic in the drinking water at home. It may be that the arsenic presenting in the urine was consumed from water sources with arsenic concentrations different from arsenic concentrations in water sources from the home or from food. Analysis are ongoing to develop predictive models that include a detailed examination of the potential impact of beverages and other food items reported in the 24 h dietary recall in the prediction of urinary arsenic concentration. Furthermore, studies should be undertaken to examine beverage consumption patterns among the various ethnicities and the role other metals (e.g., selenium or cadmium) may have on the impact of As metabolism.

5. CONCLUSION

BAsES enhanced a binational collaboration between universities in Mexico and the U.S., built research capacity, and enabled collaborations in a region experiencing rapid changes, within the same desert environment. Standardized questionnaires and recruitment methods were utilized among all recruitment sites. A standardized and common laboratory for testing total, organic, and inorganic As was utilized. Estimated arsenic concentrations consumed from water and other beverages were found to be greater among Arizonans, yet participants in Mexico had greater urinary inorganic arsenic concentrations.

ACKNOWLEDGEMENTS

Support for this research was provided by The University of Arizona Specialized Program of Research Excellence in GI Cancer (NIH/NCI CA95060). The Analytical Section of the Hazard Identification Core was supported by the Superfund Basic Research Program Grant from the National Institute of Environmental Health Sciences.

REFERENCES

1. ATDSR Toxic substance—Arsenic. Available online: http://www.atsdr.cdc.gov/substances/arsenic/index.html (accessed on 13 October 2011).
2. ATDSR Top 20 Hazerdous Substances: ATDSR/EPA Prioity List for 1997. Available online: http://www.atsdr.cdc.gov/cxcx3.html (accessed on 13 October 2011).
3. IARC Arsenic and arsenic compunds. Available online: http://www.inchem.org/documents/iarc/suppl7/arsenic.html (accessed on 13 October 2011).
4. Bates, M.N.; Smith, A.H.; Cantor, K.P. Case-control study of bladder cancer and arsenic in drinking water. *Am. J. Epidemiol.* **1995**, *141*, 523–530.
5. Goering, P.L.; Aposhian, H.V.; Mass, M.J.; Cebrian, M.; Beck, B.D.; Waalkes, M.P. The enigma of arsenic carcinogenesis: role of metabolism. *Toxicol. Sci.* **1999**, *49*, 5–14.
6. Smith, A.H.; Hopenhayn-Rich, C.; Bates, M.N.; Goeden, H.M.; Hertz-Picciotto, I.; Duggan, H.M.; Wood, R.; Kosnett, M.J.; Smith, M.T. Cancer risks from arsenic in drinking water. *Environ. Health Perspect.* **1992**, *97*, 259–267.
7. Ahsan, H.; Chen, Y.; Kibriya, M.G.; Slavkovich, V.; Parvez, F.; Jasmine, F.; Gamble, M.V.; Graziano, J.H. Arsenic metabolism, genetic susceptibility, and risk of premalignant skin lesions in Bangladesh. *Cancer Epidemiol. Biomarkers Prev.* **2007**, *16*, 1270–1278.
8. Christian, W.J.; Hopenhayn, C.; Centeno, J.A.; Todorov, T. Distribution of urinary selenium and arsenic among pregnant women exposed to arsenic in drinking water. *Environ. Res.* **2006**, *100*, 115–122.
9. Guha Mazumder, D.N.; Haque, R.; Ghosh, N.; De, B.K.; Santra, A.; Chakraborty, D.; Smith, A.H. Arsenic levels in drinking water and the prevalence of skin lesions in West Bengal, India. *Int. J. Epidemiol.* **1998**, *27*, 871–877.

10. Hopenhayn-Rich, C.; Biggs, M.L.; Smith, A.H.; Kalman, D.A.; Moore, L.E. Methylation study of a population environmentally exposed to arsenic in drinking water. *Environ. Health Perspect.* **1996**, *104*, 620–628.
11. Meliker, J.R.; Franzblau, A.; Slotnick, M.J.; Nriagu, J.O. Major contributors to inorganic arsenic intake in southeastern Michigan. *Int. J. Hyg. Environ. Health* **2006**, *209*, 399–411.
12. Xue, J.; Zartarian, V.; Wang, S.W.; Liu, S.V.; Georgopoulos, P. Probabilistic Modeling of Dietary Arsenic Exposure and Dose and Evaluation with 2003-2004 NHANES Data. *Environ. Health Perspect.* **2010**, *118*, 345–350.
13. EPA Arsenic—Safe Water. Available online: http://www.epa.gov/safewater/arsenic/index.html (accessed on 15 October 2011).
14. Zartarian, V.; Bahadori, T.; McKone, T. Adoption of an official ISEA glossary. *J. Expo. Anal. Environ. Epidemiol.* **2005**, *15*, 1–5.
15. Schoof, R.A.; Yost, L.J.; Eickhoff, J.; Crecelius, E.A.; Cragin, D.W.; Meacher, D.M.; Menzel, D.B. A market basket survey of inorganic arsenic in food. *Food Chem. Toxicol.* **1999**, *37*, 839–846.
16. Nermell, B.; Lindberg, A.L.; Rahman, M.; Berglund, M.; Persson, L.A.; El Arifeen, S.; Vahter, M. Urinary arsenic concentration adjustment factors and malnutrition. *Environ. Res.* **2008**, *106*, 212–218.
17. Vittinghoff, E. *Regression Methods in Biostatistics : Linear, Logistic, Survival, and Repeated Measures Models*; Springer: New York, NY, USA, 2005.
18. NHANES National Health and Nutrition Examination Survey. Available online: http://www.cdc.gov/nchs/nhanes.htm (accessed on 16 October 2011).
19. Popkin, B.M.; Barclay, D.V.; Nielsen, S.J. Water and food consumption patterns of U.S. adults from 1999 to 2001. *Obes. Res.* **2005**, *13*, 2146–2152.
20. Levallois, P.; Guevin, N.; Gingras, S.; Levesque, B.; Weber, J.P.; Letarte, R. New patterns of drinking-water consumption: results of a pilot study. *Sci. Total. Environ.* **1998**, *209*, 233–241.
21. Willett, W. *Nutritional epidemiology*, 2nd ed; Oxford University Press: New York, NY, USA, 1998.

CHAPTER 18

Drinking Water Quality in the Sagarmatha National Park, Nepal

Kirsten Nicholson[1], Emily Hayes[1], Klaus Neumann[1], Carolyn Dowling[1], Subodh Sharma[2]

1Department of Geological Sciences, Ball State University, Muncie, USA

2Department of Environmental Science & Engineering, Kathmandu University, Dhulikhel, Nepal

ABSTRACT

In 2014 we began the first systematic study of water quality, specifically fecal contamination of drinking water in the Khumbu Valley, Sagarmatha National Park (SNP, Mt. Everest region), Nepal. Our goal was to identify coliform bacteria and E. coli in drinking water and groundwater-fed springs to generate a data set that will function as a base for potable water supplies and further monitoring. Sampling occurred in May (pre-monsoon summer) and early November (post-mon- soon early winter) 2014. Sample sites were selected based on proximity to villages and primary use as a drinking water source. Overall, the data presented a predictable correlation between fecal contamination and both elevation and increasing population/tourist traffic. Drinking water within the study area met current World Health Organization drinking water standards for the physical properties of temperature (2.8°C - 13°C), pH (5.27 - 7.24), conductivity (14.5 - 133 μS) and TDS (7.24 - 65.5 ppm). Samples from the more populated, lower altitude areas had higher levels of E. coli. Samples collected and analyzed in May (pre-monsoon summer) had a higher level of E. coli and coliform bacteria than samples collected in November (post-monsoon early winter) suggesting a seasonal dependence overlaid on the population signature. Surface water typically had higher E. coli values than groundwater-fed springs.

Temperature, total dissolved solids and conductivity generally decreased with increasing elevation, whereas pH increased with increasing elevation. There appears to be significant presence of fecal contamination of water sources due to a combination of tourism, elevation and seasons.

Keywords: Fecal Coliform, E. coli, Mt. Everest, Drinking Water

1. INTRODUCTION

The Sagarmatha National Park (SNP; Figure 1) is located in the southeastern part of the Nepali Himalaya on the southern slope of Sagarmatha (Mt. Everest). The SNP is in the Solukhumbu district, approximately 140 km due east of Kathmandu and covers 1148 sq km. It ranges between 27°30'19"N to 27°06'45"N latitude to 86°30'53"E to 86°99'08"E longitude. The elevation of townships and communities in the SNP watersheds varies from 2610 m (Phakding) to over 5000 m at Mt. Everest base camp (EBC). The area experiences a temperate climate characterized by cold winters, warm summers and clear seasonality, with average annual precipitation ranging from 450 mm at EBC to 1800 mm at Lukla township (the town where tourists begin their trekking adventures). Maximum rainfall generally coincides with the Indian Monsoon, between June and September, and there is great variability of temperature, which ranges from 37°C in summer to −17°C in winter.

The Himalayan Mountains are the source of water for over a billion people and form the main headwaters for major river systems, such as the Ganges, Yangtze and Indus Rivers [1].

The SNP was established in 1976. It was declared a World Natural Heritage Site in 1979, and is the world's highest altitude protected area. The natural beauty and diversity in culture, flora and fauna, and the majesty of Mt. Everest make it a prime destination for adventure loving tourists [2]. Since the first ascent of Everest in 1953, around half a million trekkers have visited the SNP. The annual number of tourists in the Khumbu region was 20 in 1964, and rose to peak number 32,123 in 2009. In addition to international trekkers, a large number of people visit the SNP as porters and guides [3].

Many protected areas, such as the SNP, have promoted tourism development to improve their economic conditions [3] [4]. However, the negative effects of tourism are of significant concern and tourism operations in protected areas need to be carefully planned, managed and monitored to ensure their long-term sustainability [4].

Figure 1. Generalized map showing the major rivers, ice fields, major townships and main mountain peaks, including Mt. Everest. Inset map shows the position of the study area within the country of Nepal.

Over the past twenty years, the continuously increasing number of tourists to the SNP is causing increasing anthropogenic pressure with serious environmental consequences to the unique SNP ecosystem [5] [6]. Although the impact is visible primarily along the more popular trekking routes, where non-biodegradable solid wastes (such as water bottles, batteries and more) has resulted in environmental pollution, the effects are also felt in the waterways. Unmanaged or poorly managed solid waste disposal and open defecation have resulted in contamination of the major rivers [5]. Sewage waste is often directly discharged into nearby streams and rivers [7] resulting in considerable degradation of major rivers in the region [5]. Even though the surface water is polluted, it is still used as a potable water supply.

However, no studies of local surface water and groundwater and their suitability as reliable water sources have ever been conducted. The SNP is a well suited location to investigate surface and ground water or the effects of tourism on potable water supplies because there is access to moderate and high altitude springs that have and have not been affected by major human populations, which allows for comparative analyses. In addition, the SNP region is generally better studied than other parts of the Himalayan region in terms of geology, glaciology, and anthropology.

In natural systems, microorganisms are widely distributed whereby their diversity and abundances may be used as an indicator for suitable water sources [8]. Although there is a wide range of pathogenic microorganisms that can be transmitted to humans via water contaminated with fecal material (i.e. enteropathogenic agents such as salmonellas, shigellas, enteroviruses, and multicellular parasites as well as opportunistic pathogens like Pseudomonas aeroginosa, Klebsiella, Vibrio parahaemolyticus and Aeromonashydrophila [9]), the isolation and identification of these organisms is extremely complicated and seldom quantitative [10] [11].

As it is not practical to test water for all these organisms, the measurement of E. coli and coliform bacteria (total coliform bacteria and/or fecal coliforms) may be used as an indirect approach based on the assumption that the groups of normal enteric organisms will indicate the level of fecal contamination of the water supply [11] - [15]. The presence of coliform bacteria can be used as an indicator of potential danger to human health, as fecal contamination poses significant health risks. Finally, when combined with other indicators such as the physical characteristics of the water as well as distribution and access to water, the presence of bacteria can be used to help identify the reasons behind the contamination.

The study area begins just outside of the SNP boundary at the township of Lukla and extends through to the township of Deboche along the main trekking route to Mt. Everest base camp. It is mainly confined to the DudhKhosi drainage basin (Figure 2). The area is characterized by rugged topography, with altitudes ranging from of 2610 m at Phakding and 2774 m at Lukla airstrip, to a maximum altitude of 5364 m at Mt Everest base camp (the lowest Mt Everest base camp). Within the study area there are four main villages: Lukla, Namche Bazaar, Khumjung and Chaurikharka. The four villages are located in this lower portion of the SNP that experiences the highest numbers of tourists each year.

It is at risk from the conflicting interests between the need for economic growth (based on tourism) and the ability of the current infrastructure to handle issues such as human waste disposal [3]. The goal of this study was to characterize and monitor potable water quality close to and within the SNP using E. coli and coliform bacteria as indicators of fecal contamination.

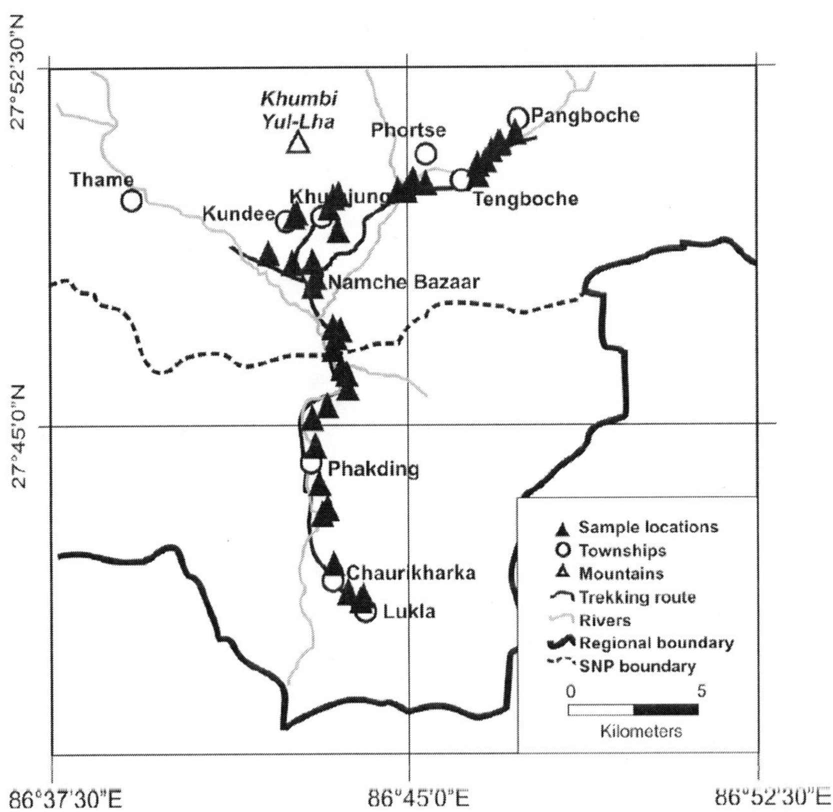

Figure 2. Map of the field study area including the major townships and sample locations (solid triangles). Sample locations correspond to data shown in Table 1. The dashed line indicates the park boundary with the SNP to the north and the buffer zone in the south.

Table 1. Data table showing all sample locations from May and October/November 2014.

Date	Sample #	Northing	Easting	Altitude (m)	Source
5/8/14	MAY 1	27.42264	086.43132	2702	spring fed
	MAY 2	27.71835	086.71604	2563	surface water
	MAY 3	27.73057	086.71380	2607	surface water
	MAY 4	27.73230	086.71326	2615	spring fed
5/9/14	MAY 5	27.75193	086.71057	2602	surface water
	MAY 6	n/a	n/a	2650	surface water
	MAY 7	27.75913	086.71003	2700	surface water
	MAY 8	27.76195	086.71467	2700	surface water
	MAY 9	27.76969	086.72402	2800	surface water
	MAY 10	27.78205	086.72275	2810	surface water
	MAY 11	27.78375	086.72253	2843	surface water
	MAY 12	27.78550	086.72172	2850	surface water
5/10/14	MAY 13	27.79960	086.71024	3350	spring fed
	MAY 14	27.80255	086.71085	3440	spring fed
	MAY 14	27.80255	86.71085	3440	spring fed
	MAY 14	27.80255	086.71085	3440	spring fed
	MAY 15	27.81528	86.7179	3440	surface water
	MAY 15	27.81528	086.71790	3440	surface water
	MAY 16	27.80636	086.70947	3440	spring fed
10/31/14	OCT 1	27.78202	086.72281	2810	surface water
	OCT 2	27.78549	086.72177	2843	surface water
	OCT 3	27.79905	086.71065	3370	surface water
11/1/14	NOV 4	27.81167	086.69595	3546	surface water
	NOV 5	27.80806	086.70241	3586	spring fed
	NOV 6	27.80634	086.70947	3512	spring fed
	NOV 7	27.80413	086.71030	3426	surface water
	NOV 8	27.81528	086.71790	3440	spring fed
	NOV 9	27.81528	086.71790	3440	spring fed
	NOV 10	27.80255	086.71085	3440	spring fed
	NOV 11	27.80255	086.71085	3440	spring fed
11/2/14	NOV 12	27.84580	086.77763	3466	surface water
	NOV 13	27.83992	086.77145	3728	spring fed
11/3/14	NOV 14	27.83969	086.77020	3762	spring fed
	NOV 15	27.83662	086.76598	3842	spring fed
	NOV 16	27.83558	086.76459	3864	spring fed
	NOV 17	27.83558	086.76459	3864	spring fed
	NOV 18	27.83511	086.76278	3812	spring fed
	NOV 19	27.83216	086.74734	3852	spring fed

	NOV 20	27.83233	086.74473	3891	surface water
	NOV 21	27.82983	086.73985	3444	spring fed
	NOV 22	27.82897	086.73801	3482	spring fed
	NOV 23	27.82503	086.73041	3521	spring fed
	NOV 24	27.82569	086.71862	3,660	spring fed
	NOV 25	27.82509	086.71645	3,764	spring fed
	NOV 26	27.82368	086.70473	3870	spring fed
	NOV 27	27.82291	086.70359	3886	spring fed
	NOV 28	27.82082	086.71576	3,891	spring fed
11/5/14	NOV 29	27.77805	086.72196	2815	spring fed
	NOV 30	27.77330	086.72262	2870	spring fed
	NOV 31	27.76991	086.72401	2908	surface water
	NOV 32	27.76742	086.72296	2896	spring fed
	NOV 33	n/a	n/a		surface water
	NOV 34	27.75175	086.71042	2688	spring fed
	NOV 35	27.73960	086.71194	2,610	spring fed
	NOV 36	27.73231	086.71326	2,610	spring fed
	NOV 37	27.68955	086.72945	2,849	spring fed
11/6/14	NOV 38	27.68955	086.72945	2,860	spring fed

"n/a" means that no data was collected.

2. METHODS

Water samples were collected in May and November 2014 (Table 1). Sites were selected based on the availability of the water that the local populations and tourist use as drinking. Sample sites include standpipes (water that is brought via tubing from higher elevations), groundwater-fed springs, creeks, and tributary streams. The major rivers and streams were omitted as they previously have been studied in detail [5].

Temperature, pH, conductivity and TDS were measured in the field using a FisherSci Ap85 pH/conductivity meter. Samples taken for bacteria analyses were collected in sterile 100 mL Whirl-pak bags containing a non-nutritive pill with 10 mg sodium thiosulfate and kept at temperatures below 20°C prior to analyses. Analysis of fecal coliform in surface and drinking water followed the EPA approved method 9222D in Standard Methods [16] by using a standard Hach® portable water test kit. Samples were filtered using a hand vacuum pump to filter sample through sterile 0.45 micron filter, placed into a petri dish treated with Hach® m-ColiBlue24® broth medium and placed in a Hach® portable field incubator at 35°C ± 0.5°C for 24 hours. Sample counts were done using a magnifying glass and a 10× geological hand lens where E. coli colonies are royal blue and coliform bacteria colonies appear to be crimson red. Two people counted bacteria samples and duplicate samples were run on approximately every tenth sample.

3. RESULTS

Samples were taken from 27 different localities in May and November 2014 (Figure 2). Of these samples, 17 were from standpipes or groundwater-fed springs and 10 were from minor creeks or tributaries. Where possible, samples were taken from local drinking water sources.

3.1. Temperature

Temperatures ranged between 2.8°C and 13°C with an average of 9.63 (±2.6)°C (Figure 3(a)). The samples collected in May were on average 2.06°C warmer than the samples collected in November. The samples in May ranged between 7.2°C and 12.9°C with an average of 11.04 (±1.5)°C, whereas the samples collected in November ranged between 2.8°C and 13°C with an average of 8.96 (±2.7)°C. It includes the November sample from Tengboche (lowest temperature recorded of 2.8°C) that was taken at 8AM when the air temperature was approximately −5°C. This sample was from shallow, slowly moving water, and the air temperature obviously affected the sample. In general, it was noted in the field that air temperature had little effect on groundwater but could have a major effect on slow moving surface waters. However, with this sample removed the November temperature data changes slightly with an increased average temperature of 9.3 (±2.4)°C. As expected, the temperatures decreased with altitude. With the exception of the Tengboche sample, both sets of our data compare well to Ghimire et al. [5] who analyzed 51 sample locations on the four major rivers (DudhKoshi, ImjaKhola, BhoteKoshi, LobucheKhola (Figure 1)) within the SNP that showed an average temperatures varying between 8°C and 13°C, with an average of 10.7 (±1.5)°C, over three years.

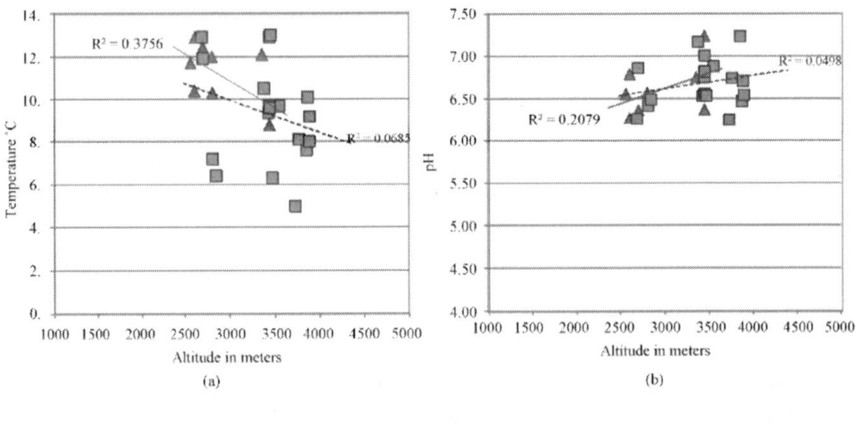

(a) (b)

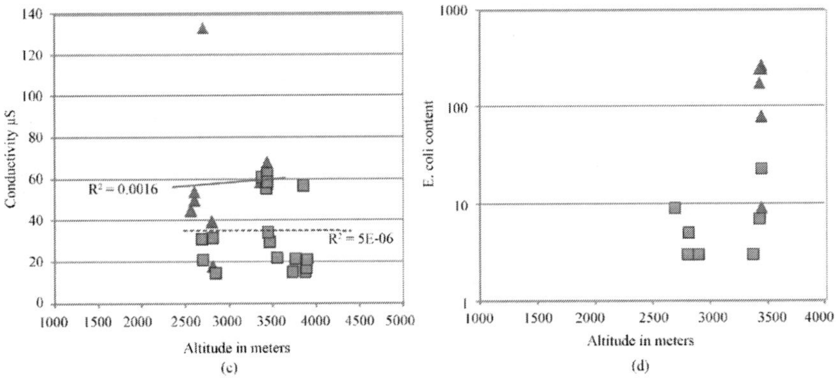

Figure 3. Bivariate plots showing the relationship between (a) temperature and altitude, (b) pH and altitude, (c) conductivity and altitude and (d) E. coli and altitude. Red triangles and corresponding red regression lines: May samples. Blue squares and corresponding dashed blue regression lines: November.

3.2. pH

The pH values ranged between 5.27 and 7.24 with an average of 6.60 (±0.4) (Figure 3(b)). Again the outlying sample is from Tengboche with a pH of 5.27. Excluding the November Tengboche sample, the samples collected in May had a pH range of 6.27 to 7.24 with an average of 6.61 (±0.3), whereas the November samples range from 6.25 to 7.23 with an average of 6.68 (±0.3). The observed pH values for drinking water are within the permissive limit pH (6.5 - 8.5), as prescribed for the drinking water by the World Health Organization (WHO) [17].

3.3. TDS and Conductivity

The range in TDS in the samples is relatively small. The average is 21.27 (±12.4) ppm, with a range between 7.24 ppm and 65.50 ppm. Samples taken in May have a slightly higher average (28.46 ± 15.5 ppm) than the samples collected in November which average 17.86 (±9.3) ppm. In the 2011 WHO [17] guidelines for drinking water, it is suggested that drinking water should contain less than 600 mg/L (which is roughly the same as 600 ppm), hence the drinking water in the SNP falls well within the recommended limit.

Conductivity was measured in the same samples and shows similar trends as TDS. The average value for both sets of data is 43.14 (±25.3) µS. Samples collected in November had lower conductivity with an average of 36.10 (±18.8) μS whereas the samples collected in May had higher values, ranging between 17.87 μS and 133 μS with an average of 57.99 (±31.5) μS (Figure 3(c)).

Figure 3 indicates that the samples show a negative correlation between altitude and the physical parameter of temperature, a positive correlation between altitude and pH, and a near neutral correlation between altitude and conductivity.

3.4. E. coli and Coliform Bacteria

A total of 41 samples from 27 locations were analyzed for E. coli and coliform bacteria (Escherichia coli and Streptococcus faecolies, respectively). Analyses were conducted proximal to the sampling sites within 20 hours of collection, with a majority conducted within 8 hours after collection. Only five samples were analyzed in May and the remaining 36 analyses conducted in November, which includes repeat sampling of the 5 location analyzed in May. The average high and low air temperatures, respectively, in Namche Bazaar are: 14°C and 4°C in May, and 9°C and 3°C in November. As anticipated, due to lack of daylight hours and colder weather conditions [18], the bacterial content was higher in May than in November. Only one sample, from the main groundwater- fed stream running through Namche Bazaar, tested positive for colonies of both E. coli and coliform bacteria. However, retesting the sample location in November showed only minor E. coli colonies (<5) and no coliform bacteria.

The five samples collected and analyzed in May all tested positive for E. coli with colony numbers ranging between 5 and 260 (Figure 3(d)). Only five out of the 36 November samples tested positive for E. coli (with greater than 3 colonies each, Figure 3(d)). Two of the samples were from Namche Bazaar: the main stream (groundwater sourced) running through town and the water supply for one of the local cafes. Three other samples tested positive for E. coli colonies, and they were collected from lower elevations: a lodge at Chaurikharka, the waterfall at Benkar, and a standpipe in Toktok (elevations between 2680 m and 2900 m). Considering the samples collected in November, there appears to be an inverse correlation between the presence of E. coli and altitude, with more E. coli detected in lower altitude townships than at higher elevations. This might be a result of several factors including: higher local population, higher numbers of tourists, higher daytime temperatures, and higher levels of surface water in the some springs.

Overall, the groundwater samples (springs and standpipes) contained fewer bacteria than surface water (streams and creeks). Typically, they have lower temperatures than the surface water samples, suggesting that the bacteria content might be related to water temperature. However, springs sampled both in May and November had similar temperatures, such as Namche Bazaar town stream, which had a May temperature of 8.8°C and a November temperature of 9.4°C. In May, the Namche Bazaar town stream had significant E. coli colonies whereas in November there were very few. There is no correlation between groundwater temperature and bacteria content.

In summary, surface water samples were more likely to contain bacteria than the groundwater samples, and water samples collected in May were more likely to contain E. coli than those collected in November. The presence of E. coli in the surface and groundwater-fed springs is a concern due to the heavy dependence on these drinking sources for the local communities and tourists.

4. DISCUSSION

This project investigated the quality of drinking water through physical field characteristics of the water and bacteria values in an effort to identify areas most at risk from fecal contaminated water. This study was necessary along the route to Mt Everest base camp (and in the SNP) because of the increasing number of tourists and lack of water and sanitation infrastructure. In the last forty years the number of visitors to the park has increased from fewer than 100 tourists per year to over 30,000 foreign tourists per year (not including their Nepalese guides and porters). As a result, the area is suffering from significant environmental degradation [5] [6].

A recent study by Salerno et al. [3] focused on tourist impressions of environmental conditions within the SNP, and their results suggest that visitors to the SNP appear to be satisfied with the current conditions, including water quality, energy use and tourist numbers [3]. Non-biodegradable waste can be seen along the major trekking routes; however, the recent addition of waste collection facilities appears to have helped alleviate this problem. Therefore, tourists do not typically see the disposal and results of biodegradable waste, such as human waste. Salerno et al. [3] suggested that there was room for further growth in tourist numbers; however, they also noted that the current environmental conditions would significantly limit growth and further development, and their research suggested that limiting tourist numbers to fewer than 15,000 would mitigate current environmental concerns [3].

Currently unmanaged and/or poorly managed solid waste disposal and open defecation throughout the SNP have resulted in fecal contamination and degradation of the major rivers [5]. In the major townships, it is possible to observe sewage and toilet waste piped into nearby streams and rivers (authors' personal observations; [7]). This study measured coliform bacteria (total coliform bacteria and/or fecal coliforms) as an indirect method to identify contamination by fecal material as fecal contamination poses significant health risks [11] - [15].

The first study of E. coli and coliform bacteria in the Sagarmatha National Park was conducted by Ghimire et al. [5] between 2008 and 2010. Their study focused entirely on the major rivers within the park, and they concluded that all of the rivers contained E. coli and coliform bacteria, although the upper reaches of most rivers did not. Their results summarized the findings from analyses of forty-five water samples that were collected and analyzed at the Central Department of Botany, Tribuvan University (Kathmandu), at Kunde Hospital in Kunde (in 2010) and in a portable laboratory in a hotel in Namche Bazaar. Total coliforms and E. coli estimations were done following APHA [19] methods.

A study published by Baghel et al. [18] of the Gangetic headwaters, beginning at the Gangotri glacier (Uttarakhand, India) and extending 27 km downstream produced similar results to this study and Ghimire et al. [5]. That investigation involved 21 sites that were sampled three times each year: the pre-monsoon summer, the monsoon, and the post-monsoon winter. The authors found that bacterial contamination increased from the upper, higher altitude, portion of the study area to the lower altitude stretches of the river [18] as shown here in the SNP. The results of Baghel et al. [18] indicate that the rising development in anthropogenic and socio-cultural activities in the lower stretch was responsible for the increased bacterial contamination. Baghel et al. [18] also found that total coliform and thermotolerant coliforms were at a maximum during summer (followed by monsoon and winter), which the researchers attributed to a larger number of pilgrims and trekkers visiting the area during the summer.

Our results are complimentary to those of Baghel et al. [18], Sood et al. [20] and Ghimirie et al. [5] [6]. In general we found that increasing altitude corresponded to decreasing temperature, and increasing pH, TDS, conductivity, and bacteria (Figure 3). Almost all of the samples collected, both in May and November, are slightly alkaline and the physical properties for all of the samples fall within the WHO [17] guidelines for safe drinking water. The bacteria content of our samples was higher in the summer (May). November, the post monsoon early winter season, has a reduced number of daylight hours (approximately 11 hours and 5 minutes daylight average), and colder conditions (average daytime highs approximately 9°C) than the pre-monsoon early summer month of May (approximately 13 hours and 20 minutes daylight, and average daytime highs of approximately 14°C) [1]. The shorter days and cooler temperatures most likely explain why samples collected in November contained fewer bacteria than the samples analyzed in May (the effects of daylight, or lack thereof, on E. coli has long been known [21]). As the samples collected in May contained more E. coli than those collected in November, there may be also a relation between E. coli and both air temperature and daylight hours in the SNP and surrounding areas.

The largest low-altitude towns (Lukla, Namche Bazaar, Khumjung and Chaurikharka) have the highest density of tourists and the greatest resident populations. Exact population data is difficult to acquire. The Chaurikharka region (including the townships of Lukla and Chaurikharka) has a population of approximately 2200. The population of Namche Bazaar township is ~1600 and the combined populations of Khumjung, Kundee and Tengboche are estimated to be 1800. Our coliform and E. coli data indicate that fecal contamination of drinking water is related to population, the number of tourists and temperature/daylight hours. These results show that field-based monitoring of physical parameters is not sufficient to assess water contamination in the region.

The results of this study also show two similar, but distinct, trends in the data based solely on altitude, which suggest that the lower altitude samples may be affected by different processes. The lower portion of the study area, between Lukla and Monjo, experiences higher tourist and local traffic and is located within the SNP buffer zone, not the SNP proper itself. The differences in field parameters and bacterial content between the buffer zone and SNP proper may relate to more humans, livestock and deforestation, in addition to altitude and tourist traffic. The higher altitude samples, between Namche Bazaar and Deboche have larger temperature ranges but overall show similar trends as the lower altitude samples. This area is completely within the SNP boundary so the difference might be the result of less deforestation, lower temperatures and/or less human impact. Alternatively, these differing trends might reflect variations in rock types, proximity to structural geology features (faulting), and/or ratios of glacial melt in the groundwater between the upper and lower reaches of the study area. Further work is needed to assess the cause of the different trends.

Fecal pollution indicators, such as E. coli and coliform bacteria, can be used to evaluate the health of aquatic ecosystems and the potential for health effects among individuals within those environments [22]. The presence of such indicator organisms may provide evidence of water-borne problems and direct threats to human and animal health. Our study clearly suggests that there is a significant presence of bacterial indicators of fecal pollution in this study area. This work suggests that groundwater-fed springs may be safer to drink during the cold winter months. However, further work is needed to ascertain the bacteria content of the drinking water during the warmer summer months when bacteria levels are likely to be higher.

Importantly, more research is needed to ascertain the ultimate water sources, such as glacial melt, to surface water or the shallow groundwater. The quality and quantity of these sources are likely to change with the current global warming trends and accelerated melting of glacial ice, which could have a significant impact on the potable water in the region. Continued monitoring of microbial contamination in the SNP should be an essential component of the environmental protection strategy in the SNP.

The World Tourism Organization (WTO) states that tourism operations in protected areas need to be carefully managed and monitored to ensure their long-term sustainability. Otherwise, such operations will have negative environmental and cultural consequences, and tourism will contribute to the further deterioration [4]. Although the negative effects of tourism are of significant concern, many protected areas have promoted tourism development to improve their economic conditions, particularly to generate revenue to finance other social and economic development activities and to provide direct income and employment opportunities for the local population [3] [4]. In the SNP there are conflicting interests between the need for economic growth based on tourism and the ability of the current infrastructure to handle issues such as human waste disposal. As such, continued monitoring of water quality, specifically fecal contamination, can be used as a key indicator of the effects of both continued economic growth and climate change.

5. CONCLUSIONS

Our study shows that:

1) Drinking water within the SNP region currently meets current WHO [17] drinking water standards for the physical properties of temperature (2.8°C - 13°C), pH (5.27 - 7.24), conductivity (14.50 - 133 mS) and TDS (7.24 - 65.5 ppm).

2) Samples collected and analyzed in May (pre-monsoon summer) had a higher level of E. coli and coliform bacteria than samples collects in November (post-monsoon early winter) suggesting a seasonal dependence.

3) Samples from surface water sources have higher concentrations of E. coli and coliform bacteria than groundwater-fed springs.

4) Samples from lower altitude drinking water sources were higher in E. coli and coliform bacteria than samples from higher altitude sources.

5) Field-based measurements of physical parameters, temperature, pH, TDS and conductivity alone are not sufficient to detect or monitor drinking water contamination in the SNP.

ACKNOWLEDGEMENTS

The authors would like to thank the following people and organizations for their support: Troy Risk Inc. (Indianapolis), Indiana Geologists, Ball State University, AngNorbu Sherpa (Holiday Namche), AngPema Sherpa (Valley View Khumjung) and the Sagarmatha National Park Service.

REFERENCES

1. Balestrini, R., Polesello, S. and Sacchi, E. (2014) Chemistry and Isotopic Composition of Precipitation and Surface Waters in Khumbu Valley (Nepal Himalaya): N Dynamics of High Elevation Basins. Science of the Total Environment, 485-486, 681-692. http://dx.doi.org/10.1016/j.scitotenv.2014.03.096
2. Salerno, F., Viviano, G., Thakuri, S., Flury, B., Maskey, R.K., Khanal, S.N., Bhuju, D., Carrer, M., Bhochhibhoya, S., Melis, M.T., Giannino, F., Staiano, A., Carteni, F., Mazzoleni, S., Cogo, A., Sapkota, A., Shrestha, S., Pandey, R.K. and Manfredi, E.C. (2010) Energy, Forest, and Indoor Air Pollution Models for Sagarmatha National Park and Buffer Zone, Nepal: Implementation of a Participatory Modeling Framework. Mountain Research and Development, 30, 113-126. http://dx.doi.org/10.1659/mrd-journal-d-10-00027.1
3. Salerno, F., Viviano, G., Manfredi, E.C., Caroli, P., Thakuri, S. and Tartari, G. (2013) Multiple Carrying Capacities from a Management-Oriented Perspective to Operationalize Sustainable Tourism in Protected Areas. Journal of Environmental Management, 128, 116-125. http://dx.doi.org/10.1016/j.jenvman.2013.04.043
4. WTO (World Tourism Organization) (2005) Tourism's Potential as a Sustainable Development Strategy. World Tourism Organization, Madrid.
5. Ghimirie, N.P., Jha, P.K. and Caravello, G. (2013) Physico-Chemical Parameters of High-Altitude Rivers in the Sagarmatha (Everest) National Park, Nepal. Journal of Water Resource and Protection, 5, 761-767. http://dx.doi.org/10.4236/jwarp.2013.58077
6. Ghimirie, N.P., Jha, P.K. and Caravello, G. (2013) Water Quality of High-Altitude Lakes in the Sagarmatha (Everest) National Park, Nepal. Journal of Environmental Protection, 4, 22-28. http://dx.doi.org/10.4236/jep.2013.47A003
7. Caravello, G.U., Boselli, A.M., Bertollo, P. and Baroni, A. (2007) Assessing Ecosystem Health: An Analysis of Tourism Related Change and Impact in Khumbu Valley. International Journal of Ecology, 14, 45-64.
8. Okpokwasili, G.C. and Akujobi, T.C. (1996) Bacteriological Indicators of Tropical Water Quality. Environmental Toxicology and Water Quality, 11, 77-81. http://dx.doi.org/10.1002/(SICI)1098-2256(1996)11:2<77::AID-TOX1>3.0.CO;2-5
9. Hodegkiss, I.J. (1988) Bacteriological Monitoring of Hong Kong Marine Water Quality. Environment International, 14, 495-499. http://dx.doi.org/10.1016/0160-4120(88)90410-2
10. Cairneross, S., Carruthers, I., Curtis, D., Feachem, R., Bradley, D. and Baldwin, G. (1980) Evaluation for Village Water Supply Planning. Wiley, Chichester, 277.
11. World Health Organization (WHO) (2008) Guidelines for Drinking Water Quality [Electronic Resource]: Third Edition, Incorporating the First and Second Addenda, Volume 1, Recommendations. World Health Organization, Geneva, 668.
12. Kistemann, T., Claben, T., Koch, C., Dangendorf, F., Fischeder, R., Gebel, J., Vacata, V. and Exner, M. (2002) Microbial Load of Drinking Water Reservoir Tributaries during Extreme Rainfall and Runoff. Applied and Environmental Microbiology, 68, 2188-2197. http://dx.doi.org/10.1128/AEM.68.5.2188-2197.2002

13. Pathak, S.P. and Gopal, K. (2001) Rapid Detection of Escherichia coli as an Indicator of Faecal Pollution in Water. Indian Journal of Microbiology, 41, 139-151.
14. Harwood, V.J., Brownell, M., Perusek, W. and Whitelock, J.E. (2001) Vancomycin-Resistant Enterococcus sp. Isolated from Waste Water and Chicken Feces in the United States. Applied and Environmental Microbiology, 67, 4930-4933. http://dx.doi.org/10.1128/AEM.67.10.4930-4933.2001
15. Vaidya, S.Y., Vala, A.K. and Dube, H.C. (2001) Bacterial Indicators of Faecal Pollution and Bhavnagar Coast. Indian Journal of Microbiology, 41, 37-39.
16. APHA (1998) Standard Method for the Examination of Water and Wastewater, 19th ed. American Public Health Association, Washington DC.
17. WHO (2011) "Guidelines for Drinking Water Quality" [Electronic Resource]: 4th Edition. World Health Organization, Geneva, 541.
18. Baghel, V.S., Gopalb, K., Dwivedia, K. and Tripathi, R.D. (2005) Bacterial Indicators of Faecal Contamination of the Gangetic River System Right at Its Source. Ecological Indicators, 5, 49-56. http://dx.doi.org/10.1016/j.ecolind.2004.09.002
19. American Public Health Association (APHA) (2005) Standard Method for the Examination of Water and Wastewater. 21st Edition, American Public Health Association, Washington DC.
20. Sood, A., Singh, K.D., Pandey, P. and Sharma, S. (2008) Assessment of Bacterial Indicators and Physicochemical Parameters to Investigate Pollution Status of Gangetic River System of Uttarakhand (India). Ecological Indicators, 8, 709-717. http://dx.doi.org/10.1016/j.ecolind.2008.01.001
21. 21. Gameson, A.L.H. and Saxon, J.R. (1967) Field Studies on Effect of Daylight on Mortality of Coliform Bacteria. Water Research, 1, 279-295. http://dx.doi.org/10.1016/0043-1354(67)90004-8
22. 22. McLellan, S.L., Daniels, A.D. and Salmore, A.K. (2001) Clonal Populations of Thermotolerant Enterobacteriaceaes in Recreational Water and Their Potential Interference with Foecal Escherichia coli Counts. Applied and Environmental Microbiology, 67, 4934-4938. http://dx.doi.org/10.1128/AEM.67.10.4934-4938.2001

CHAPTER 19

Problematic of Drinking Water Access in Rural Area: Case Study of the Sourou Valley in Burkina Faso

Savadogo Boubacar[1], Kaboré Aminata[2], Zongo Dramane[1], Poda Jean Noel[1], Bado Hortense[3], Rosillon Francis[4], Dayeri Dianou[1*]

[1]*National Center for Scientific and Technological Researches, Institute for Health Sciences Research, Ouagadougou, Burkina Fas*

[2]*Research Center for Biological, Alimentary and Nutritional Sciences, Research and Training Unit, Life and Earth Sciences, University of Ouagadougou, Ouagadougou, Burkina Faso*

[3]*Convention for the Promotion of a Sustainable Development, Non-Governmental Organization, Ouagadougou, Burkina Faso*

[4]*Water, Environment, Development Unit, Arlon Campus, University of Liège, Arlon, Belgique.*

ABSTRACT

Safe drinking water access for rural populations in developing countries remains a challenge for a sustainable development. The study aims to investigate the drinking water quality and the factors affecting this quality in the Sourou valley in Burkina Faso. A total of 135 water samples were collected in sterile glass bottles during the dry seasons 2007, 2008, and 2012 from 10 drillings and 5 wells. Fifteen physicochemical parameters and two fecal pollution indicators (*Escherichia coli* and fecal Coliforms) were monitored based on laboratory standard methods. Datas were analyzed, using the Student's t-test and XLSTAT 7.5.2 statistical software. From results obtained, water quality was related to water source and sampling period as well ($p < 0.0001$). 30% of drillings provided water with nitrates concentration over the World Health Organization (WHO) guideline value. High turbidity was also observed for some drillings. Moreover, 90% of drillings showed water total hardness largely over the WHO threshold value. Water from drillings were exempt of fecal pollution, contrasting with the wells one which appeared uniformly polluted with concentrations exceeding sometimes 10^3 and 10^4 CFU/100 ml for *E. coli* and fecal Coliforms, respectively. Field investigations showed a preference of wells as drinking water source, and that appeared related to the lack of self-management of drillings and to cultural considerations. Overall, this study highlighted that a regular survey of water quality, management of protection zones around drinking water sources, sensitization on water resources self-management, hygiene and health issues, and providing appropriate household disinfection methods could help advancing to reach an effective safe drinking water access for rural populations in the country.

Keywords: Drinking Water; Chemistry; Bacteriology; Pollution; Sourou; Burkina Faso

1. INTRODUCTION

Access to safe drinking-water is important as a health and development issue at national, regional and local levels. In some regions, it has been shown that investments in water supply and sanitation can yield a net economic benefit, since the reductions in adverse health effects and health care costs outweigh the costs of undertaking the interventions. This is true for major water supply infrastructure investments through to water treatment in the home. Experience has also shown that intervenetions in improving access to safe water favor the poor in particular, whether in rural or urban areas, and can be an effective part of poverty alleviation strategies [1].

However, safe drinking water access for rural populations in developing countries (DC) remains a challenge to overcome for a sustainable development. Despite appreciable efforts undertaken to achieve the Millennium Development Goals (MDG), many of these countries are still suffering from a lack of drinking water access [2,3]. In view of the increase in water demand, measures undertaken generally focused the quantitative aspect to meet the needs of populations. However, beyond the quantitative aspect, it is advisable to pay attention on the quality of water consumed by the populations. Unfortunately, in most DC, analytical data of water quality are missing [4], although it is well known that the control of water in its different components is essential for the socioeconomic development of a country, and that also determine the implementation of the MDG in the other sectors [5,6].

Considering water access for populations in Burkina Faso, the National Action Plan for Integrated Water Management Resources [7] depicted the country as in situation of hydric stress, the mean water resources available and accessible being 850 m^3 per year and per inhabitant, compared to the threshold water shortage generally evaluated to 1000 m^3. The situation is partially related to the fact that in most parts of the country, the aquifer system is located in a sedimentary zone with hard stones represented by sandstones and limestone-dolomites with a thickness estimated to a hundred meters [8,9] and that limits strongly the possibility to realize drillings with a high flow and at low cost. As a consequence, the urban hydraulic resorts quite exclusively to surface waters to feet populations demand (i.e. dams of Ziga and Loumbila which actually aliment the capital of the country, Ouagadougou) [10].

At national level, the regulation on the quality of water resources is sustained by a decree which fixes the standards of pollutants in surface waters and drinking water as well with regards to WHO standards and the specific situations in the country [11]. Specifically for drinking water quality, the guideline values currently referred by the national office of water are the ones recommended by WHO [12].

According to the United Nations Development Program [13], the access to safe drinking water in Burkina Faso clearly improved these years with a national rate of water access passing from 18.3% in 1993 to 66.3% in 2007. These good performances are the consequence of the efforts undertaken by the country to achieve the Millennium Development Goals (MDG). These efforts led to the reinforcement of the infrastructures of water supply. The network of drinking water adduction which was of 881 kilometers in 1986 reached 3129 kilometers in 2004 while between 2006 and 2007, the projects and programs allowed the realization of approximately 1882 drillings [13]. However, the situation is undoubtedly variable from one place to another in the country, the urban environment being privileged compared to the rural one [14].

Although Burkina Faso already reached the MDG for the access to safe drinking water [13], the situation is not therefore satisfactory, in particular in rural environment where the populations are confronted with the optimal management of the water supply points. If in urban environment, water distributed is subjected to regular control, in rural area the indicators of drinking water quality are missing due to the lack of analytical data [15-17]. An improvement of knowledge is however essential to make the water services more efficient and to reinforce the policy for an effective access to safe drinking water in the country. From this view point, one of the recommendations of the National Action Plan for Integrated Water Management Resources (APIWRM), consists to the installation of networks for water analysis [7,18].

To meet this recommendation, a network for surface waters and groundwaters quality survey was initiated since 2006 within the framework of the Contract of the Sourou River [19]. Following a first investigation on the quality of surface waters [17], the present study examines the general physicochemical and bacterial parameters of water from wells and drillings consumed by populations in the Sourou valley.

After the presentation of the context of the study and the methodology used, the results obtained will be presented. The quality of water will be discussed with regard to the national and WHO standards [11,12,20-22], and the environmental factors and socio-behavioral attitudes impacting the access to safe drinking water before proposing some issues in order to improve the situation.

2. STUDY CONTEXT

As underlined in our previous works [15-17], the study was realized within the framework of the implementation of a project of river contract in the Sourou' watershed. In 2003, through cooperation with the Walloon Region of Belgium, a river contract based on the Walloon model was initiated [19]. This model is an approach of integrated and participatory management which aims at gathering within a river committee the representatives of all the users of water in order to define and implement a restoration actions program of the water resources, waterways and their accesses. This program was elaborated according to a consensual approach which takes care to integrate the concerns of each user while improving the protection of environment. In Burkina Faso, it was proved that the river contract could also be an issue to fight against desertification and poverty [19].

This project which has been developed on nearly ten years was framed locally by a NGO, the Convention for the promotion of a sustainable development (COPROD) which dealt with the animation and the coordination of the activities. The Department of Environment of the University of Liege in Belgium ensured the general coordination and the scientific expertise for the account of the Walloon Region in collaboration with the Institute for Health Sciences Research (IRSS) of the Scientific and Technological National Research Center (CNRST) of Ouagadougou.

Considering the operational characteristic of the river contract, many field activities were performed [23]. The activities were divided into five sets of themes: 1) Coordination, animation and dialog between actors of water; 2) Improvement of knowledge through the data acquisition and the organization of many formations for the users of water and the local collectivities; 3) Communication, information and sensitization of schools and water users; 4) Restoration of the water resources and the environment which is concretized by an improvement of water access and environment protection (waterways protection and fight against desertification); and 5) Income-generating activities in a context of fighting against poverty.

The present study is in line with an improvement of knowledge related to the quality of water resources in the Sourou basin. It also meets one of measures recommended by the APIWRM in the actions field No. 2 "water information system": action 2.2, the implementation of national networks to monitor the water quality, water uses, water requests and the risks [7].

3. STUDY ZONE

The Sourou valley as previously described [15-17] is located in the North-West of Burkina Faso, in the area of the Mouhoun loop (Figure 1). The Sourou River takes its source in Mali at the level of Baye. It makes border between Burkina Faso and Mali, by then crossing the Burkina Faso from north to south before joining Mouhoun River at Léry. The Sourou' watershed covers a surface of 16,200 km^2 but only the central part located on the left bank of the river (approximately 5000 km^2) is the object of this study.

The Sourou valley is especially known for its hydroagricultural installations following the erection of dam valves at the junction of Sourou and Mouhoun rivers in 1984. The realization allowed to increase significantly the level of water of the Sourou River (600,000,000 m^3) through the valley [15]. This availability of water thus allowed the creation of irrigated perimeters, making the Sourou valley an important agricultural production zone [15].

Currently, the irrigated perimeters extend on a surface of 3200 ha (which will be soon increased with a new additional zone of 2033 ha) and are managed under the direction of the Sourou Valley Development Authority (SVDA). The Sourou valley thus constitutes an important agrarian production zone benefitting the whole country [14].

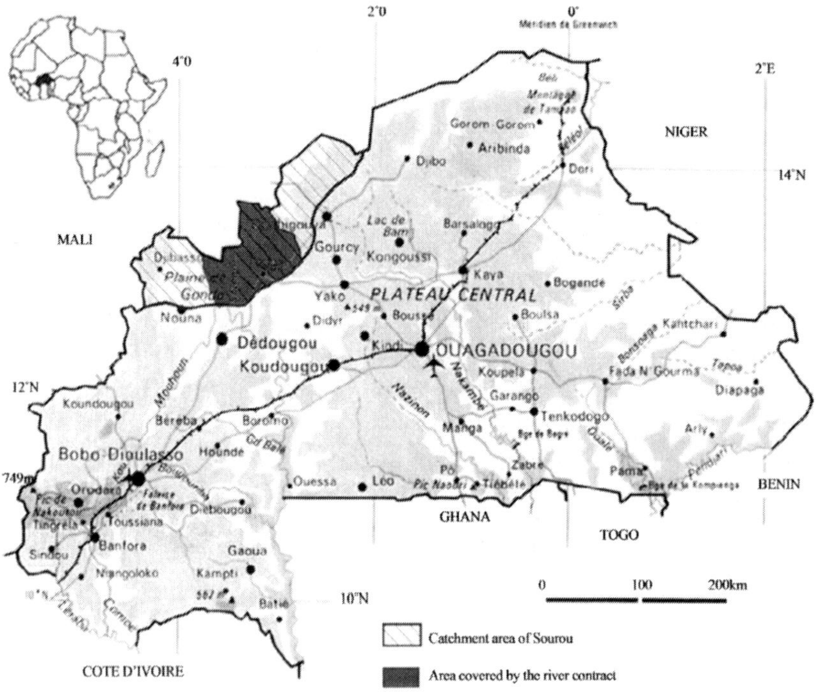

Figure 1. Map showing the catchment area of the Sourou valley and the zone covered by the river contract [19].

Drinking water access for Sourou populations is possible from various sources of supply, mainly from groundwaters. Except for some more important localities like Tougan, chief town of the Sourou Province, which profits from a partial system of water adduction, the rural populations generally feed on from traditional wells to a relatively dense network of drillings installed by NGO within the framework of cooperative projects or programs supported by the Government, notably the second Soils Management National Plan (SMNP 2); however, populations of insular villages or close to the Sourou river use surface water for their alimentation (Figure 2).

The availability of groundwater is rather stable. The Sourou aquifer system is located in a sedimentary zone and consists of hard stones represented by sandstones and limestone-dolomites which can be crossed by faults. The thickness of the aquifers' sandstones is estimated to a hundred meters [24]. The depth of drillings is about a sixty meters while the water level in the traditional wells is variable from one site to another, the depth being of approximately 10 to 20 m in the zone of the study. The refill of the aquifer can be established through a slow infiltration in the subsoil. This diffuse refill could be supplemented by a preferential water flow through fractured zones [14].

Figure 2. Some aspects of drinking water access in the Sourou valley. (a) Getting water from Niassan-AMVS drilling (January 2012); (b) Aspects of water sampled from drillings (November 2007); (c) Getting water from a modern well at Kouy (January 2012); (d) Getting water from the traditional well of Kouy Mosque (March 2008); (e) Getting drinking water from Sourou River at Toma-island (January 2012); (f) Transportation of drinking water collected from drilling to home at Kiembara (January 2012).

Within the framework of the Sourou river contract, an inventory of wells and drillings were carried out [19]. This inventory identified many non-functional works which were the object of repair within the framework of this river contract. More than 100 works were thus given back in activity.

4. MATERIAL AND METHODS

4.1. Water Sampling and Field Information Gathering

The groundwater resources focused in this study consisted of 10 drillings and 5 wells (traditional or modern, large diameter wells) located throughout the valley (Figure 3).

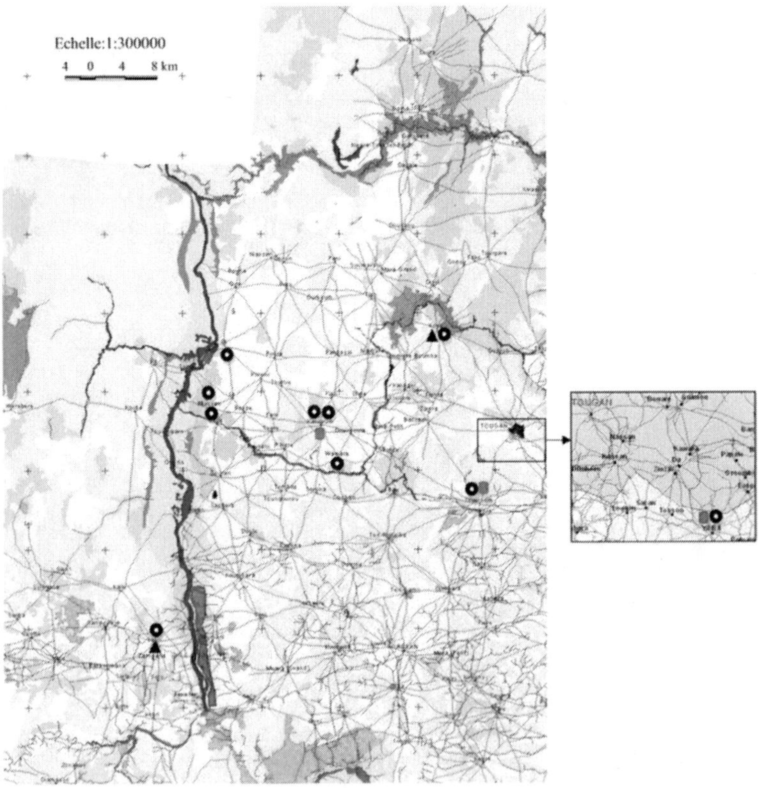

Figure 3. Location of sampling sources within the zone covered by the river contract in the Sourou valley [19].■ Modern well (3); ▲ Traditional well (2); ● Drilling (10).

Sampling campaigns were carried out in 2007, 2008 and 2012 during the dry season for which water demand and pressure on water sources were particularly high.

Water samples were collected aseptically in triplicate into sterile glass bottles at the field (Figure 2(b)). A total of 45 samples were taken and analyzed during each campaign. Samples collected at the field were preserved at 4°C in cool boxes, carried to the National Laboratory for Water Analysis in Ouagadougou and stored in a refrigerator at 4°C before analysis. Some samples were blocked by the addition of mercuric chloride and analyzed in Belgium for nitrates at the laboratory of water resources, University of Liège.

During sampling, field observations, supplemented by information collected from water users and local authorities fed the reflection in order to try to identify the sources of constraints to safe drinking water access.

4.2. Sample Analysis

A total of 15 physicochemical parameters were determined. Parameters such as pH and conductivity were monitored on-site with a portable Hatch Multimeter and kit. Other parameters were determined from samples preserved at 4°C in cool boxes by the National Laboratory for Water Analysis of the Ministry of Environment and Sustainable Development in Ouagadougou, in the 2 - 3 days following sampling, using standard methods [25, 26].

For nitrates/nitrites determination, proportioning was carried out by molecular absorption spectrophotometry through nitrates reduction in nitrites by cadmium (spectro Hach DR2400 method 8171).

Iron in the water samples was analyzed using Atomic Absorption Spectrophotometer (AAS) after pre-concentration [25,26].

Two bacterial indicators of fecal contamination namely *Escherichia coli,* and fecal Coliforms were enumerated 3 - 4 hours following sampling at the Laboratory of Tougan hospital, using the membrane filtration technique [25]. Bacterial cells were concentrated on a 0.2 µm Millipore Membrane Filter followed by culture on the chromogenic RAPID E. COLI 2 AGAR (BIO RAD) medium which contains 2 substrates specific to the β-DGlucuronidase (Gluc) and β-D-Galactosidase (Gal) enzymes, respectively. Incubation was performed at 44.5°C for 24 h. Colonies of E. coli ($Gal^+/Gluc^+$) appears violet to pink while other coliforms colonies stain blue.

4.3. Statistical Method

Data obtained were analyzed for water source and sampling period variations using the Student's t-test and XLSTAT 7.5.2 statistical software. Mean parameters concentrations were compared according to the Newman Keuls' test.

5. RESULTS AND DISCUSSION

5.1. Water Physicochemical Characteristics

The results of the statistical analysis on the data obtained are presented in Tables 1(a) and (b) while the mean characteristics are shown in Tables 2(a) and (b).

The Student's t-test revealed that all the characteristics of water were significantly related to the sampling site ($p < 0.0001$) and period as well ($p < 0.0001$), except calcium, magnesium and total hardness (Tables 1(a) and (b)). The joined effects of site and period affected also significantly ($p < 0.0001$) these characteristics except total hardness ($p = 0.291$). The spatio-temporal variation in response to water source and sampling period implied that water samples were collected from sources of different physicochemical characteristics which are more influenced by the period of water collection.

Mean turbidity values recorded for some water samples were higher than the national and WHO guideline value of 5 NTU for drinking water [11,12,20]. According to the sampling period, high turbidity of water was recorded for the drillings of Kassoum-CEG (12 NTU in 2007), Yaba (14.6 NTU in 2007) and especially Sono (104-180 NTU in 2007-2008) (Table 2(a)). For the later, the turbidity seemed related to the oxidation of iron in contact of air as indicated by the reddish-brown color observed after a few minutes subsequent to water collection into flasks (Figure 2(b)). All samples from wells showed uniformly turbidity over the WHO threshold value. Since wells are not protected, they can receive suspended material leading to turbidity increase as confirmed our field observations. As also underlined in studies, excessive turbidity may also be associated with unpleasant tastes and odors [27]. Turbidity also correlates with iron content of water samples [28] as observed at some sampling periods for water samples of Sono and Yaba drillings (Tables 2(a) and (b), Figure 2(b)). As a consequence, the water source can be rejected by populations and that corroborated the case of Sono drilling.

The mean pH of the water samples ranged from slightly acid (pH 5.96) to slightly basic (pH 7.70) (Table 2(a)). Although pH usually has no direct impact on consumers, high pH can affect the palatability. No healthbased guideline value has been proposed for pH; however, an acceptable range for drinking water pH is from 6.5 to 8.5 [20- 22]. Corrosion effects may also become significant below pH 6.5 and that corroborated the case of Sono drilling, and at a less extent Yaba drilling!

Mean water conductivity values recorded ranged from 45 - 2465 $\mu S{\cdot}cm^{-1}$ (Table 2(b)). Although no health hazard for populations was found associated to conductivity, classification of potability based on electrical conductivity ascribes <325 $\mu S{\cdot}cm^{-1}$ for fresh and potable water [29]. From the results obtained, 60% of the water sources showed mean values of water electrical conductivity over the ascribed potability value of 325 $\mu S{\cdot}cm^{-1}$. Mean values were particularly high throughout the 3 sampling periods for the drillings of Wawara (2215 - 2465 $\mu S{\cdot}cm^{-1}$), Diouroum (837 - 1168 $\mu S{\cdot}cm^{-1}$); Di (764 - 906 $\mu S{\cdot}cm^{-1}$), Niassan-AMVS (541 - 606 $\mu S{\cdot}cm^{-1}$) and Niassan-Clinic (557 - 600 $\mu S{\cdot}cm^{-1}$). The high conductivity of these ground-waters may be related to the bedrock they flow through as suggested [21].

Total Hardness values in the water samples ranged from 16.0 - 1447.3 mg/l $CaCO_3$. This parameter varied significantly with only the sampling source ($p < 0.0001$) but not with the sampling period ($p = 0.383$) or both joined factors ($p = 0.291$). Water hardness was particularly remarkable for all the water sources studied, except the drilling of Kouy, the modern well of Yaba and the traditional well of Sono. A high and practically constant value throughout the study periods was recorded for the water of Wawara drilling (1446 - 1447 mg $CaCO_3$/l). Hardness in water comprises the determination of calcium and magnesium as the main constituents and their widespread abundance in rock formations leads often to very considerable hardness levels in surface waters or groundwaters. One of several arbitrary classifications of waters by hardness includes: Soft, up to 50 mg/l; Moderately Soft, 51 - 100 mg/l; Slightly Hard, 101 - 150 /l; Moderately Hard, 151 - 250 mg/l; Hard, 251 - 350 mg/l; Excessively Hard, over 350 mg/l [30]. The degree of hardness of drinking water is important for aesthetic acceptability by consumers.

From the values recorded (Table 2(b)), waters were classified as Soft for the drilling of Kouy (18 - 20 mg $CaCO_3$/l), the modern well of Kassoum-Market (25 - 35.3 mg $CaCO_3$/l), and the traditional well of KouyMosque (16 - 22.6 mg $CaCO_3$/l).

Table 1. (a) Variance of the physicochemical characteristics of water from drillings and wells with regard to sampling site and sampling period. (b) Variance of the physicochemical characteristics of water from wells and drillings with regard to sampling site and period (continued).

(a)

Source of variation	df	Nitrates (mg NO_3^-/l)		Nitrites (mg NO_2^-/l)		pH		Potassium (mg K^+/l)		Sodium (mg Na^+/l)		Sulfates (mg SO_4^{2-}/l)		Turbidity (NTU)	
		MS	p	MS	p	MS	p	MS	p	MS	p	MS	p	MS	p
Site	14	312192.8	$< 0.0001^{**}$	4882.4	$< 0.0001^{**}$	0.982	$< 0.0001^{**}$	13572.61	$< 0.0001^{**}$	195.65	$< 0.0001^{**}$	868184.6	$< 0.0001^{**}$	5202.21	$< 0.0001^{**}$
Period	2	5512.7	$< 0.0001^{**}$	0.002	$< 0.0001^{**}$	0.128	$< 0.004^{**}$	540.16	$< 0.0001^{**}$	41.27	$< 0.0001^{**}$	1371.13	$< 0.0001^{**}$	24.24	$< 0.0001^{**}$
Site*Period	28	4882.4	$< 0.0001^{**}$	0.001	$< 0.0001^{**}$	0.101	$< 0.0001^{**}$	371.42	$< 0.0001^{**}$	3.25	$< 0.0001^{**}$	1052.77	$< 0.0001^{**}$	1600.09	$< 0.0001^{**}$

MS means square; **significant $p < 0.01$.

(b)

Source of variation	df	Ammonium (mg NH_4^+/l)		Conductivity (µS/cm)		Arsenic (µg As/L)		Calcium (mg Ca^{2+}/l)		Calcium hardness (mg $CaCO_3$/l)		Total hardness (mg $CaCO_3$/l)		Total Iron (mg Fe/l)		Magnesium (mg Mg^{2+}/l)	
		MS	p	MS	p	MS	p	MS	p	MS	p	MS	p	MS	p	MS	p
Site	14	1.993	$< 0.0001^{**}$	3572554.2	$< 0.0001^{**}$	1.548	$< 0.0001^{**}$	89167.4	$< 0.0001^{**}$	553279	$< 0.0001^{**}$	1136052.2	$< 0.0001^{**}$	27.17	$< 0.0001^{**}$	7238.99	$< 0.0001^{**}$
Period	2	0.035	$< 0.0001^{**}$	6154.45	$< 0.0001^{**}$	2.118	$< 0.0001^{**}$	2.164	0.754^{ns}	74.63	0.018^{*}	7329.80	0.0383^{ns}	61.15	0.0001^{**}	14.92	0.038^{*}
Site*Period	28	0.020	$< 0.0001^{**}$	16955.9	$< 0.0001^{**}$	0.239	$< 0.0001^{**}$	40.929	$< 0.0001^{**}$	233.61	$< 0.0001^{**}$	8792.11	0.291^{ns}	15.38	$< 0.0001^{**}$	31.45	$< 0.0001^{**}$

MS means square; *significant $p < 0.05$; **significant $p < 0.01$; ^{ns}not significant $p < 0.05$.

Table 2. (a) Mean* physicochemical characteristics of water from wells and drillings during the dry season in 2007, 2008 and 2012 (*mean of 3 replicates); (b) Mean* physicochemical characteristics of water from wells and drillings during the dry season in 2007, 2008 and 2012 (*mean of 3 replicates).

(a)

Sampling Site	Period	Nitrates ($mg\ NO_3^-/l$)	Nitrites ($mg\ NO_2^-/l$)	pH	Potassium ($mg\ K^+/l$)	Sodium ($mg\ Na^+/l$)	Sulfates ($mg\ SO_4^{2-}/l$)	Turbidity (NTU)
	2007	115^{g}	0.027^{defghi}	6.9^{bcdefgh}	55^{f}	7^{g}	38^{e}	1^{h}
Di-CP^{D}	2008	126.5^{ef}	0.013^{ghi}	6.8^{cdefghi}	120^{c}	2.2^{klmn}	40^{e}	1^{h}
	2012	128^{ef}	0.017^{fghi}	7.0^{bcde}	90^{d}	5^{hi}	42^{e}	1.16^{gh}
	2007	860^{a}	0.200^{b}	6.46^{hijklm}	6^{klm}	14^{c}	43^{e}	1^{h}
Dioroum-COPROD^{D}	2008	610^{c}	0.187^{b}	6.46^{hijklm}	5.03^{klm}	9.1^{e}	45^{e}	1^{h}
	2012	751.7^{b}	0.290^{a}	6.10^{no}	5.13^{klm}	12^{d}	42^{e}	1^{h}
	2007	3^{q}	0.040^{cdef}	6.6^{efghijkl}	3^{lm}	0.86^{no}	5.66^{ij}	12^{ef}
Kassoum-CEG^{D}	2008	2.23^{q}	0.014^{ghi}	6.6^{efghijkl}	4.2^{klm}	0.53^{o}	7^{hij}	2.63^{gh}
	2012	3^{q}	0.011^{ghi}	6.43^{ijklmn}	2.8^{lm}	0.83^{no}	6^{ij}	3^{gh}
	2007	19.3^{lmn}	0.040^{cdef}	7.13^{bc}	6^{klm}	2^{klmno}	1.6^{j}	1^{h}
Kassoum-School^{D}	2008	23^{lm}	0.035^{defg}	7.0^{bcde}	2.6^{lm}	1.5^{klmno}	2.3^{j}	1.3^{gh}
	2012	24^{l}	0.026^{efghi}	6.96^{bcdef}	5.7^{klm}	1.8^{klmno}	2^{j}	1^{h}
	2007	19.5^{lmn}	0.004^{hi}	6.43^{ijklmn}	0.20^{m}	0.73^{no}	1.7^{j}	2.76^{gh}
Kouy-COPROD^{D}	2008	19.9^{lmn}	0.004^{hi}	6.50^{ghijklmn}	0.25^{m}	0.8^{no}	1.5^{j}	2.73^{gh}
	2012	13.8^{lmnopq}	0.002^{i}	6.63^{efghijk}	0.23^{m}	0.86^{no}	1.76^{j}	2.3^{gh}
	2007	12.1^{lmnopq}	0.020^{fghi}	7.0^{bcde}	4^{klm}	8.3^{ef}	13.5^{fghi}	1^{h}
Niassan-AMVS^{D}	2008	12.6^{lmnopq}	0.016^{fghi}	7.0^{bcde}	2.5^{lm}	4.7^{hi}	14.16^{fghi}	1.3^{gh}
	2012	18.3^{lmn}	0.030^{defgh}	7.0^{bcde}	3.1^{lm}	9^{e}	15^{efghi}	1.2^{gh}
	2007	11^{mnopq}	0.010^{ghi}	7.1^{bcd}	3^{lm}	9^{e}	13^{fghi}	1^{h}
Niassan-Clinic^{D}	2008	15.3^{lmnop}	0.010^{ghi}	7.7^{a}	4.4^{klm}	4.9^{hi}	14^{fghi}	1.5^{gh}
	2012	12.1^{lmnopq}	0.010^{ghi}	7.0^{bcde}	5^{klm}	7^{g}	16^{defgh}	4^{gh}
	2007	3^{q}	0.040^{cdef}	6.26^{klmno}	2^{m}	3^{jk}	6^{ij}	180^{a}
Sono-Clinic^{D}	2008	4^{pq}	0.023^{efghi}	6.16^{mno}	1.9^{m}	1.73^{klmno}	5.33^{ij}	104^{b}
	2012	4.23^{pq}	0.012^{ghi}	6.66^{defghijk}	2.5^{lm}	2.76^{jkl}	8^{ghij}	2.26^{gh}
	2007	122.3^{f}	0.013^{ghi}	6.03^{o}	15^{j}	9^{e}	24^{de}	14.66^{de}
Yaba-Clinic^{D}	2008	134^{e}	0.018^{fghi}	6.20^{lmno}	5.2^{klm}	5.2^{hi}	22^{def}	2.8^{gh}
	2012	82.7^{i}	0.027^{defghi}	5.96^{o}	6^{klm}	7.5^{fg}	25^{d}	4^{gh}
	2007	5.23^{opq}	0.010^{ghi}	6.73^{cdefghij}	130^{ab}	17^{b}	1175^{b}	1.2^{gh}
Wawara-COPROD^{D}	2008	5.20^{opq}	0.013^{ghi}	6.70^{cdefghijk}	132^{a}	17^{b}	1175^{b}	1.2^{gh}
	2012	2.3^{q}	0.042^{cdef}	6.60^{efghijk}	120^{c}	10^{e}	1301.66^{a}	4^{gh}
	2007	37.3^{k}	0.021^{fghi}	7.20^{b}	32.3^{i}	2^{klmno}	16^{defgh}	5^{gh}
Dioroum-PNGT^{MW}	2008	96.7^{h}	0.022^{fghi}	7.20^{b}	80^{e}	1^{mno}	17^{defg}	2.5^{gh}
	2012	45.2^{jk}	0.025^{efghi}	6.93^{bcdefg}	30^{i}	2^{klmno}	18^{def}	5.16^{gh}

Continued

Kassoum-Market^{MW}	2007	15^{lmnop}	0.010^{ghi}	$6.7^{cdefghijk}$	1^{m}	3^{jk}	2^{j}	4^{gh}
	2008	13.7^{lmnopq}	0.010^{ghi}	6.8^{cdefgh}	0.9^{m}	1.3^{lmn}	2^{j}	2.4^{gh}
	2012	10.4^{nopq}	0.010^{ghi}	6.8^{cdefgh}	3^{lm}	1.5^{klmn}	2^{j}	6.43^{g}
Yaba-Clinic^{TW}	2007	240^{d}	0.050^{cd}	6.30^{klmno}	50^{g}	5.1^{hi}	39^{c}	2.5^{gh}
	2008	93^{h}	0.020^{fghi}	6.30^{klmno}	50.7^{g}	5.1^{hi}	40^{c}	3^{gh}
	2012	97.3^{h}	0.060^{c}	$6.63^{efghijk}$	44.7^{h}	5.8^{h}	43^{c}	4.9^{gh}
Kouy-Mosque^{TW}	2007	16^{lmno}	0.030^{defgh}	$6.7^{cdefghijk}$	1^{m}	4^{ij}	1.63^{j}	23.66^{c}
	2008	18.5^{lmn}	0.025^{efghi}	$6.9^{bcdefgh}$	0.6^{m}	2.8^{jkl}	1.8^{j}	22^{c}
	2012	20.5^{lmn}	0.047^{cde}	6.20^{lmno}	0.8^{m}	3^{jk}	2^{j}	23.36^{c}
Sono-Centre^{TW}	2007	46^{jk}	0.010^{ghi}	$6.70^{cdefghijk}$	9^{k}	2^{klmn}	6^{ij}	15^{de}
	2008	50.5^{j}	0.018^{fghi}	6.80^{cdefgh}	4.3^{klm}	1.66^{klmn}	5^{ij}	11.16^{f}
	2012	37.3^{k}	0.010^{ghi}	$6.53^{fghijklm}$	8^{kl}	2.53^{klm}	8^{ghij}	16.86^{d}
Guideline values [20-22]		50	3	6.5 - 8.5	12	≤200	≤500	≤5

Means with a same letter within a column are not significantly different according to Newman Keuls' test $p < 0.05$. ^{D}drilling; ^{MW}modern well; ^{TW}traditional well.

(b)

Sampling Site	Period	Ammonium ($mg\ NH_4^+/l$)	Conductivity (µS/cm)	Arsenic (µg As/l)	Calcium ($mg\ Ca^{2+}/l$)	Calcium hardness ($mg\ CaCO_3/l$)	Total hardness ($mg\ CaCO_3/l$)	Total Iron (mg Fe/l)	Magnesium ($mg\ Mg^{2+}/l$)
Di-CP^{D}	2007	0.030^{nop}	865^{f}	2^{bc}	68^{d}	171^{def}	323.6^{bcde}	0.09^{jk}	37^{e}
	2008	0.030^{nop}	906^{e}	1^{d}	70^{d}	174^{de}	325^{bcde}	0.1^{jk}	38^{e}
	2012	0.030^{nop}	764^{h}	2^{bc}	66^{d}	176^{d}	326^{bcde}	0.09^{jk}	36^{e}
Dioroum-COPROD^{D}	2007	0.430^{e}	1079.3^{d}	2^{bc}	92^{c}	231^{c}	490^{bc}	0.03^{k}	62^{d}
	2008	0.553^{d}	1168.6^{c}	1.33^{cd}	100^{b}	249^{b}	535.3^{b}	9.8^{a}	70^{c}
	2012	0.380^{f}	837.6^{g}	1.66^{bcd}	95^{c}	235^{c}	501.6^{bc}	0.02^{k}	65^{d}
Kassoum-CEG^{D}	2007	0.050^{mnop}	207.6^{vwx}	2^{bc}	20^{kl}	50^{mn}	107.6^{ef}	0.05^{jk}	14^{hijkl}
	2008	0.130^{ijklm}	257.6^{u}	1^{d}	26^{hij}	63.6^{l}	128.6^{ef}	0.2^{ijk}	16^{hij}
	2012	$0.080^{jklmnop}$	260^{u}	1^{d}	23.33^{ijk}	55^{m}	111.6^{ef}	0.02^{k}	15^{hijk}
Kassoum-School^{D}	2007	$0.080^{jklmnop}$	440.3^{n}	1.66^{bcd}	44^{e}	110.3^{i}	240^{def}	0.17^{jk}	31.3^{f}
	2008	0.070^{klmnop}	206.6^{vwx}	1^{d}	22^{kl}	55^{m}	117^{ef}	0.3^{ijk}	15^{hijk}
	2012	0.063^{lmnop}	332.3^{r}	1^{d}	35^{fg}	80^{jk}	202^{ef}	2.5^{g}	18^{h}
Kouy-COPROD^{D}	2007	0.020^{op}	45^{z}	1^{d}	4^{n}	10.3^{qr}	18.6^{f}	1.3^{h}	2.5^{m}
	2008	0.003^{p}	46^{z}	1^{d}	4^{n}	11.6^{qr}	18^{f}	1.5^{h}	2.23^{m}
	2012	0.002^{p}	54.6^{z}	1^{d}	5^{n}	11.3^{qr}	20^{f}	0.29^{ijk}	2.26^{m}
Niassan-AMVS^{D}	2007	0.050^{mnop}	580^{j}	1.93^{bc}	63^{d}	157.6^{g}	282.6^{cdef}	0.02^{k}	39^{e}
	2008	0.160^{h}	606.3^{i}	1^{d}	69^{d}	170.6^{def}	318.6^{bcde}	0.35^{ijk}	26^{g}
	2012	0.143^{ijkl}	541.3^{j}	1^{d}	67^{d}	165^{efg}	271^{cdef}	0.03^{k}	30^{f}

Continued

Niassan-Clinic^{D}	2007	0.053^{mnop}	582.3^{j}	1.93^{bc}	66^{d}	163^{g}	315^{bcde}	0.02^{k}	37^{c}
	2008	0.060^{lmnop}	600^{j}	1.06^{d}	66.6^{d}	170^{def}	320^{bcde}	7^{c}	37^{c}
	2012	0.080^{klmnop}	557.3^{k}	2^{bc}	65^{d}	150^{h}	310^{bcde}	0.02^{k}	35^{c}
Sono-Clinic^{D}	2007	0.553^{d}	217.3^{v}	1.66^{bcd}	17^{klm}	42.6^{pq}	91^{ef}	3.5^{e}	12^{ijkl}
	2008	0.703^{c}	204.6^{vwx}	1^{d}	15^{lm}	38^{q}	90.3^{ef}	2.8^{f}	13^{hijkl}
	2012	0.700^{c}	147.3^{y}	1^{d}	16^{lm}	40.3^{pq}	90.3^{ef}	0.25^{ijk}	15^{hijk}
Yaba-Clinic^{D}	2007	$0.087^{jklmnop}$	422.6^{o}	2^{bc}	33^{fg}	82.3^{jk}	146^{ef}	0.023^{k}	15^{hijk}
	2008	0.150^{ijk}	394^{pq}	2^{bc}	32^{fgh}	79^{jk}	138^{ef}	5.35^{d}	
	2012	$0.083^{jklmnop}$	383^{q}	2.33^{ab}	34^{fg}	82.3^{jk}	142^{ef}	0.09^{jk}	14^{hijkl}
Wawara-COPROD^{D}	2007	1.95^{a}	2465a	1^{d}	407^{a}	1015^{a}	1447^{a}	0.5^{i}	105^{a}
	2008	1.95^{a}	2465a	1^{d}	410^{a}	1015^{a}	1447.3^{a}	0.5^{i}	105^{a}
	2012	1.65^{b}	2215b	1^{d}	405^{a}	1012^{a}	1446.3^{a}	0.4^{i}	100^{b}
Dieroum-PNGT^{MW}	2007	0.020^{op}	370.3^{rs}	1^{d}	32^{fgh}	80^{jk}	113.6^{ef}	0.02^{k}	9^{j}
	2008	0.013^{p}	407.6^{p}	1^{d}	30^{fgh}	75^{k}	453^{bcd}	0.3^{ijk}	11^{jkl}
	2012	0.020^{op}	360^{s}	1^{d}	28^{ghi}	81^{jk}	116^{ef}	0.02^{k}	10^{kl}
Kassoum-Market^{MW}	2007	0.020^{op}	71.3^{z}	1^{d}	11^{mn}	28^{p}	35.3^{f}	0.04^{k}	2^{m}
	2008	0.323^{e}	60^{z}	1^{d}	8no	20^{s}	27^{f}	0.2^{ijk}	2^{m}
	2012	0.020^{op}	51^{z}	1^{d}	15^{lm}	18^{qr}	25^{f}	0.08^{jk}	2^{m}
Yaba-Clinic^{MW}	2007	0.110^{ijklmn}	467^{m}	1^{d}	35^{fg}	87.3^{j}	157.3^{ef}	9.5^{h}	17^{hi}
	2008	0.113^{ijklmn}	468^{m}	1^{d}	35.66^{f}	87.3^{j}	157.6^{ef}	9.5^{h}	17^{hi}
	2012	0.100^{jklmno}	442^{n}	1^{d}	32^{fgh}	85^{jk}	155^{ef}	0.1^{jk}	17^{hi}
Kouy-Mosque^{TW}	2007	0.250^{h}	46^{z}	2.66^{a}	4.16^{p}	10^{qr}	21^{f}	0.08^{jk}	3^{m}
	2008	0.250^{h}	50^{z}	2^{bc}	3.1^{o}	8^{r}	16^{f}	0.15^{jk}	3^{m}
	2012	0.500^{d}	54^{z}	2^{bc}	5^{o}	11^{qr}	22.6^{f}	0.32^{ijk}	2.3^{m}
Sono-Centre^{TW}	2007	$0.083^{jklmnop}$	172^{wx}	1^{d}	19^{kl}	48^{mn}	64^{ef}	0.1^{jk}	4^{m}
	2008	0.130^{ijklm}	179^{wvx}	1^{d}	19^{kl}	47^{mno}	63^{ef}	0.4^{ij}	4^{m}
	2012	0.180^{i}	191^{vw}	1^{d}	18^{kl}	46^{mno}	62^{ef}	0.3^{ijk}	4^{m}
Guideline values [20-22]	-	<350	10	-	-	-	≤ 0.30	-	-

Means with a same letter within a column are not significantly different according to Newman Keuls' test $p < 0.05$. ^{D}drilling; ^{MW}modern well; ^{TW}traditional well.

Moderately Soft for the drilling of Sono-Clinic (90.3 - 91 mg $CaCO_3$/l) and the traditional well of Sono-Centre (62 - 64 mg $CaCO_3$/l), Slightly Hard for the drillings of Kassoum-CEG (107.6 - 128.6 mg $CaCO_3$/l) and Yaba-Clinic (142 - 146 mg $CaCO_3$/l); Slightly-to-Moderately Hard for the drilling of Kassoum-School (117 - 240 mg $CaCO_3$/l), Moderately Hard for the modern well of Yaba-Clinic (155 - 157.3 mg $CaCO_3$/l), Hard for the drillings of Di (323.6 - 326 mg $CaCO_3$/l), Niassan-AMVS (271 318.6 mg $CaCO_3$/l) and Niassan-Clinic (310 - 320 mg $CaCO_3$/l); Slightlyto-Excessively Hard for the modern well of Diouroum (113.6 - 453 mg $CaCO_3$/l) and Excessively Hard for the drillings of Diouroum (490 - 535.3 mg $CaCO_3$/l) and Wawara

(1446.3 - 1447.3 mg $CaCO_3$/l). Although there is evidence from epidemiological studies for a protective effect of magnesium, calcium or hardness on cardiovascular mortality, the evidence is being debated and does not prove causality. There are insufficient data to suggest either minimum or maximum concentrations of minerals, as adequate intake will depend on a range of other factors. Therefore, no guideline values are proposed [20-22].

Out of the 15 water sources examined, 3 drillings (Di, Diouroum and Yaba) and one modern well (Yaba) provided water with Nitrates concentrations exceeding the WHO and USEPA guideline value (50 mg NO_3^-/l) [15, 16]. The mean values for all the study periods were 123, 740, 113, and 143 mg NO_3^-/l for these water sources, respectively. The highest concentrations of nitrates were recorded in 2007 for the drilling of Diouroum and the modern well of Yaba (860 and 240 mg/l, respectively), in 2008 for the drilling of Yaba (134 mg/l), and in 2012 for the drilling of Di (128 mg/l) (Table 2(a)). Nitrates concentration in drinking water is more focused because high level can be hazardous to infants. The nitrate itself is not a direct toxicant but is a health hazard because of its conversion to nitrite, which reacts with blood haemoglobin to cause methaemoglobinaemia. Hence, 50 mg NO_3^-/l nitrate is set as Guideline standard for nitrate in drinking water [20-22]. Concerning the study zone, concentrations higher than 100 mg NO_3^-/l were observed by FAO [31] in the areas of Mouhoun and Sourou. Moreover, Nabayaogo [32] within the framework of a study on the impact of agricultural management on the water resources and the ecosystems of the Sourou valley, found nitrates contents of 2.7 to 37.2 mg/l for some wells and drillings located in Niassan village and on the riverside. These values reported by the author correspond approximately to the range of values recorded in this same village for drillings during the present study (11 to 18.3 mg/l). Dugué [33] concluded that the pollution risk of the aquifer by nitrates of agricultural origin is nearly zero in the Sourou valley (zone of Di). However, concentrations reaching 41 mg NO_3^-/l were evidenced at the beginning of rainy season in water of the Sourou River, particularly downstream Di village, within the framework of this project and these concentrations could be due to the cultivation practices in line with the production of tomato and onions [17]. For the nitrates concentrations in groundwater exceeding sometimes at a large extent the WHO standard in the Sourou valley, investigations were performed to elucidate the origin [15]. Several tracks of contamination were thus evoked near the works, in relation with the anthropic activities: animal and human wild defecation, waste discharges, wastewaters rejections and so on. It is particularly the case at Diouroum village where the ground is strewn with excrements. In addition, specific organic matter deposits such as dunghills and composting areas can also generate rejections of nutrients. Latrines, although not very widespread in the Sourou villages can also punctually influence the quality of water (case of Yaba village), particularly in fractured zones where a fast contamination of the aquifer is possible. To these possible causes of nitrates pollution, the use of dynamite for digging on the groundwater quality was suggested since many explosives contain in their structure a nitrate radical which could remain in water after drilling. However, according to the authors, the precise diagnosis is not obvious to establish and it requires more in-depth investigation. Moreover, bacterial oxidation and fixation of nitrogen by plants can both produce nitrates [31]. Concerning nitrates originating from agricultural inputs, no connection between the fertilizers used and the contamination of groundwaters could be rigorously established, since most of the water supply points are far away from the hydro-agricultural perimeters. Moreover, samples collected from Niassan village, next to the irrigated perimeters presented nitrates concentrations lower than the guideline value for drinking water [20-22].

Throughout the study periods, mean potassium concentrations ranged from 0.2 - 132 mg K^+/l (Table 2(a)). Rather high potassium concentrations were observed for 2 (13%) of the water sources examined, namely Wawara (128 - 132 mg/l) and Di (55 - 120 mg/l) drillings. As for nitrates, the variation in concentration, was related to the sampling period ($p < 0.0001$), the highest values being observed in 2008 for these water sources (132 and 120 mg/l, respectively). Potassium is an essential element in humans and is seldom, if ever found in drinking water at levels that could be a concern for healthy humans. The recommended daily requirement is greater than 3000 mg [20]. Potassium occurs widely in the environment, including all natural waters. Currently, there is no evidence that potassium levels in municipally treated drinking water, even water treated with potassium permanganate, are likely to pose any risk for the health of consumers [20]. Therefore, it is not considered necessary to establish a health-based guideline value for potassium in drinking water. Although potassium may cause some health effects in susceptible individuals, potassium intake from drinking water is well below the level at which adverse health effects may occur. Health concerns would be related to the consumption of drinking water treated by potassium-based water treatment (principally potassium chloride for regeneration of ion exchange water softeners), affecting only individuals in high-risk groups (i.e. individuals with kidney dysfunction or other diseases, such as heart disease, coronary artery disease, hypertension, diabetes, adrenal insufficiency, pre-existing hyperkalemia, people taking medications that interfere with normal potassium-dependent functions in the body, and older individuals or infants) [20].

For all the 15 water sources examined, the mean Sodium concentrations ranged from 0.5 to 19 mg Na^+/l (Table 2(a)). Although concentrations of sodium in potable water are typically less than 20 mg/l, they can greatly exceed this in some countries. No firm conclusions can be drawn concerning the possible association between sodium in drinking water and the occurrence of hypertension. Therefore, no health-based guideline value is proposed. However, concentrations in excess of 200 mg/l may give rise to unacceptable taste [20].

The mean Magnesium and Calcium concentrations ranged from 2 - 105 mg Mg^{+}/l and 4 - 410 mg Ca^{2+}/l (Table 2(b)). Both parameters varied significantly with the water source ($p < 0.01$) and the sampling period ($p < 0.05$) (Table 1(b)). Garzon and Eisenberg [34] first showed that there is a large variation in mineral contents of commercially available bottled drinking waters. The magnesium content of bottled water available in North America ranged from 1 to 120 mg/l, and the calcium content ranged from 1 to 240 mg/l, whereas concentration in bottled waters that are commercially available in Europe ranged from 0 to 546 mg/l for calcium and from 1 to 126 mg/l for magnesium. Comprehensive follow-up studies [35-37] on a wide range of commercially available bottled waters suggested that the mineral levels varied tremendously in bottled waters within countries and around the world. As summarized by WHO [38] from the above sources, magnesium and calcium concentrations found in water ranged from 0 - 29 mg Mg^{2+}/l and 2 - 83 mg Ca^{2+}/l for surface water sources, 2 - 48 mg Mg^{2+}/l and 26 - 85 mg Ca^{2+}/l for groundwater sources, 1-130 mg Mg^{2+}/l and 3 - 310 mg Ca^{2+}/l for Mineral water. Therefore, out of the 15 water sources studied, only the drilling of Wawara (7% of the water sources) with concentrations of 100 - 105 mg Mg^{2+}/l and 405 - 410 mg Ca^{2+}/l can be classified as highly mineralized, while the other water sources (93%) crossed the values found for groundwater sources. Magnesium and Calcium are naturally occurring in surface or groundwater from erosion and weathering of soils, minerals or ores. There is no evidence of adverse health effects from calcium or magnesium in drinking water; both ions contribute to water hardness. Therefore, guideline values are not proposed [20,22,39].

Total Iron in water samples ranged from 0.02 - 9.8 mg Fe/l (Table 2(b)). These concentrations varied significantly with the sampling site and the sampling period as well ($p < 0.0001$; Tables 1(b) and 2(b)). Concentrations over the WHO and USEPA Guideline value (≤0.3 mg Fe/l) were found in water samples from the modern well of Yaba and the drillings of Diouroum, Niassan-clinic, Sono, Yaba and Kouy (6.3, 3.3, 2.3, 2.2, 1.8, and 1 mg Fe/l, respectively). Iron is one of the most abundant metals in Earth's crust. It is found in natural fresh waters at levels ranging from 0.5 to 50 mg/l [20]. Toxic effects have resulted from the ingestion of large quantities of iron, but there is no evidence to indicate that concentrations of iron commonly present in drinking water constitute any hazard to human health; hence, a maximum acceptable concentration has not been set. However; at concentrations above 0.3 mg/l (drinking water standard), iron can stain laundry and plumbing fixtures and produce undesirable tastes in beverages. The precipitation of excessive iron impacts an objectionable reddish-brown color to water and may also promote the growth of certain microorganisms (i.e. Iron-Reducing Bacteria), leading to the deposition of a slimy coating in water distribution pipes [20,31,39].

Ammonium concentrations in the water samples ranged from 0.002 - 1.95 mg NH_4^+ /l (Table 2(b)). Concentrations recorded in the water samples varied significantly with the sampling site and the sampling period as well ($p < 0.0001$, Table 1(b)). The term ammonia includes the non-ionized (NH_3) and ionized (NH_4^+) species. Ammonia in the environment originates from metabolic, agricultural and industrial processes and from disinfection with chloramine. Natural levels in groundwater and surface water are usually below 0.2 mg/l. Anaerobic groundwaters may contain up to 3 mg/l. Intensive rearing of farm animals can give rise to much higher levels in surface water. Ammonia in water is an indicator of possible bacterial, sewage and animal waste pollution. It represents a major component of the metabolism of mammals. Exposure from environmental sources is insignificant in comparison with endogenous synthesis of ammonia. Toxicological effects are observed only at exposures above about 200 mg/kg body weight. Ammonia in drinking-water is not of immediate health relevance, and therefore no health-based guideline value is proposed [20,22]. However, ammonia can compromise disinfecttion efficiency, result in nitrite formation in distribution systems; it can also cause the failure of filters for the removal of manganese, taste and odor problems [20].

Mean Sulfates concentrations in the water samples ranged from 1.5 - 1301.66 mg SO_4^{2-} /l (Table 2(a)). Concentrations were significantly related to the sampling site and the sampling period as well ($p < 0.0001$, Table 1(a)). Sulfates occur naturally in numerous minerals and are used commercially, principally in the chemical industry. The highest levels usually occur in groundwater and are from natural sources. In general, the average daily intake of sulfates from drinking water, air and food is approximately 500 mg, food being the major source. However, in areas with drinking water supplies containing high levels of sulfates, drinking water may constitute the principal source of intake [20-22]. No health-based guideline is proposed for sulfate. However, because of the gastrointestinal effects (diarrhea or dehydration) resulting from ingestion of drinking water containing high sulfates levels, it is recommended that health authorities be notified of sources of drinking water that contain sulfate concentrations in excess of 500 mg/l. The presence of sulfates in drinking water may also cause noticeable taste over 250 mg/l (threshold taste) and may contribute to the corrosion of distribution systems [20-22]. From the data recorded, out of the 15 water sources examined, only the drilling of Wawara (1175 - 1301.66 mg/l) appears to be of health concern regarding the defined standard (≤500 mg/l). The high sulfates levels leading to noticeable taste may partially explain the rejection by the village populations of this drilling for their alimentation as revealed our field investigations.

The mean Arsenic concentrations recorded for all the water sources studied were in the range of 1 - 2.66 μg/l and crossed the guideline value for the maximum acceptable concentration of Arsenic in drinking water [20-22]. Arsenic is found widely in Earth's crust in oxidation states of −3, 0, +3 and +5, often as sulfides or metal arsenides or arsenates. In water, it is mostly present as arsenate (+5), but in anaerobic conditions, it is likely to be present as arsenite (+3). Levels in natural waters generally range between 1 and 2 μg/l, although concentrations may be elevated (up to 12 mg/l) in areas containing natural sources [20]. Arsenic has not been demonstrated to be essential in humans. The acute toxicity of arsenic compounds in humans is predominantly a function of their rate of removal from the body. Arsine is considered to be the most toxic form, followed by the arsenites, the arsenates and organic arsenic compounds. Acute arsenic intoxication associated with the ingestion of well water containing very high concentrations (21.0 mg/l) of arsenic has been reported [20]. Numerous epidemiological studies have examined the risk of cancers associated with arsenic ingestion through drinking water. The International Program on Chemical Safety (IPCS) concluded that long-term exposure to arsenic in drinking-water is causally related to increased risks of cancer in the skin, lungs, bladder and kidney, as well as other skin changes, such as hyperkeratosis and pigmentation changes. In view of the practical difficulties in removing arsenic from drinking water, particularly from small supplies, and the practical quantification limit for arsenic (1 - 10 μg/l), the guideline value of 10 μg/l is retained as a goal [20].

5.2. Water Microbiology

The results of the microbiological examination of the water sources samples showing Escherichia coli and fecal Coliforms cells count throughout the sampling periods are presented in Tables 3 and 4. All the drillings were exempt of fecal pollution and crossed the recommended guideline values [20-22]. By contrast, water from wells appeared uniformly polluted with concentrations exceeding sometimes 10^3 and 10^4 CFU/100 ml for E. coli and fecal Coliforms, respectively (Table 4). For the latters, water pollution by both indicators was significantly ($p < 0.0001$) related to the water source location and the sampling period as well (Table 3). Since wells are not protected from environmental contaminations (Figures 3(c) and (d)), they may receive several depositions with regard to the absence of latrines in rural area, animals frequentation and the atmospheric conditions leading to water microbial pollution [15,17]. These potential sources or amplifying factors of microbial pollution may vary with the sampling source location and/or the sampling period as evidenced the data recorded (Table 3).

Field investigations showed a preference of wells as source of water for the alimentation of populations in the Sourou Valley, and that appeared mainly related to the high water hardness or sulfates content (noticeable taste) of water provided by some drillings (i.e. Wawara), poverty and cultural considerations. Therefore, these populations are exposed to health risks [20-22] as underlined previous investigations in the zone [40-42].

5.3. Drinking Water Quality

Physico-chemical and microbiological parameters were analyzed to identify the physical status, impurities, other dissolved substances and microorganisms that affect water used for domestic purposes in the Sourou valley.

From the 15 water sources examined, water from 4 (26%) drillings was of health concern in view of nitrates concentration (Di, Yaba and Diouroum), sulfates content (Wawara) and the guideline values for these parameters, respectively [11,12,20-22]. The 6 other drillings (40%) can be classified as safe sources for drinking water with regards to the physico-chemical and microbiological qualities, guideline standards and health risks.

All water samples from wells (33% of water sources studied) appeared uniformly polluted from fecal contaminations with concentrations of *E. coli* and fecal Coliforms largely over the guideline values [11,12,20-22].

Overall, 9 (60%) of the water sources examined do not provide safe water for populations alimentation. As a consequence of such unhygienic water quality, waterborne diseases have proven to be the biggest health threat worldwide and they contribute between 70% - 80% of health problems in developing countries [1,28]. These diseases continue to be a major cause of human mortality and morbidity. Diarrheal diseases remain a leading cause of illness and death in the developing world which alone causes 2.2 million of the 3.4 million water-related deaths per year, 90% of these deaths involving children less than five years [1,20,28].

As revealed studies [17,40] and the present investigations, populations in the Sourou valley are preferentially getting drinking water from wells and even do from the surface water of the Sourou River, although presenting unsafe quality. What could be the reasons undergoing such situation in the Sourou valley?

Field investigations (group discussions and interviews) were performed to elucidate the situation and to propose some issues for improving safe drinking water access for populations in the Sourou valley.

Table 3. Variance of bacterial fecal pollution indicators in water from drillings and wells with regard to sampling site and sampling period.

Source of variation	df	*Escherichia coli* (CFU/100 ml)		Fecal Coliforms (CFU/100 ml)	
		MS	p	MS	p
Site	14	2414336.67	$<0.0001^{**}$	68372495.25	$<0.0001^{**}$
Period	2	3171524.12	$<0.0001^{**}$	76018114.69	$<0.0001^{**}$
Site*Period	28	1440370.22	$<0.0001^{**}$	33793213.73	$<0.0001^{**}$

^{MS}mean square; **significant $p < 0.01$.

Table 4. Means* concentrations of E. coli and fecal Coliforms in water from drillings and wells during the dry season in the Sourou valley in 2007, 2008, and 2012 (*mean of 3replicates).

Sampling Site	Period	*E. coli* (CFU/100 ml)	Fecal Coliforms (CFU/100 ml)
	2007	0^{l}	0^{m}
Di-CP^{D}	2008	0^{l}	0^{m}
	2012	0^{l}	0^{m}
	2007	0^{l}	0^{m}
Dioroum-COPROD^{D}	2008	0^{l}	0^{m}
	2012	0^{l}	0^{m}
	2007	0^{l}	0^{m}
Kassoum-CEG^{D}	2008	0^{l}	0^{m}
	2012	0^{l}	0^{m}
	2007	0^{l}	0^{m}
Kassoum-School^{D}	2008	0^{l}	0^{m}
	2012	0^{l}	0^{m}
	2007	0^{l}	0^{m}
Kouy-COPROD^{D}	2008	0^{l}	0^{m}
	2012	0^{l}	0^{m}
	2007	0^{l}	0^{m}
Niassan-AMVS^{D}	2008	0^{l}	0^{m}
	2012	0^{l}	0^{m}
	2007	0^{l}	0^{m}
Niassan-Clinic^{D}	2008	0^{l}	0^{m}
	2012	0^{l}	0^{m}

Yaba-Clinic^{PM}	2007	60.3^{k}	1381.6^{g}
	2008	314.6^{g}	840^{i}
	2012	398.3^{f}	21050^{a}
Kouy-Mosque^{TW}	2007	599.3^{e}	800^{i}
	2008	890^{c}	10722^{c}
	2012	2801.6^{b}	11993.3^{b}
Sono-Centre^{TW}	2007	220^{i}	407.3^{k}
	2008	780^{d}	1140^{h}
	2012	200^{i}	670^{j}
Guideline values [20-22]		0	0
Sono-Clinic^{D}	2007	0^{l}	0^{m}
	2008	0^{l}	0^{m}
	2012	0^{l}	0^{m}
Yaba-Clinic^{D}	2007	0^{l}	0^{m}
	2008	0^{l}	0^{m}
	2012	0^{l}	0^{m}
Wawara-COPROD^{D}	2007	0^{l}	0^{m}
	2008	0^{l}	0^{m}
	2012	0^{l}	0^{m}
Dioroum-PNGT^{MW}	2007	274^{h}	4136^{e}
	2008	75.3^{k}	334.6^{k}
	2012	800^{d}	2203.3^{f}
Kassoum-Market^{MW}	2007	127^{j}	3167.3^{e}
	2008	140^{j}	234^{l}
	2012	4400^{a}	7033.3^{d}

*Means with a same letter within a column are not significantly different according to Newman Keuls' test $p < 0.05$. ^{D}drilling; ^{MW}modern well; ^{TW}traditional well.

5.4. Problematic of Safe Drinking Water Access in the Sourou Valley

Although according to the United Nations Development Program, Burkina Faso already reached the MDG for the access to safe drinking water [13], the situation is nottherefore satisfactory, in particular in rural environment as underlined the Ministry in charge of water management [14].

The improvement these years in the access to safe drinking water in the country, focused mainly on the quantitative aspect through the reinforcement of the infrastructures of water supply [13]. However, beyond the quantitative aspect, it is also advisable to take consideration on the quality of water consumed by populations.

In the Sourou valley, populations fed on water from drillings, modern and traditional wells, and from the surface water of the Sourou River as well (Figures 2(a)-(e)) [17,40]. As revealed the data recorded, if some drillings (60% of those examined) provide safe drinking water, in contrast, other drillings (40%) deliver water with either high nitrates [10] or sulfates concentrations over the guideline values [11,12,20-22]. Moreover, water from all the wells examined (100%) was microbiologically polluted (Table 4).

Field investigations revealed some factors contributing to the non-achievement in the access to safe drinking water for populations in the Sourou valley.

5.4.1. The Non-Regular Control of Water Quality

If in urban environment, distributed water is the object of regular control, it is not the case in rural environment where the indicators of drinking water quality are missing due to the lack of analytical data. For all the drillings examined in the study, no report attesting the quality of water provided by the hydraulic realization prior to populations' utilization was found. This situation illustrates clearly a lack of control during the realization of drillings.

As a consequence, some drillings are providing unsafe drinking water to populations. Moreover, although the APIWRM recommended the installation of national networks to monitor water quality, water uses, water requests and health risks, such operational structures remain to be created.

In line with this gap and to meet the recommendation of APIWRM [7,18], the present study intends to improve the knowledge related to the quality of drinking water resources in the Sourou basin [15-19].

5.4.2. The Lack of Self-Management of Infrastructures for Drinking Water Supply

During field investigations, several drillings were found broken down and not repaired. According to Ouédraogo [43], this apparent indifference in the water resource management, could be related to the fact that water is free and in addition, the hydraulic realization is a donation of government or NGO. Therefore, populations are just waiting for the donators to repair.

Such situation was well illustrated during the campaign of drillings restoration by the COPROD, one action of the Sourou river contract. Within this framework, a hundred broken down drillings were repaired but some were broken down again after a few months and were never been repaired by the populations [16]. This negative perception of self-water resource management by populations could be in line with a lack of education and sensitization. These gaps can be overcome, since primary school through the sensitization for an efficient integrated water resources management [16]. Unfortunately, such pedagogic activities are relatively scarce in Burkina Faso, although the intention is underlined in the APIWRM action plan [7].

5.4.3. Socio-Cultural Considerations

During samples collection, one remarkable observation was that water sources are mainly frequented by women and children (Figures 2(a) and (d)). This situation is related to the socio-cultural context, for which these fractions of the population are mainly in charge of household chores [19].

Moreover, wells appeared much more frequented than drillings for drinking water gathering in some villages (Figure 2(c)). As revealed investigations from water users, collecting water from wells gives opportunity to exchange on social and current events than drillings can afford [23].

For populations of villages located on the Sourou River, only the surface water is used for their alimentation (Figure 2(e)). All efforts from local authorities encouraging these populations to leave the unsafe environment were unsuccessful. This attitude in line with traditional believes leads these populations to consume unsafe water exposing them to waterborne diseases as revealed previous investigations in the zone [17,40-42].

5.4.1. The Lack of Hygiene and Basic Sanitation

In developing countries, the lack of hygiene and basic sanitation is still among the main factors limiting the access to safe drinking water for rural populations, despite efforts at international and national levels to provide water facilities. In addition to the factors limiting safe drinking water access evoqued above, the absence of disinfecting methods of drinking water in household should be also evoqued [44-47]. It has also been indicated that even do for water distribution networks, the duration of water storage in pipe, temperature variations and pipe materials can lead to bacterial proliferation in water collected in household from water pump [48,49].

Although we didn't examine the quality of drinking water in households, field observations supported a poor microbial quality of water from drillings in households as revealed the aspects of water collecting and storing materials (Figure 2(a)) and the method of water transportation to household (Figure 2(f)). The conesquences of such practices and attitudes on the quality of drinking water were highlighted in previous studies [44,50].

5.5. Some Issues to Improve Drinking Water Quality and Access in the Sourou Valley

In line with the limited safe drinking water access for populations in the Sourou valley, the OXFAM Belgique in collaboration with the Water Unit of Liège University and the Rotary Club Burkina Faso undertook to equip school classes with a local material-based technology (Figure 4) for preserving drinking water safety for school children. Over improving drinking water quality, these actions contributed to the fight against poverty through a financial support to local associations in charge of setting the technology design. Although the analysis of water issued from the disposal showed a rather good quality at microbiological level (data not shown), some improvement of the technology is needed to obtain a safe drinking water in line with the guideline standards [11,12,20-22].

Investigations using local plant material (grains of Moringa oleifera) to improve the quality of some surface waters and water from wells which feed the populations in the Sourou valley are currently performed [47]. These investigations, the ones in RDC Congo [45,46] and the outputs from previous studies in Burkina Faso [44,50] may help improving the quality of drinking water for rural populations in Sub-Saharan developing countries.

Beyond these direct actions intending to improve the quality of drinking water, it is also advisable within the framework of an integrated approach, to develop preventive measures at land management and soil occupation nearby water sources, and at hygiene and basic sanitation levels as well.

Therefore, the construction of protection zones around drinking water sources, the promotion of hygiene-based rules and the installation of bottom-tight latrines could help improving drinking water quality for rural populations in the Sourou valley and in developing countries.

Figure 4. Technology based on local material experienced in schools of the Sourou valley for the preservation of safe drinking water quality.

6. CONCLUSIONS AND PROSPECTS

Access to safe drinking water for rural populations in the Sourou valley is still facing the optimization of water resources management. The lack of an efficient network of water quality survey and a self-management of drinking water sources facilities, the inertia of mentalities and traditional believes appear the main constraints governing this problematic in the Sourou valley. To overcome these constraints, information and sensitization actions on water resources management and on water and health relationships are required. This approach involves pedagogic aspect and should therefore be performed in collaboration with school teachers, focusing children and women the main actors at the center of the problematic.

In addition, scientific expertise, through improving knowledge on water resources quality and providing appropriate methods for water disinfection in households could help solving the problematic. Moreover, an integrated approach which associates water and land manaments could allow setting preventive measures for the protection of drinking water sources in the Sourou valley.

Overall, these proposed issues for improving the access to safe drinking water for populations must be provided to the institutions in charge of the water policy and to the local collectivities, which within the framework of the decentralization, were seen entrusted the responsibility for the natural resources management. It is also advisable that local participating structures as the Local Water Committees (LWC) catch these problems in order to improve the water services in response to the populations needs with respect to the national and international standards of drinking water.

ACKNOWLEDGEMENTS

Savadogo Boubacar, Kaboré Aminata, Zongo Dramane, Poda Jean Noel, Bado Hortense, Rosillon Francis, Dayeri Dianou The authors would like to express profound gratitude to the Walloon Region of Belgique, University of Liège, CNRST/IRSS, INERA-Di, COPROD, LNAE and Tougan Hospital for providing financial support, laboratory and other facilities.

REFERENCES

1. World Health Organization, "Guidelines for Drinking Water Quality," 3rd Edition, World Health Organization, Geneva, 2008
2. United Nations Development Programme, "Burkina Faso: Access to Safe Drinking Water Has Improved a Lot," 2010.
3. United Nations Environment Program, UNEP, "The Problematic of Water in the Democratic Republic of Congo," Technical Report, 2011.
4. K. G. R. Kouam, H. G. Mpakam, S. A. Ndonwy, S. L. D. Bopda and G. E. Ekodeck, "Integrated Management of Water Resources and Millennium Development Goals for the Development of Africa: Case study of Cameroun," Vertigo, Vol. 7, No. 2, 2006, pp. 1-9.
5. F. Julien, "Water Control and Sustainable Development in West Africa: A Need of a Regional Cooperation on the Trans-Boundaries Hydrologic Systems," Vertigo, Vol. 7, No. 2, 2006, pp. 1-18.
6. D. J. H. Phillips, M. Daoudy, S. C. McCaffrey and A. J. R. Öjendal, "Trans-Boundary Water Cooperation as a Tool for Conflict Prevention and Broader Benefit Sharing," Swedish Ministry of Foreign Affairs, Stockholm, 2006, p. 249.
7. Ministry of Agriculture, "National Action Plan for Integrated Water Resources Management (APIWRM)," Hydraulics and Fishing Resources, Ouagadougou, 2003, p. 62.
8. Ministry of Water, "Leaflet of the Hydrogeological Map of Burkina Faso, Scale 1:500,000, Sheet Bobo-Dioulasso," Direction of Hydraulics Resources Inventory, 1993, p. 40.
9. Ministry of Agriculture,"Conception Plan and Implementation of the National Water Information System," Direction of Hydraulics Resources Inventory, Hydraulics and Fishing Resources, 2004, p. 135.

10. Ministry of Agriculture, "National Program of Drinking Water and Sanitation Supply for the 2015 year issue (NPDWSS2015)," Direction of Hydraulics Resources Inventory, Hydraulics and Fishing Resources, 2006, p. 43.
11. Ministry of Water and Environment, "Decree n 2001-185/ PRES/PM/MEE, Fixing the Standardsof Pollutants in Air, Water and Soil," 7 May 2001.
12. World Health Organization, "Guidelines for Drinking Water Quality," Third Edition, Geneva, 2004.
13. United Nations, United Nations Development Programme, "Access to Safe Drinking Water," 2010.
14. Ministry of Agriculture, "IWRM 2010 and 2011 Perspectives," Hydraulics and Fishing Resources, Ouagadougou, 2011, p. 29.
15. F. Rosillon, B. Savadogo, A. Kaboré, H. Bado-Sama and D. Dianou, "Attempts to Answer on the Origin of the High Nitrates Concentrations in Groundwaters of the Sourou Valley in Burkina Faso," Journal of Water Resource and Protection, Vol. 4, No. 8, 2012, pp. 663-673. doi:10.4236/jwarp.2012.48077
16. F. Rosillon, B. Savadogo, A. Kaboré, H. Bado-Sama and D. Dianou, "Estimation of the Nitrates Contents in Waters by Using Reagent Strips: An Environment Education Exercise in the Sourou Valley in Burkina Faso," Vertigo, Vol. 12, No. 2.
17. D. Dianou, B. Savadogo, D. Zongo, T. Zougouri, J. N. Poda, H. Bado and F. Rosillon, "Surface Waters Quality of the Sourou Valley: The Case of Mouhoun, Sourou, Debe and Gana Rivers in Burkina Faso," International Journal of Biological and Chemical Sciences, Vol. 5, No. 4, 2011, pp. 1571-1589.
18. Ministry of Agriculture, "IWRM 2010 and 2011 Perspectives," Hydraulics and Fishing Resources, Ouagadougou, 2011, p. 29.
19. F. Rosillon, P. Vander Borght and H. Bado-Sama, "River Contract in Wallonia (Belgium) and Its Application for Water Management in the Sourou Valley (Burkina Faso)," Water Sciences and Technology, Vol. 52, No. 9, 2005, pp. 85-93.
20. World Health Organization, "Guidelines for Drinking Water Quality," Fourth Edition, World Health Organization, Geneva, 2011.
21. United States Environmental Protection Agency "Drinking Water Standards and Health Advisories," 2012, p. 12.
22. Health Canada, "Guidelines for Canadian Drinking Water Quality Summary Table," Water, Air and Climate Change Bureau, Healthy Environments and Consumer Safety Branch, Health Canada, Ottawa, p. 22.
23. H. Bado-Sama and F. Rosillon, "River Contract of Sourou, Outcome of the Actions Program 2003-2008," Report COPROD/ULG," 2009, p. 76.
24. Direction of Hydraulics Resources Inventory, Ministry of Water and Burkina Faso, "Leaflet of the Hydrogeological Map of Burkina Faso Scale 1:500,000 Sheet Bobo-Dioulasso," 1993, p. 40.
25. American Public Health Association, American Water Works Association and World Peace Choral Festival, "Standard Methods for the Examination of Water and Wastewater," 20th Edition, by American Public Health Association, American Water Works Association and Water Pollution Control Federation, Washington, 1998.
26. J. Rodier, "Water Analysis: Natural Waters, Wastewaters, Sea Water," Ed. Dunod, Paris, 2005, p. 1384.
27. G. I. Oyhakilome, A. F. Aiyesanmi and F. C. Akharaiyi, "Water Quality Assessment of the Owena Multi-Purpose Dam, Ondo State, Southwestern Nigeria," Journal of Environmental Protection, Vol. 3, No. 1, 2012, pp. 14-25. doi:10.4236/jep.2012.31003 [Citation Time(s):1]
28. L. J. Bina, T. Prasai, A. Singh and K. D. Yami, "Assessment of Drinking Water Quality of Madhyapur-Thimi and Study of Antibiotic Sensitivity Against Bacterial Isolates," Nepal Journal of Science and Technology, Vol. 10, 2009, pp. 167-172. [Citation Time(s):1]
29. D. McKelvie, "In Situ Measurement of Physicochemical Water Quality Parameters," In: F. R. Burden, D. Donnert, T. Godish and I. D. McKelvie, Eds., Environmental Monitoring Handbook, McGraw-Hill Co., Boston, 2004, pp. 3.1-3.21.

30. Ireland Environmental Protection Agency, "Parameters of Water Quality: Interpretation and Standards," Environmental Protection Agency, Johnstown, 2001, p. 133.
31. FAO, "Aquastat, Water and Agriculture FAO Information System: Burkina Faso," 2005.
32. D. Nabayaogo, "Agricultural Activity in the Sourou Valley: Impact on Water Resources and Ecosystems," Memory, University of Ouagadougou, 2006, p. 42.
33. P. Dugué, "Environmental Assessment Study and Development of Sustainable Production Systems in Projects to Support Food Production in Mali and Burkina Faso," Foundation for Agriculture and Rural Life in the World, CIRAD, 2009, p. 97.
34. P. Garzon and M.J. Eisenberg, "Variation in the Mineral Content of Commercially Available Bottled Waters: Implications for Health and Disease," The American Journal of Medicine, Vol. 105, No. 2, 1998, pp. 125-130. doi:10.1016/S0002-9343(98)00189-2
35. Von Wiesenberger, "The Pocket Guide to Bottled Water," Contemporary Books, Chicago.
36. Azoulay, P. Garzon, and M.J. Eisenberg, "Comparison of the Mineral Content of Tap Water and Bottled Waters," Journal of General Internal Medicine, Vol. 16, No. 3, 2001, pp. 168-175. doi:10.1111/j.1525-1497.2001.04189.x
37. Rosborg, B. Nihlgard, L. Gerhardsson, M. L. Gernersson, R. Ohlin and T. Olsson, "Concentrations of Inorganic Elements in Bottled Waters on the Swedish Market," Environmental Geochemistry and Health, Vol. 27, No. 3, 2005, pp. 217-227. doi:10.1007/s10653-004-1612-8
38. World Health Organization, "Calcium and Magnesium in drinking water: Public Health Significance, World Health Organization, Geneva, 2009.
39. European Union's Drinking Water Standards, "Council Directive 98/83/EC on the Quality of Water Intended for Human Consumption," Geneva, 1998.
40. D. Dianou, J. N. Poda, L. G. Savadogo, H. Sorgho, S. P. Wango and B. Sondo, "Intestinal parasites in the zone of the hydro-agricultural complex of Sourou in Burkina," Vertigo, Vol. 5, No. 2, 2004, pp. 1-8.
41. N. Poda, J. R. Mwanga, D. Dianou, A. Garba, F. C. Ouédraogo, D. Zongo and K. B. Sondo, "Parasite Infections that Undermine the Pole of Development in Burkina Faso: Case Study of Schistosomiasis and Geohelminthiasis in the Hydro-Agricultural Complex of Sourou," Vertigo, Vol. 7, No. 2, 2006, pp. 1-7.
42. Koukounari, A. F. Gabrielli, S. Toure, E. Bosque-Oliva, Y. Zhang, B. Sellin, C. A. Donnelly, A. Fenwick and J. P. Webster, "Schistosoma haematobium Infection and Morbidity before and after Large-Scale Administration of Praziquantel in Burkina Faso," Journal of Infectious Diseases, Vol. 196, No. 5, 2007, pp. 659-669. doi:10.1086/520515]
43. Ouedraogo, "Water Management and Environmental Education in the Pedagogic Documents in Burkina Faso," Vertigo, 25 December 2003. http://vertigo.revues.org/1968
44. Dianou, J. N. Poda, L. Thiombiano and H. Sorgho, "Evaluation of Contamination Tracks and Some Disinfecting Methods of Households' Drinking Water in Rural Area: Case Study of Thion village in Burkina Faso," Annals University of Ouagadougou, Vol. C, No. 2, 2004, pp. 168-181.
45. M. Sunda, F. Rosillon and K. M. Taba, "Contribution to Water Disinfection through Photosensitization with Plant Extracts," Environment Journal of Water Quality, Vol. 39, No. 2, 2008, pp. 199-209. doi:10.1051/water/2008006
46. M. Sunda, "Contribution to Water Disinfection through Photosensitization with Plant Extracts," Ph.D. Dissertation, University of Liège, Faculty of Sciences, 2012.
47. Kaboré, B. Savadogo, F. Rosillon, A. S. Traoré and D. Dianou, "Optimization of theEffectiveness of Moringa Oleifera Seeds in the Treatment of Drinking Water in Sub-Saharan Africa: Case Study of Waters of Burkina Faso," Annals University of Ouagadougou, Vol. 9, 2011, pp. 880-889.
48. Lautenschlager, N. Boon, Y. Wang, T. Egli and F. Hammes, "Overnight Stagnation of Drinking Water in Household Taps Induces Microbial Growth and Changes in Community Composition," Water Research, Vol. 44, No. 17, 2010, pp. 4868-4877. doi:10.1016/j.watres.2010.07.032
49. S. Patrick, A. Roland, C. Anani1 and S. B. Dampare, "Nutrient-Induced Growth of Coliform and HPC Bacteria in Drinking Water Pipes," Journal of Environmental Protection, Vol. 3, No. 6, 2012, pp. 508-517 doi:10.4236/jep.2012.36061

50. D. Dianou, J. N. Poda, H. Sorgho and L. Thiombiano, "Quality of Households Drinking Water at Thion, Blé- dougou and Kangoura in Burkina Faso," South Sciences and Technologies, Vol. 9, 2002, pp. 25-33.

INDEX